MARINE CHEMISTRY

Water Science and Technology Library

VOLUME 25

Editor-in-Chief
V. P. Singh, *Louisiana State University,*
Baton Rouge, U.S.A.

Editorial Advisory Board

M. Anderson, *Bristol, U.K.*
L. Bengtsson, *Lund, Sweden*
A. G. Bobba, *Burlington, Ontario, Canada*
S. Chandra, *New Delhi, India*
M. Fiorentino, *Potenza, Italy*
W. H. Hager, *Zürich, Switzerland*
N. Harmancioglu, *Izmir, Turkey*
A. R. Rao, *West Lafayette, Indiana, U.S.A.*
M. M. Sherif, *Giza, Egypt*
Shan Xu Wang, *Wuhan, Hubei, P.R. China*
D. Stephenson, *Johannesburg, South Africa*

The titles published in this series are listed at the end of this volume.

MARINE CHEMISTRY

An Environmental Analytical Chemistry Approach

edited by

ANTONIO GIANGUZZA

*Department of Inorganic Chemistry,
University of Palermo,
Palermo, Italy*

EZIO PELIZZETTI

*Department of Analytical Chemistry,
University of Torino,
Torino, Italy*

and

SILVIO SAMMARTANO

*Department of Chemistry,
University of Messina,
Messina, Italy*

KLUWER ACADEMIC PUBLISHERS
DORDRECHT / BOSTON / LONDON

A C.I.P. Catalogue record for this book is available from the Library of Congress.

ISBN 0-7923-4622-X

Published by Kluwer Academic Publishers,
P.O. Box 17, 3300 AA Dordrecht, The Netherlands.

Sold and distributed in the U.S.A. and Canada
by Kluwer Academic Publishers,
101 Philip Drive, Norwell, MA 02061, U.S.A.

In all other countries, sold and distributed
by Kluwer Academic Publishers Group,
P.O. Box 322, 3300 AH Dordrecht, The Netherlands.

Printed on acid-free paper

All Rights Reserved
© 1997 Kluwer Academic Publishers
No part of the material protected by this copyright notice may be reproduced or
utilized in any form or by any means, electronic or mechanical,
including photocopying, recording or by any information storage and
retrieval system, without written permission from the copyright owner.

Printed in the Netherlands

TABLE OF CONTENTS

Preface — vii

List of Contributors — viii

Environmental Analytical Chemistry in Marine Science
E.Pelizzetti — 1

The Effect of Ionic Interactions on Thermodynamic and Kinetic Processes in Natural Waters
F.J.Millero — 11

Light and Chemically Driven Reactions and Equilibria in the Presence of Organic and Inorganic Colloids
C.Minero — 39

Chemical Speciation of Some Classes of Low Molecular Weight Ligands in Seawater
A.De Robertis, C.Foti, S.Sammartano, and A.Gianguzza — 59

Computer Tools for the Speciation of Natural Fluids
C.De Stefano, S.Sammartano, P.Mineo, and C.Rigano — 71

Metal Ions and Organometallic Compounds in Sea Water and in Sediments: Biogeochemical Cycles
P.J.Craig and D.Miller — 85

Nutrients in the Sea
M.Ribera d'Alcalà, V.Saggiomo, and G.Civitarese — 99

Sampling Techniques for Sea Water and Sediments
G.Capodaglio — 115

Trace and Ultratrace Metal Analysis: Matrix Removal and Sample Preconcentration
C.Sarzanini — 131

Hyphenated Instrumental Methods for the Detection of Heavy Metals in Marine Environment
R.Frache — 149

Analysis and Speciation of Organometallic Compounds in the Marine Environment - General Considerations
P.J.Craig and D.Miller — 161

Chromatographic Analysis of Organic Micropollutants in Marine Environments
R.Fuoco and M.P.Colombini   173

Application of Mass Spectrometric Techniques to the Detection of Natural and Anthropogenic Substances in the Sea
M.Vincenti   189

Techhiques of Extraction and Analytical Methods for Humic Substances in Sea Water and Sediments
B.M.Petronio   211

Certified Reference Materials for Chemical Analysis in Marine Ecosystems
R.Morabito and P.Quevauviller   225

Distribution Models of Pollutants in the Marine Environment
R.Cecchi and G.Ghermandi   237

Sewage and Nutrients in the Marine Environment: Stimulants for Good or Vectors for Harm?
M.R.Preston   259

Arsenic in the Marine Environment
M.E.Farago   275

The Introduction of Some Selected Persistent Organic Pollutants to the Marine Environment and Aspects of their Subsequent Behaviour
M.R.Preston   293

The Interface Air-Water: Oil Spills and Tropospheric Chemistry. Techniques and Remediation
E.Bolzacchini, S.Meinardi, M.Orlandi, and B.Rindone   309

Interactions of Marine Biogeochemical Cycles and the Photodegradation of Dissolved Organic Carbon and Dissolved Organic Nitrogen
R.G.Zepp   329

Effects of Light on the Biological Availability of Trace Metals
B.Sulzberger   353

The Influence of Iron on Carbon Dioxide in Surface Seawater
F.J.Millero   381

Subject Index   399

PREFACE

This volume documents the scientific activity at the "International School on Marine Chemistry" held from September 14 to September 20, 1996 in the island of Ustica, Palermo, Italy. In the island the first Italian Marine Reserve is operating since 1987.

Twenty-three contributions are collected in the book presenting the more recent developments on the Chemistry of the Sea, with particular emphasis on the environmental analytica chemistry approach.

The School was organized under the auspices and financial support of the Marine Reserve of Ustica Island, Regione Sicilia, Italian Research Council, and University of Palermo. The scientific advice of the Analytical Chemistry Division of the Italian Chemical Society is gratefully acknowledged.

The ninety participants, including the lecturers, are grateful to the generous support of the agencies.

The Editors also thank the high quality and creative contributions of the lecturers. It is they who made this volume a reality.

Antonio Gianguzza
Ezio Pelizzetti
Silvio Sammartano

LIST OF CONTRIBUTORS

Bolzacchini, E.	309
Capodaglio, G.	115
Cecchi, R,	237
Civitarese, G.	99
Colombini, M.P.	173
Craig, P.J.	85, 161
De Robertis, A.	59
De Stefano, C.	71
Farago, M.E.	275
Foti, C.	59
Frache, R.	149
Fuoco, R.	173
Ghermandi, G.	237
Gianguzza, A.	59
Meinardi, S.	309
Miller, D.	85, 161
Millero, F.J.	11, 381
Mineo, P.	71
Minero, C.	39
Morabito, R.	225
Orlandi, M.	309
Pelizzetti, E.	1
Petronio, B.M.	211
Preston, M.R.	259, 293
Quevauviller, P.	225
Ribera d'Alcalà, M.	99
Rigano, C.	71
Rindone, B.	309
Saggiomo, V.	99
Sammartano, S.	59, 71
Sarzanini, C.	131
Sulzberger, B.	353
Vincenti, M.	189
Zepp, R.G.	329

ENVIRONMENTAL ANALYTICAL CHEMISTRY IN MARINE SCIENCE

Ezio PELIZZETTI
*Dipartimento di Chimica Analitica, Università di Torino
via Pietro Giuria 5, 10125 Torino, Italy*

1. Introduction

Two elements are at the basis of the music: the musical instruments and the idea. They do not represent the form and the content, but melt organically into the music [1]. Similarly in the environmental analytical chemistry, the analytical instrumentation and methodologies combine with theories and models to gain insight the concert of the nature.

The assessment of methodologies for the characterization of the chemical and physicochemical properties of environmental samples represents in fact the domain in which the environmental analytical chemistry operates [2]. The correct understanding of environmental processes implies to unravel the extreme physical and chemical complexity of the different environmental compartments and their interconnections. It is recognized that the determination of global parameters, such as total concentrations or average constants, can only be a first step in the correct representation of environmental properties. The environmental processes are affected by many factors still poorly evaluated or not evaluated at all. New approaches and methodologies are needed such as the combination of analytical techniques or the use of mathematical procedures to extract detailed information, and finally a critical compilation of data.

The IUPAC Commission on Environmental Analytical Chemistry [2] listed the following "Principal Activities" in which the evaluation of concepts and methodologies are of primary importance:

Particles and colloids
- Characterization of nonliving particles and colloids
- Kinetic and equilibrium aspects of exchange reactions on nonliving particles and colloids (complexation, adsorption, acid-base effects)
- Processes related to the transfer of matter by particles and colloids
- Role of the living particles (plankton and bacteria)

Transfer of matter at interfaces
- Air-water interface studies and methodologies
- Transfer between saturated and nonsaturated zones of soils

Other problems
- Factors affecting biological productivity: Methods and interpretation
- Speciation of soluble compounds by specific analysis of species (organic compounds, N- and P-containing compounds, organometallic species)
- Speciation of soluble compounds by kinetic and equilibrium studies of exchange reactions: metal complexation, acid-base, interactions between organics
- Redox processes and eutrophication

Chemical measurements in seawater involve the peculiarity related to its distinctive composition, the large spatial and temporal scales over which the measurements are made and the frequent need to perform analysis in difficult conditions [3].
Improvements in methodologies of chemical analysis are requested for studies on the biological, physical and chemical processes that control the flow of chemical through the ocean and its linkage with the atmosphere. Increasing attention has been focused on marine chemistry because of the introduction in the ocean of chemicals as a consequence of human activities.
In the following some aspects in which environmental analytical chemistry contributes to marine science are briefly outlined.

2. Particles and colloids

The ocean and seas represent the largest colloidal system on the planet. The study of particles and colloids in the marine environment is a subject that cannot be separated from studies of dissolved chemicals [4-6]. Many ions and compounds are adsorbed onto the particles, whereas others are actively taken up and transformed by the biota. Many of the particles are living organisms and much of the dissolved organic carbon (DOC) is humic material, a poorly characterized substance of high molecular weight [7]. Inorganic particles are also present affecting chemical cycling and in surface seawater photoassisted processes [8].
Nonliving particles and colloids can significantly influence the migration behavior of metals (toxic and radioactive). Adsorption is different between complexes and free metals and the knowledge of these properties may enable better predictions of their behavior in marine system [9].
Colloids have been shown to affect also the bioavailability and transport of organic contaminants. Marine sediments have been reported to be enriched in anthropogenic toxic compounds, the majority of which is associated with the particulate phase. Several techniques have been used [10,11] but recently ultrafiltration and reverse-phase chromatography were evaluated for isolating colloids and the environmentally associated contaminants. These two techniques were applied to natural colloids and in-place pollutants from an environmentally contaminated site [12].

3. The carbon cycle

The processes occurring at the air-ocean interface are closely linked to the Earth's climate [13]. Despite their importance there are large gaps in their understanding. For example, most of the increasing CO_2 in the atmosphere will enter in the ocean, but the rate of CO_2 absorption in seawater is not yet well known [14].

Direct measurements of the increase of total inorganic carbon are not yet feasible because its spatial and temporal variability requires a large number of samples. Measurements of the variation of $^{13}C/^{12}C$ ratio with time demonstrate however that the ocean is receiving fossil fuel carbon [15].

There are substantial difficulties in the routine measurements of inorganic carbon as well as the related parameters alkalinity and pH in seawater with errors < 0.1%. A coulombic titration procedure with precision of 0.1%, usable aboard ship, has been reported [16] and a spectrophotometric determination of pH [17] and of CO_2 partial pressure [18] could be the simplest procedure and represents the basis for an optical sensor for observe changes in the oceanic inorganic carbon.

4. Speciation of organic carbon

The measurement of DOC in seawater is matter of great controversy. The classical methods have been questioned by high-temperature combustion methods, although an interlaboratory comparison failed to resolve the discrepancies among the different methods for DOC analysis [19,20]. It is worth to note that the difference between the low estimate (30-100 μM) and the high (300 μM) amounts to a pool of organic carbon that is greater in mass than the total amount of carbon stored on land in soils and vegetation [21].

The concentration and its variation in time and space of individual organic compounds are important in the biogeochemical cycles. Only <10% of the dissolved organic carbon has been identified.

Some compounds are useful tracers of ocean circulation (chlorofluorocarbons and ^{14}C) [22,23], ancient ocean temperature [24], global sulfur cycle (COS and dimethylsulfide) [21,25,26].

The identification and quantitation of individual organic compounds present in seawater at nano to picomolar concentration represents a challenge for the analytical chemists and an obstacle in the studies of carbon cycling. Current methodologies require large samples, time-consuming chromatographic separation, purging, preconcentration [27], many of which are difficult to maintain on ship cruise [28]. Advances in mass spectrometry (MS) that provide detection limits in picomolar range [29] and fast derivatization techniques for hydrophilic compounds [30] promise to expand the number of compounds that can be detected.

5. Trace elements and their speciation

The above mentioned large spatial and temporal scales over which chemical measurements are required in seawater have also to face a concentration range of 15 orders of magnitude in the elemental composition. Seven ions (present at concentration > 1 mM) constitute >99.5% of the dissolved chemicals in seawater and dramatically affect rates and equilibria of chemical reactions in the ocean. Many of the remaining ions, despite their low concentration, may have a significant influence on global cycling as micronutrients in the enzymes and electron transport system present in the living organisms.

The accurate measurements of the trace elements at nanomolar to picomolar concentrations represent a formidable challenge for analytical chemists [31]. The analytical methods most commonly used are atomic absorption spectrometry (AAS) with graphite furnace or inductively coupled plasma mass spectrometry [32,33]. The development of sampling procedures (avoiding contamination), of preconcentration and separation steps led to dramatic decrease (in some cases several orders of magnitude) in the reported trace elements concentration [34].

However, measurements of total element concentration are not sufficient to define its environmental behavior. The chemical speciation will determine its biological availability, toxicity and geochemical reactivity in seawater. Electrochemical methods are employed most often in these studies [35-37].

As for the analysis and speciation of organic compounds, much of the focus is on the development of methods that can be used aboard ships to produce large data sets. Since AAS and MS are actually sensitive to constant vibrations, a variety of methods (GC, flow injection analysis with chemiluminescence and fluorometric detection, atomic fluorescence, electrochemical techniques)[38-40] have been exploited. To avoid risk of contamination , in situ measurements represent the preferred option and future work in this direction is needed.

6. Dissolved Gases

Dissolved gases play an important role in marine environment. Deviation from equilibrium concentrations are common and reflect biological activity and physical processes acting on the system. On the three most concentrated gases in aerobic waters, N_2 and O_2 are affected by both biological and physical processes [41] whereas Ar is affected strictly by physical processes. In some cases only small fractional (<1%) deviations from equilibrium occur and is necessary to measure dissolved gases with high precision.

The principal methods of analysis include gas-chromatography (GC) or mass-spectrometry (MS). GC provides direct concentration while MS provides gas ratio data with high precision (in the order of 0.05%) which can be converted to concentration data with independent measurement of one of the component gases (e.g. O_2). Recent advancements involving the use of a membrane inlet mass spectrometer [42] allows the direct analysis of a seawater sample, thus avoiding a separate time-consuming degassing step which can also limit the overall precision of the gas analysis [43].

7. Radioisotopes

Radioisotopes have the unique feature of allowing oceanographer to elucidate rates of mixing and chemical transformations in the oceans. Among the sixty radionuclides occurring naturally and the fission products of thermonuclear power generation and weapons detonation, the radioactive isotopes ^{234}Th (predominantly present as hydrolytic species, strongly interacting with particles) and its ^{238}U parent (present as negatively charged carbonate complex) can be used to calculate the rate at which thorium is scavenged from the solution to the particulate phase and the rate at which particles are removed from the water column.
The determination of ^{234}Th requires several steps before alpha counting, i.e. coprecipitation, ion-exchange, electrodeposition [44]. Recently extraction through a cartridge followed by gamma detection can be operated still at sea [45]. Thermal ionization mass spectrometry is now being used to detect the longer lived Th and U series isotopes, improving the detection limits [46].

8. Marine resources

The ocean is a precious commons and was the place that first nourished life and still provides much of our sustenance. We must protect mother ocean [47].
The ocean has been viewed by humans as an infinitely broad highway on which to transport people and things, as a source of food, as a protective shield or battle zone, and recently as a receiver for waste disposal [48,49], including CO_2 from fossil-fueled power plants [50]. Addition of only 1 nM of iron to surface seawater has been found to cause a significant increase of carbon fixation by phytoplankton. This "Iron Hypothesis" has received much attention [51] and is subject of scientific debate [52].
In addition to food, researchers have found new compounds from marine organisms to be used as drugs (anticancer and anti-AIDS) or in cosmetic formulations [53]. Extraction procedures and structural investigations are the challenge for analytical chemists in this area.

9. Marine pollution

Increasing attention has been focused on ocean chemistry because of man-made effects and impacts due to the introduction of chemicals into the sea from which they were previously absent or less prevalent.
Environmental research, policy and management decisions related to the consequences of harmful substances on humans and, on more broadly basis, on the survival of living

species, need some quantitative bases of analytical data accounting for the location and the duration of the events. Environmental analytical chemistry is then an important frontier, including from one side the invention of basic measurement tools and their specific applications and, from the other side, the investigation of the fate of the contaminants (transport, degradation, accumulation in marine organisms).

Main classes of pollutants include: sewage and nutrients with the related eutrophication phenomena; persistent organic compounds, such as halogenated aromatics, pesticides [54]; oil spills and petroleum derivatives [55,56]; metals and organometallic compounds [57]. The concern about the protection of the world's marine environment from threats as sewage, pesticides and other persistent organic pollutants has committed representatives of 102 countries, under the sponsorship of the United Nations Environment Program, to control, reduce and/or eliminate emissions, discharge and use of persistent organic substances. The program (called the Global Program of Action for the Protection of Marine Environment from Land-Based Activities) will first focus of 12 pollutants (PCB's, dioxins, furans, aldrin, dieldrin, DDT, endrin, chlordane, hexachlorobenzene, mirex, toxaphene, and heptachlor), but it will be expanded to cover other persistent organic substances in future years [58].

10. Analytical challenges associated with monitoring programs in marine science

Some of the analytical problems associated with measurements in seawater have already been mentioned [59]. Since long-term series of chemical relevations are required to identify the processes that lead to natural variability and to assess the human impacts on biogeochemical cycles, sample collection and analysis on a ship should be desirable. High costs make this solution prohibitive and long-term studies are then done at very few location [60].

Automated systems could represent an interesting possibility, once drift and biofouling will not compromise their use. Recent progress has been made [61] and new generation for long-term (1-12 months) and deep sea operations should be developed.

Sensor systems (electrodes, fiber-optic sensors, chemical field transistors) promise to play an important role in the future [62]. They are simple and may be placed in a remote location [63]. The only chemical sensors regularly used are oxygen and pH electrodes. To be suitable for environmental monitoring of other chemicals at concentration of 0.1 µM, these systems should be adequately selective and sensitive toward the target analyte and to be of low cost and high reliability. It is also possible to determine the concentration of dissolved chemicals in situ by using unsegmented continuous flow analyzers. These submersible chemical analyzers are based on the principle of flow injection analysis, although without the injection valve [64].

Finally in order to plan monitoring studies with a clear scientific value for validate models and understand processes, the reliability of the data is crucial. Common analytical problems associated with monitoring programs are:
- Sample collection, conservation, contamination
- Matrix problems

- Extraction procedures. Hydrophilic compounds derivatization
- Improper standardization. Absence of standards for metabolites and degradation products
- Improper use or lack of methods for the analyte
- Chemical form to be determined
- Incorrect reporting of analytical statistics

The solutions to many of the above problems require to intercalibrate methods, develop reference materials [65,66], increase the research of new methods, train the chemical analysts, improve quality assurance/quality control plans.

The school and the book that represents its outcome are not pretending to give a complete treatment of this complex subject. The aim was to offer to the analytical chemist a sense of the enormous challenge facing the chemical oceanographer. Thus it will be possible to avoid that monitoring becomes a goal in itself and to set the conceptual basis on which analytical protocols and campaigns are programmed. Then, it should hopefully emerge the power of environmental analytical chemistry to give a contribution to the understanding, preservation and proper use of the ocean and the Earth as a whole.

ACKNOWLEDGEMENTS

The author thanks MURST and CNR, Progetto Sistema Lagunare Veneziano for support.

REFERENCES

1. U.Michels, *Atlas zur Musik*, Deutscher Taschenbuch Verlag, Munchen, 1977.
2. J.Buffle, H.P.van Leeuwen, *Environ.Sci.Technol.*, 1989, 23, 931.
3. K.S.Johnson, K.H.Coale, H.S.Jannash, *Anal.Chem.*, 1992, 64, 1065A.
4. D.C. Hurd and D.W.Spencer, Eds., *Marine Particles: Analysis and Characterization*, Geophysical Monograph no. 63, American Geophysical Union, Washington DC, 1991.
5. J.Buffle and H.P.Van Leeuwen, Eds., *Environmental Particles*, vol.1 and 2, IUPAC Ser. on Environmental Analytical and Physical Chemistry, Lewis Publ., Chelsea MI, 1992.
6. J.Buffle, G.G. Leppard, *Environ. Sci Technol.*, 1995, 29, 2169, and 2176.
7. J.Buffle, *Complexation Reactions in Aquatic Systems: An Analytical Approach*, Ellis Horwood, Chichester, 1988.
8. E.Pelizzetti, C.Minero, V.Maurino, *Adv.Colloid Interface Sci.*, 1990, 32, 271.
9. J.E.Szecsody, J.M.Zachara, P.L.Bruckhart, *Environ.Sci.Technol.*, 1994, 28, 1706.
10. J.Kukkonen, J.Pellinen, *Sci. Total Environ.*, 1994, 152, 19.

11. G.A.Harkley, P.F.Landrum, S.J.Klaine, *Chemosphere*, 1994, 28, 583.
12. R.M.Burgess, R.A.Mckinney, W.A.Brown, J.G.Quinn, *Environ.Sci.Technol.*, 1996, 30, 1923.
13. J.L.Sarmiento, *Chem.&Eng.News*, 1993, May 31, p. 30.
14. J.L.Sarmiento, E.T.Sundquist, *Nature*, 1992, 356, 589.
15. P.D.Quay, B.Tilbrook, C.S.Wong, *Science*, 1992, 256, 74.
16. K.M.Johnson, J.M.Sieburth, P.J.L.Williams, L.Braendstroem, *Mar.Chem.*, 1987, 21, 117.
17. G.Robert-Baldo, M.J.Morris, R.H.Byrne, *Anal.Chem.*, 1985, 57, 2564.
18. M.D.DeGrandpre, *Anal. Chem.*, 1993, 65, 331.
19. Y.Sugimura., Y.Suzuki, *Mar.Chem.*, 1988, 24, 105.
20. J.H.Martin, S.E.Fitzwater, *Nature*, 1992, 356, 699.
21. R.G.Zepp and Ch.Sonntag, Eds., *The Role of Nonliving Organic Matter in the Earth's Carbon Cycle*, Wiley, New York, 1995.
22. M.Stuiver, P.D.Quay, H.G.Ostlund, *Science*, 1983, 219, 849.
23. R.F.Weiss, J.L.Bullister, R.H.Gammon, M.J.Warner, *Nature*, 1985, 314, 608.
24. S.C.Brassell, G.Eglinton, I.T.Marlowe, U.Pflaumann, M.Sarnthein, *Nature*, 1986, 320, 129.
25. M. O.Andreae, *Mar.Chem.*, 1990, 30, 1.
26. C.Lee, S.G.Wakeham, in *Chemical Oceanography*, J.P.Riley Ed., Academic, London, 1988, vol. 9, p. 1.
27. X.H. Yang, C. Lee, M.I.Seranton, *Anal. Chem.*, 1993, 65, 572.
28. M.O.Andreae, W.Barnard, *Anal.Chem.*, 1983, 55, 608.
29. A.L.Burlingame, D.S.Millington, D.L.Norwood, D.H.Russell, *Anal.Chem.*, 1990, 62, 268R.
30. C.Minero, M.Vincenti, S.Lago, E.Pelizzetti, *Fresenius J.Anal.Chem.*,1994, 350, 403.
31. K.W.Bruland, in *Chemical Oceanography*, J.P.Riley and R.Chester Eds., Academic, London, 1983, vol. 8, p. 157.
32. K.J.Orians, E.A.Boyle, K.W.Bruland, *Nature*, 1990, 348, 322.
33. L.C.Alves, L.A. Allen, R.S. Houk, *Anal. Chem.*, 1993, 65, 2468.
34. H.L.Windom, J.T.Byrd, R.G.Smith, F.Huan, *Environ.Sci.Techmol.*, 1991, 25, 1137.
35. C.M.G.Van Den Berg, in *Chemical Oceanography*, J.P.Riley Ed., Academic, London, 1988, vol. 9, p. 197.
36. J.R.Donat, K.W.Bruland, in *Trace Metals in Natural Waters*, E.Steinnes and B.Salbu Eds., CRC Press, Boca Raton, FL, 1993, ch. 12.
37. P.J. Brendel, G.W. Luther III, *Environ. Sci. Technol.*, 1995, 29, 751.
38. J.A. Resing, C.I. Measures, *Anal.Chem.*, 1994, 66, 4105.
39. J.L.Nowicki, K.S.Johnson, K.H.Coale, V.A.Elrod, S.H. Lieberman, *Anal. Chem.*, 1994, 66, 2732.
40. R. Garcia-Monco Carra, A.Sanchez-Misiego, A. Zirino, *Anal. Chem.*,1995, 67, 4484.
41. H.Craig, T.Hayward, *Science*, 1987, 235, 199.

42. T.Kotiaho, F.R.Lauritsen, T.K.Choudhury, R.G.Cooks, G.T.Tsao, *Anal.Chem.*, 1991, 63, 875.
43. T.M.Kana, C.Darkangelo, M.D.Hunt, J.B.Oldham, G.E.Bennett, J.C.Cornwell, *Anal.Chem.*, 1994, 66, 4166.
44. K.H.Coale, K.W.Bruland, *Limnol.Oceanogr.*, 1985, 30, 22.
45. K.O.Buesseler, J.K.Cochran, M.P.Bacon, H.D.Livingston, S.A.Casso, D.Hirschberg, M.C.Hartman, A.P.Fleer, *Deep-Sea Res.*, 1992, 39, 1103.
46. J.H.Chen, R.L.Edwards, G.J.Waaserburg, *Earth Planet.Sci.Lett.*, 1986, 80, 241.
47. W.H.Glaze, *Environ.Sci.Technol.*, 1993, 27, 419.
48. R.P Eganhouse, *Am.Chem.Soc. Environ. Chem. Div.*, 1996, 36(1), 145.
49. *Wastes in the Oceans*, vol 1-6, Environ. Sci Technol. Ser., Wiley, N.Y.
50. J.Rose, *Environ.Sci.Technol.*, 1993, 27, 1282.
51. J.A .Martin, R.M.Gordon, S.Fitzwater, W.W.Broenkow, *Deep-Sea Res.*, 1989, 36, 649.
52. F.J.Millero, *this book*, p.381
53. A.M.Rouhi, *Chem.&Eng.News*, 1995, Nov. 20, p. 42.
54. M.R.Preston, in *Chemical Oceanography*, J.P.Riley Ed., Academic, London, 1988, vol. 9, p. 53.
55. J.R.Payne, C.R.Philipps, *Environ.Sci.Technol.*, 1985, 19, 569.
56. R.Bongiovanni, E.Borgarello, E.Pelizzetti, *Chim.Ind.(Milan)*, 1989, 71(12), 12.
57. P.J.Craig, *this book*, p.85
58. J.Long, *Chem.&Eng.News*, 1995, November 13, p. 8.
59. *Chemical Measurements Technologies for Ocean Sciences*, National Research Council, Washington DC, 1993.
60. D.M.Karl, C.D.Winn, *Environ.Sci.Technol.*, 1991, 25, 1977.
61. D.W.R.Wallace, C.D.Wirick, *Nature*, 1992, 356, 694.
62. C.Goyet, D.R.Walt, P.G.Brewer, *Deep-Sea Res.*, 1992, 39, 1015.
63. J.Wang, D. Larson, N.Foster, S.Armalis, J.Lu, X.Rongrong, K.Olsen, A. Zirino, *Anal. Chem.*, 1995, 67, 1481.
64. H.W.Jannasch, K.S.Johnson, C.M.Sakamoto, *Anal. Chem.*, 1994, 66, 3352.
65. S.A.Wise, M.M.Schantz, B.A.Benner, M.J.Hays, S.B.Schiller, *Anal. Chem.*, 1995, 67, 1171.
66. R.Morabito, *this book*, p.225

THE EFFECT OF IONIC INTERACTIONS ON THERMODYNAMIC AND KINETIC PROCESSES IN NATURAL WATERS

DR. FRANK J. MILLERO
Rosenstiel School of Marine and Atmospheric Science
4600 Rickenbacker Causeway
Miami, FL 33149
Tel: 1 305 361 4707
Fax: 1 305 361 4144
Email: fmillero@rsmas.miami.edu

Abstract

Ionic interactions can affect the thermodynamic and kinetic behavior of metals in natural waters. A quantitative treatment of these interactions requires an appropriate, self consistent model describing the variation of activity coefficients with ionic strength and composition. One would also like to know the form or speciation of metals in natural waters of different composition. The estimation of the activity coefficients of ions in natural waters can be determined by using an ion pairing model and the specific interaction model. In this paper we will describe a chemical equilibrium model which can be used to characterize metals in natural waters from 0 to 50°C and high ionic strengths (6 m). The model considers the ionic interactions in solutions of the major sea salts (H - Na - K - Mg - Ca - Sr - Cl - Br - OH - HCO_3 - $B(OH)_4$ - HSO_4 - SO_4 - CO_3 - CO_2 - $B(OH)_3$ - H_2O). The estimated activity coefficients and infinite dilution constants have been used to determine the dissociation constants of acids (H_3CO_3, $B(OH)_3$, H_2O, etc.) that can form anions that interact with metals. The speciation and activity coefficients of divalent and trivalent metals have been added to the model. The application of the model is demonstrated by examining the speciation of iron in natural waters and its affect on the reduction of Fe(III) with sulfite in cloud and rain waters.

1. Introduction

Ionic interactions have been shown to affect the thermodynamic and kinetic behavior of metals in natural waters [1,2]. In recent years we have been interested in understanding how ionic interactions affect the redox reactions of metals. These effects can be divided into the interactions of anions (Cl^-, SO_4^{2-}, OH^-, CO_3^{2-}, etc.) on the rates of the reactions of cations and the interactions of cations (H^+, Na^+, Mg^{2+}, etc.) on the rates of the reactions of anions. Examples of the effect of anions on the rates of cations are the formation of ion pairs

$$Cu^+ + Cl^- \rightarrow CuCl^° \qquad (1)$$

$$Fe^{2+} + OH^- \rightarrow FeOH^+ \qquad (2)$$

The formation of $CuCl^°$ causes the rates of oxidation of Cu(I) to decrease [3], while the formation of $FeOH^+$ causes the rates of oxidation of Fe(II) to increase [4]. Examples of the effect of cations on the rates of reactions of anions are

$$Fe^{2+} + H_2S \rightarrow FeHS^+ + H^+ \qquad (3)$$

$$Zn^{2+} + H_2S \rightarrow ZnHS^+ + H^+ \qquad (4)$$

The formation of $FeHS^+$ causes the rates of oxidation of H_2S to increase; while, the formation of $ZnHS^+$ causes the rates to decrease [5].

Our major interest has been on the effects of ionic interactions on the rates of redox reactions. These redox processes occur at oxic-anoxic interfaces and in the surface waters of the oceans due to photochemical processes. The formation of reactive HO_2 and H_2O_2 in surface waters [6] is thought to occur by the following reaction scheme

$$C + h\nu \rightarrow C^* \qquad (5)$$

$$C^* \rightarrow C^+ + e^-(aq) \qquad (6)$$

$$O_2 + e^-(aq) \rightarrow O_2^- \qquad (7)$$

$$O_2^- + 2H^+ \rightarrow H_2O_2 \qquad (8)$$

The O_2^- can react with the oxidized forms of metals to produce reduced species

$$Fe^{3+} + O_2^- \rightarrow Fe^{2+} + O_2 \qquad (9)$$

$$Cu^{2+} + O_2^- \rightarrow Cu^+ + O_2 \qquad (10)$$

The residence time of these reduced metals will be influenced by the rates of oxidation with O_2 and H_2O_2.

A quantitative treatment of these effects requires an appropriate, self consistent model describing the variation of activity coefficients with ionic strength and composition. One would also like to know the form or speciation of a metal in natural waters of different composition since the redox reactions are strongly dependent on the speciation. The estimation of the activity coefficients of ions in natural waters can be determined by using an ion pairing model and the specific interaction model. The ion pairing model can be used to estimate activity coefficients for a number of major and minor ions to 1 m and at 25°C [7]. Extensions to higher ionic strengths and other temperatures is complicated by the requirement of experimental data for the large number of ion pairs - 50 in the case of the major components of seawater. The Pitzer model [8] for the same components requires stability constants for only 6 ion pairs. Stability constants at high ionic strengths and temperatures other than 25°C are not readily available. Reliable extensions to higher ionic strength are, thus, difficult due to our lack of knowledge of the activity coefficients of the ion pairs of various charge types.

The specific interaction model as formulated by Pitzer [8] has made a large impact on our ability to estimate the activity of ionic and nonionic solutes in natural waters. The model was extensively applied to seawater by Harvie et al. [9]. The present model [10,11,12] can be used to make reliable estimates of the activity coefficients of the major components (H^+, Na^+, K^+, Mg^{2+}, Ca^{2+}, Cl^-, OH^-, HSO_4^-, SO_4^{2-}) of natural waters over a wide range of temperatures (0 to 250°C) to high ionic strengths (<6 m). The extension of the model to include the speciation of divalent [13] and trivalent [14] metals with OH^-, HCO_3^-, CO_3^{2-}, $H_2PO_4^-$, and HPO_4^{2-} have been added to the model at 25°C. More recently, Millero and Roy [12] have developed a Pitzer model [8] for the major sea salts that considers the ionic interactions of H - NH_4 - Na - K - Mg - Ca - Sr - Cl - Br - OH - HS - HCO_3 - $B(OH)_4$ - HSO_4 - H_2PO_4 - HPO_4 - SO_4 - CO_3 - PO_4 - CO_2 - $B(OH)_3$ - H_2S - NH_3 - H_2O from 0 to 50°C and ionic strengths from 0 to 6 m. As pointed out by Whitfield [15], the Pitzer model [8] can serve as a solid thermodynamic for the speciation of trace constituents in natural waters over a wide range of ionic strength and composition. We have attempted to use this approach to examine the interactions of the major components of seawater with minor metals [14] and non-metals [16]. The single ion activity coefficients calculated from the specific interaction model are used for the extension of the ion pairing model. The activity coefficients of ion pairs are treated in a similar manner as single ions or uncharged solutes. The specific interaction model accounts for the ionic interactions such as Na-SO_4, Na-Cl, K-Cl, K-SO_4, Ca-Cl, Ca-SO_4, Mg-Cl, Mg-SO_4 which are difficult to characterize experimentally using the ion-pairing concept and usually are the major components of natural waters. The strong cation-anion interactions of trace metals with the major and minor anions can be accounted for by using the ion pairing model. We have attempted to unite these two models over the last few years. Although the combination of the two models has been used to determine activity coefficients of the major components of natural waters, the extension of this approach to the speciation of trace components has not been widely attempted. Millero [14] has recently used the Pitzer model [8] combined with the ion pairing model to examine the speciation of rare earth's in natural waters as a function of ionic strength.

2. Iron as a Test Case for Understanding Ionic Interactions

Iron has been suggested to be a potential factor in limiting phytoplankton production in high nutrients, low chlorophyll areas of the oceans [17,18]. The form or interaction of iron in natural waters with the major and minor ligands affects the thermodynamic solubility, the rates of redox processes, and the interaction of iron with organisms. Although a number of computer codes are available to calculate the speciation of elements in natural waters, they are generally limited because the data base is confined to low ionic strengths. A quantitative treatment of the interactions of iron in natural waters requires an appropriate, self consistent model describing the variation of activity coefficients with ionic strength and composition. The limitations of these calculations are due to the lack of reliable stability constants for the formation of iron complexes valid over a wide range of temperatures and ionic strengths. This makes it difficult to determine the activity coefficient of the ion pairs and complexes as a function of ionic strength and composition. Our recent attempt to circumvent some of these limitations for Fe(III) are discussed below.

3. Thermodynamics of Fe(III) in Natural Waters

Ionic interactions affecting the non ideal behavior of Fe^{3+}, in an electrolyte solution are related to the stoichiometric or total activity coefficients, $\gamma_T(Fe)$, which are related to the activity, a_{Fe}, and stoichiometric or total concentration, $[Fe]_T$, by

$$a_{Fe} = \gamma_T(Fe) [Fe]_T \tag{11}$$

The value of $\gamma_T(Fe)$ is controlled by the composition of the natural water at a given temperature and pressure. It can be estimated in an electrolyte solution by using various extensions of the Debye-Hückel equation. More reliable estimates can be made by using the specific interaction model that uses parameters derived from experimental measurements made at the same ionic strength. Although reliable parameters are available for the major and some minor components of natural waters [8], values are not available for the strong interactions of Fe^{3+} with the minor anions of natural waters (e.g. OH^-, CO_3^{2-}). The strong interactions of Fe^{3+} with minor anions (e.g. Fe^{3+} with CO_3^{2-} and OH^-) can be treated by considering the formation of ion pairs or complexes. When using the ion pairing model the activity of Fe^{3+} is given by

$$a_{Fe} = \gamma_F(Fe) [Fe]_F \tag{12}$$

where $[Fe]_F$ is the free concentration, and $\gamma_F(Fe)$ is the activity coefficient of the free or uncomplexed Fe(III). The value of $\gamma_F(Fe)$ is typically assumed to be only a function of the ionic strength and independent of the relative composition. However, if the Pitzer model [8] is used to estimate $\gamma_F(Fe)$, one can account for the influence of medium composition on $\gamma_F(Fe)$. The activity of Fe can be related to the total and free Fe by

$$a_{Fe} = \gamma_T(Fe) [Fe]_T = \gamma_F(Fe) [Fe]_F \qquad (13)$$

which gives the expression relating the two activity coefficients

$$\gamma_T(Fe) = \{ [Fe]_F / [Fe]_T \} \gamma_F(Fe) \qquad (14)$$

If the values of $\gamma_F(Fe)$ are determined for the major components of the solution (e.g., Na-Mg-Cl-SO$_4$) using the Pitzer equation [8], the ion pairing model can be used to account for strong interactions between Fe^{3+} and minor anions (OH^-, HCO_3^-, CO_3^{2-}, and PO_4^{3-}).

The formation of an ion pair or complex between a metal (Fe^{3+}) and a ligand (CO_3^{2-}) can be characterized by

$$Fe^{3+} + CO_3^{2-} \rightarrow FeCO_3^+ \qquad (15)$$

The stoichiometric stability constant, K^*_{FeCO3}, for the formation of this ion pair is given by

$$K^*_{FeCO3} = K_{FeCO3} \gamma_{Fe} \gamma_{CO3} / \gamma_{FeCO3} \qquad (16)$$

where K_{FeCO3} is the thermodynamic constant in pure water and the values of γ_i are the activity coefficients of the ions and the ion pair. The fraction of the free Fe is given by

$$[Fe]_F / [Fe]_T = (1 + \sum K^*_{FeXi} [X_i]_F)^{-1} \qquad (17)$$

Similarly, the fraction of a free anion X is given by

$$[X]_F / [X]_T = (1 + \sum K^*_{MiX} [M_i]_F)^{-1} \qquad (18)$$

where the summations are made over all the complexes of each cation M_i and anion X_i in the solution (including the major components, Mg^{2+}, Ca^{2+}, Cl^-, and SO_4^{2-} when appropriate). The fraction of a given ion pair can be determined from

$$[FeX]/[Fe]_T = K^*_{FeXi} [X_i]_F / (1 + \sum K^*_{FeXi} [X_i]_F) \qquad (19)$$

$$[MX]/[X]_T = K^*_{MX} [M]_F / (1 + \sum K^*_{MiX} [M_i]_F) \qquad (20)$$

To solve Eqs. (17) to (20) and determine the speciation of ions in a natural water one needs to know the values of K^*_{MX} over a wide range of ionic strengths in different ionic media.

The value of K^*_{MX} in a given natural water can be estimated by using an experimental value measured in an ionic medium like NaCl and NaClO$_4$ at the same ionic strength of the natural water or by estimating the activity coefficients of ions and ion pairs in the natural water. The first method is convenient to use if the experimental data are available, but it has some limitations [7]. The reliability of the literature values of K^*_{MX} at a given ionic strength may be questionable due to the methods used in their determination. The second method requires thermodynamic constants that have been extrapolated to infinite dilution by the same method used to estimate the activity coefficients. Frequently infinite dilution constants in the literature have been extrapolated by methods which do not give reliable activity coefficients above 0.1 m [7]. Reinterpretation of thermodynamic constants in older literature is often needed [14] using up to date theory [8]. For the major components of natural waters Pitzer equations [8] are available to calculate reliable total or stoichiometric activity coefficients from 0 to 50°C and 0 to 6 m [12]. Unfortunately, the experimental data necessary to extend this treatment to trace metal ions are not always available. It is possible, however, to use the limited measurements of stability constants in various media to extend the Pitzer [8] ionic interaction model to divalent [13] and trivalent metals [14]. This method has an advantage in that differences in the specific interactions occurring in an ionic medium can be considered in a manner consistent with the major components of natural waters. The application of the Pitzer model [8] to determine the activity coefficients and the speciation of Fe(III) in natural waters [19] is described in the next section.

4. Construction of the Model

Figure 1 shows the main steps used to calculate the speciation of trace metals for natural water. First, the major cations and anions of the medium are inputted to determine the ionic strength and the total concentration of the ligands in the solution that can interact with Fe. The activity coefficients of the ions and ion pairs are then estimated for the given medium using Pitzer's equations. The dissociation and association constants in the medium (pK^*_{HX} and pK^*_{MX}) are determined from the estimated activity coefficients (γ_M, γ_X, γ_{MX}, γ_{HX}) and the thermodynamic stability constants (pK_{HX} and pK_{MX}). Next, the pH and total concentration of minor ligands (such as H$_2$S and organic ligands) are input and the free ligand concentrations $[X]_F$ are determined (those not interacting with the major cations). Finally, the speciation of a metal is determined

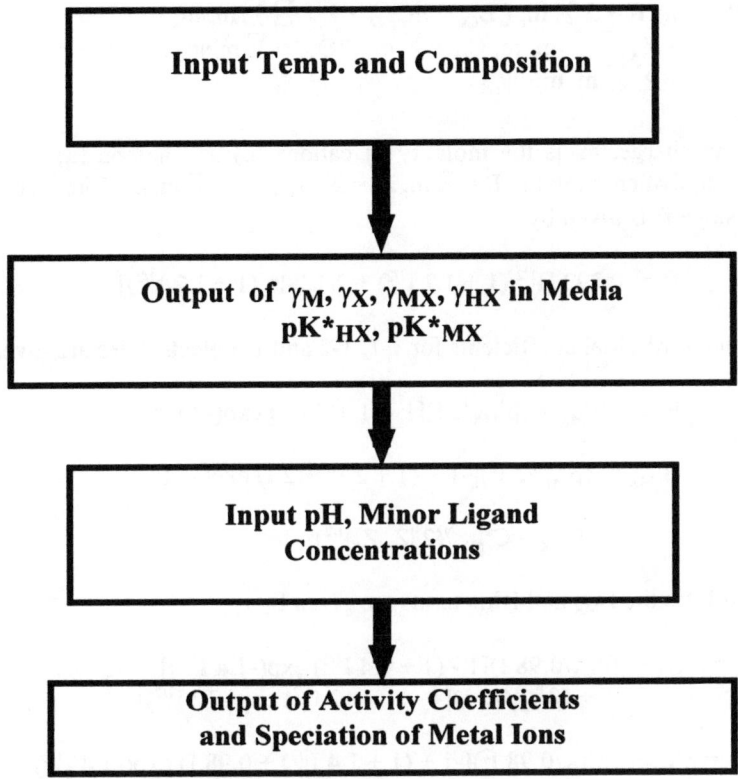

Figure 1. Sketch of speciation model used for trace metals in natural waters.

(fraction of free $[M]_F$ and $[MX_i]$) from the values of $[X]_F$ and pK^*_{MX}. This latter calculation can be made without iterations if the concentration of the ligand is significantly greater than the trace metal (e.g., CO_3^{2-}). A brief discussion of the Pitzer equations used to estimate the activity coefficients of ions is given in the next section.

5. Pitzer's Equations and Activity Coefficients of Ions

Pitzer [8] has described the general equations that can be used to estimate the activity coefficients of electrolytes (MX) or ions (M, X) in mixed salt solutions. In their simplest form [14,20], the activity coefficients of cation M and anion X are given by

$$\ln \gamma_M = Z_M^2 f^\gamma + 2 \sum m_a (B_{Ma} + E\, C_{Ma}) + Z_M^2 \sum \sum m_c m_a B'_{ca}$$
$$+ Z_M \sum \sum m_c m_a C_{ca} + \sum m_c (2\Theta_{Mc} + \sum m_a \psi_{Mca})$$
$$+ \sum \sum m_a m_{a'} \psi_{aa'M} \quad (21)$$

$$\ln \gamma_X = Z_X^2 f^\gamma + 2 \sum m_c (B_{cX} + EC_{cX}) + Z_X^2 \sum \sum m_c m_a B'_{ca}$$
$$+ Z_X \sum \sum m_c m_a C_{ca} + \sum m_a (2\Theta_{Xa} + \sum m_c \psi_{Xca})$$
$$+ \sum \sum m_c m_{c'} \psi_{cc'X} \qquad (22)$$

where Z_i is the charge, m_i is the molality of cations (c) and anions (a) in the mixed solution. The equivalent molality $E = \sum m_c Z_c = \sum m_a Z_a = \frac{1}{2} \sum m_i Z_i$. The Debye-Hückel limiting law slope f^γ is given by

$$f^\gamma = -0.392 \left[I^{1/2}/(1 + 1.2 I^{1/2}) + 2/1.2 \ln(1 + 1.2 I^{1/2}) \right] \qquad (23)$$

The second and third virial coefficients for 1-1, 1-2 and 1-3 electrolytes are given by

$$B_{MX} = \beta^0_{MX} + (\beta^1_{MX}/2 I)[1 - (1 + 2 I^{1/2}) \exp(-2 I^{1/2})] \qquad (24)$$

$$B'_{MX} = (\beta^1_{MX}/2 I^2)[-1 + (1 + 2 I^{1/2} + 2 I) \exp(-2 I^{1/2})] \qquad (25)$$

$$C_{MX} = C^\phi_{MX} / (2 |Z_M Z_X|^{1/2}) \qquad (26)$$

For 2-2 electrolytes the B_{MX} and B'_{MX} terms are given by

$$B_{MX} = \beta^0_{MX} + (\beta^1_{MX}/0.98 I)[1 - (1 + 1.4 I^{1/2}) \exp(-1.4 I^{1/2})]$$
$$+ (\beta^2_{MX}/72 I)[1 - (1 + 12 I^{1/2}) \exp(-12 I^{1/2})] \qquad (27)$$

$$B'_{MX} = \beta^1_{MX} + (\beta^1_{MX}/0.98 I^2)[-1 + (1 + 1.4 I^{1/2}) + 0.98 I) \exp(-1.4 I^{1/2})]$$
$$+ (\beta^2_{MX}/72 I^2)[-1 + (1 + 12 I^{1/2} + 72 I) \exp(-12 I^{1/2})] \qquad (28)$$

Values of β^0_{MX}, β^1_{MX}, β^2_{MX}, and C^ϕ_{MX} are tabulated elsewhere [8,20]. The term Θ_{ij} is related to the interactions of similarly charged ions (Na^+-Mg^{2+}). The term Ψ_{ijk} is related to the triple ionic interactions of two similarly charged ions with an ion of opposite charge (Na^+-Mg^{2+}-Cl^-). These mixing parameters are determined from mixtures of two electrolytes with a common ion.

The Pitzer equations described above can be used to estimate the activity coefficients of the major components of seawater [12]. For those divalent and trivalent metals where Pitzer coefficients [8] are available for chlorides and sulfates [13,14], the equations can also be used to estimate the "free" trace metal activity coefficients needed in the ion pairing model. The dissociation constants for various acids can also be estimated using these activity coefficients and thermodynamic constants in pure water.

The measured stability constants (K^*_{FeX}) for the formation of Fe(III) complexes can be used to estimate the activity coefficient of the ion pairs (γ_{FeX}) using

$$\ln(K^*_{FeX}/\gamma_{Fe}\gamma_X) = \ln K_{FeX} - \ln(\gamma_{FeX}) \qquad (29)$$

where γ_{Fe} and γ_X are the activity coefficients of Fe and X and K_{FeX} is the thermodynamic constant [14]. In the earlier paper by Millero and Hawke [13] the values of γ_M and γ_X for divalent metals were estimated from the values of the ions in pure chloride (MCl_2) or sodium solutions (NaX). More reliable values of γ_M and γ_X can be determined using the Pitzer equations [14]. Since the stability constants are normally determined in a simple medium like $NaClO_4$ or NaCl, the values of γ_{Fe} and γ_X can be determined from

$$\ln \gamma_{Fe} = Z^2_{Fe} f' + 2 m_{Cl} (B_{FeCl2} + m_{Cl} C_{FeCl2}) + m^2_{Cl} (Z^2_{Fe} B'_{NaCl} + Z_{Fe} C_{NaCl}) + m_{Na} (2\Theta_{FeNa} + m_{Cl} \Psi_{FeNaCl}) \tag{30}$$

$$\ln \gamma_X = Z^2_X f' + 2 m_{Na} (B_{NaX} + m_{Na} C_{NaX}) + m^2_{Cl} (Z^2_X B'_{NaCl} + Z_X C_{NaCl}) + m_{Cl} (2\Theta_{XCl} + m_{Cl} \Psi_{XClNa}) \tag{31}$$

The values of $\ln \gamma_{FeX}$ for the ion pairs can be determined from

$$\ln \gamma_{FeX} = Z_{FeX}^2 f' + 2 m_{Cl} (B_{FeX,Cl} + m_{Cl} C_{FeX,Cl}) + m^2_{Cl} (Z^2_{FeX} B'_{NaCl} + Z_{FeX} C_{NaCl}) \tag{32}$$

For an uncharged ion pair equation (32) simplifies to

$$\ln \gamma_{FeX} = 2 m_X B_{FeX} = 2 (m_{Na} \lambda_{Na, FeX} + m_{Cl} \lambda_{Cl, FeX}) \tag{33}$$

where the values of λ_{ij} are parameters that account for the interactions of neutral species i with ions j [8]. Pitzer coefficients [8] (B_{FeX}, C_{FeX}) for Fe(III) ion pairs and the thermodynamic constants (K_{FeX}) can be estimated by using equations (29)-(33) from the values of K^*_{FeX} at different ionic strengths. The choice of which parameters (β^0_{FeX}, β^1_{FeX}, and C^ϕ_{FeX}) that are needed to fit the experimental values of K^*_{FeX} is evaluated using an F test. When limited data is available only the value of β^0_{FeX} can be determined and if no concentration data is available one must estimate the value of β^0_{FeX} using an ion pair of similar charge [13,14,19].

6. Stability Constants for the formation of Fe(III) Complexes

The thermodynamic stability constants for the formation of Fe(III) ion pairs are tabulated elsewhere [19]. The data is quite limited and there is a need for more reliable measurements in various ionic media ($NaClO_4$ and NaCl) over a wide range of ionic strengths (I = 0.1 to 6m) and temperature (0 to 50°C). When new experimental data becomes available the Pitzer parameters [8] for Fe(III) ion pairs and salts can be recalculated. From the measurements in $NaClO_4$ solutions as a function of ionic strength estimates can be made for the activity coefficients of ion pairs. Since the Pitzer coefficients [8] for Fe(III) salts are not available, it is not possible to determine the activity coefficients of Fe(III) in a $NaClO_4$ or NaCl medium. Recently Zhu et al. [21] used the activity coefficient of Ga^{3+} in NaCl as an estimate of the $\gamma(Fe^{3+})$. In our recent paper [19], we have taken a different approach. Since the hydrolysis constant for the formation of $FeOH^{2+}$ are known in $NaClO_4$ solutions over a wide range of ionic strength (**Figure 2**), we have estimated the Pitzer coefficients [8] for Fe^{3+} from this data.

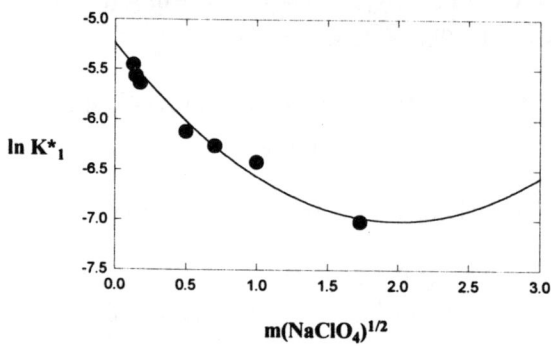

Figure 2. The hydrolysis constants for the formation of $FeOH^{2+}$ in $NaClO_4$ solutions at 25 °C [19].

The experimental values of K^*_1 for the first hydrolysis of Fe(III)

$$Fe^{3+} + H_2O \rightarrow FeOH^{2+} + H^+ \tag{34}$$

are given by

$$K^*_1 = K_1 \gamma_{Fe} a_{H2O}/(\gamma_{FeOH} \gamma_H) \tag{35}$$

where a_{H2O} is the activity of water in $NaClO_4$ [14]. Rearranging this equation we have

$$\ln \{K^*_1 \gamma_{FeOH} \gamma_H / a_{H2O}\} = \ln K_1 + \ln \gamma_{Fe} \tag{36}$$

The values of γ_{FeOH} needed to evaluate K_1 and γ_{Fe} have been estimated from the values of γ_{LnF} complexes [14]. Since the Pitzer parameters for the rare earth ion pairs of the same charge type are equal and $\gamma_{LnOH} \cong \gamma_{LnF}$, this is a reasonable assumption to use to approximate γ_{FeOH}. The Pitzer coefficients for Fe^{3+} and FeX ion pairs are given elsewhere [19]. The value of γ_{Fe} in 0.7 m $NaClO_4$ (0.08) is much lower than the value for other trivalent ions ($\gamma_{Al} = 0.23$, $\gamma_{La} = 0.22$, $\gamma_{Ga} = 0.24$). The activity coefficient of Fe^{3+} in seawater is assumed to be equal to the value in $NaClO_4$ at the same ionic strength and Fe^{3+} is considered to form F^-, Cl^-, OH^-, and SO_4^{2-} ion pairs. The speciation of Fe(III) as a function of pH determined from the model is shown in **Figure 3**. Over most of the pH range of natural water the hydrolyzed species $FeOH^{2+}$, $Fe(OH)^+$, and $Fe(OH)_3$ dominate. The reliability of the model can be demonstrated by examining the solubility of Fe(III) in seawater [22,23,24] and is discussed in the next section.

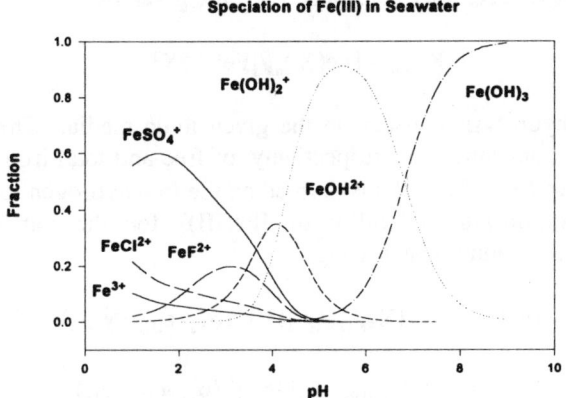

Figure 3. The speciation of Fe(III) in seawater (S = 35) as a function of pH [19].

7. Solubility of Fe(III) in Seawater

The solubility of Fe(III) of various solid phases in seawater can be estimated from the calculated speciation of seawater [19,21]. The solubilities of the minerals Fe_2O_3 (s), $Fe(OH)_3$ (s) and FeO(OH) (s) as a function of pH are given by:

$$Fe_2O_3 \ (s) + 6H^+ \rightarrow 2Fe^{3+} + 3H_2O \tag{37}$$

$$Fe(OH)_3 \ (s) + 3H^{3+} \rightarrow Fe^{3+} + 3H_2O \tag{38}$$

$$FeOOH \ (s) + 3H^+ \rightarrow Fe^{3+} + 2H_2O \tag{39}$$

At equilibrium, the thermodynamic equilibrium constants are given by:

$$K_{Fe2O3(s)} = a_{Fe}^2 \ a_{H2O}^3 / a_H^6 \tag{40}$$

$$K_{Fe(OH)3(s)} = a_{Fe} \ a_{H2O}^3 / a_H^3 \tag{41}$$

$$K_{FeOOH(s)} = a_{Fe} \ a_{H2O}^2 / a_H^3 \tag{42}$$

where $a_i = [i] \gamma_i$ are the activities of species i. If one uses the ion pairing model, the value of a_{Fe} is given by:

$$a_{Fe} = \gamma_{Fe} [Fe^{3+}] = \gamma_{Fe} [Fe(III)]/(1 + \sum K^*_{x,n} [X]^n + \sum K^*_j [H^+]^{-n}) \quad (43)$$

where the cumulative stability constants $K^*_{x,n}$ are given by:

$$K^*_{x,n} = [Fe(X_i)_n]/[Fe^{3+}] [X]^n \quad (44)$$

and K^*_j is the hydrolysis constant in the given ionic media. The values of $[Fe^{3+}]$ and $[Fe(III)]$ are the concentrations, respectively, of free and total iron and γ_{Fe} is the activity coefficient of free Fe^{3+}. The $[H^+]$ is defined on the free hydrogen ion molality scale.

The equilibrium solubility of $[Fe(III)]$ for the various minerals can be determined for any natural water using

$$[Fe(III)] = \{K_{Fe2O} \gamma_H^6 [H^+]^6/(\alpha_{Fe}^2 a_{H2O}^3 \gamma_{Fe}^2)\}^{0.5} \quad (45)$$

$$[Fe(III)] = K_{Fe(OH)3} \gamma_H^3 [H^+]^3/(\alpha_{Fe} a_{H2O}^3 \gamma_{Fe}) \quad (46)$$

$$[Fe(III)] = K_{FeO(OH)} \gamma_H^3 [H^+]^3/(\alpha_{Fe} a_{H2O}^2 \gamma_{Fe}) \quad (47)$$

At a fixed ionic strength and temperature the solubility of the minerals as a function of pH can be determined from

$$[Fe(III)] = \{K^*_{Fe2O3} [H^+]^6/\alpha_{Fe}^2\}^{0.5} \quad (48)$$

$$[Fe(III)] = K^*_{Fe(OH)3} [H^+]^3/\alpha_{Fe} \quad (49)$$

$$[Fe(III)] = K^*_{FeO(OH)} [H^+]^3/\alpha_{Fe} \quad (50)$$

where $K^*_{Fe2O3} = K_{Fe2O3} \gamma_H^6/(a_{H2O}^3 \gamma_{Fe}^2)$, $K^*_{Fe(OH)3} = K_{Fe(OH)3} \gamma_H^3 /a_{H2O}^3$, and $K^*_{FeO(OH)} = K_{FeO(OH)} \gamma_H^3 /a_{H2O}^2$ are the stability constants for the media. The fraction of free Fe in the solutions can be calculated from

$$\alpha_{Fe(III)} = [Fe^{3+}]/[Fe(III)] = 1/(1 + \sum K^*_{x,n} [X]^n + \sum K^*_j [H^+]^{-n}) \quad (51)$$

The stability and hydrolysis constants at a given ionic strength in seawater can be determined from the equations given by Millero et al. [19]. The reliability of the model can be demonstrated by calculations of the solubility of Fe(III) in seawater.

The solubility of $Fe(OH)_3(s)$ has been made in seawater by Byrne and Kester [22] and Kuma et al. [23]. The values of the log [Fe(III)] determined in these studies are shown as a function of pH in **Figure 4**. Byrne and Kester [22] used four methods to separate the solid and liquid phases: 1) Filtration through a 0.05 μm filter; 2) Ultrafiltration (150 to 350 D cutoff), 3) Acidification followed by filtration thorough a 0.05 μm filter, and 4) Dialysis rates. Kuma et al. [23], also, determined the solubility by dialysis rate measurements. From a pH of 3.3 to 7.0 the 0.05 μm filtration and the Ultrafiltration method gave similar results (open circles). The dialysis results of Kuma et al. [23] from pH = 5.5 to 8.06 (solid circles) are in good agreement with the Byrne and Kester [22] measurements determined by dialysis and acidification followed by filtration (open triangles). The measured solubilities between a pH of 7 to 9 determined by dialysis (solid circles and open triangles) are higher than the ultrafiltration measurements (open circles).

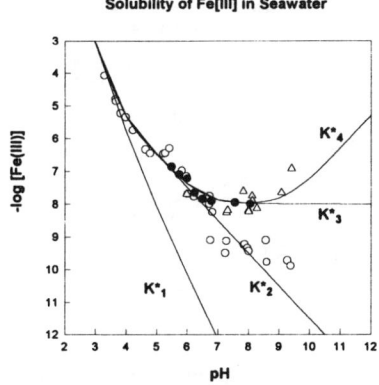

Figure 4. The solubility of Fe(III) in seawater as a function of pH determined by Byrne and Kester [22] and Kuma et al. [23]. The curves are the calculated solubilities using successively the K^*_1, K^*_2, K^*_3, and K^*_4 hydrolysis constants [19].

The solubility product for $Fe(OH)_3(s)$ at given salinity in seawater (I = 0.72 M) can be estimated from equation (52) where $K^*_{Fe(OH)_3}(s) = 10^{4.5}$ is the solubility of $Fe(OH)_3(s)$ [19]. The fraction of free Fe^{3+} and solubility in seawater at a given salinity is only affected by the hydrolysis of Fe^{3+} (**Figure 3**). To fit the dialysis results [22,23], one needs to consider the formation of $FeOH^{2+}$, $Fe(OH)_2^+$ and $Fe(OH)_3$ (K^*_1, K^*_2, and K^*_3). The filtration (0.05 μm) solubilities of Byrne and Kester [22], however, can be predicted by only considering the formation of $FeOH^{2+}$ and $Fe(OH)_2^+$ (K^*_1 and K^*_2). The differences are quite significant with changes in the solubility of Fe(III) going from 10 nM using the dialysis experiments to 0.3 nM for the filtration results at a pH = 8. More recently Kuma et al. [24] have measured the solubility of Fe(III) in open ocean and coastal waters as a function of pH with and without UV irradiation and after filtration through a 0.025 μm filter. These results are discussed below.

8. Effect of Organic Ligands on the Solubility of Fe(III) in Seawater

The solubility of Fe(OH)$_3$(s) without UV irradiation [24] after 5 weeks of equilibration are shown in **Figure 5**. The solubilities in the coastal waters (1.7 nM at pH = 8.16) are much higher than in open-ocean waters (0.6 nM at pH = 8.16). This effect has been attributed to higher concentrations of organic ligands in the coastal waters. The model calculations for seawater (using K^*_1, K^*_2) are in better agreement with the coastal values than the open ocean values. By adding a value of K^*_3, it is possible to get a reasonable fit of the coastal data; while, the open ocean data requires changes K^*_2 and K^*_3 (see **Figure 5**). These results, if correct, would indicate that the organic ligands present in the coastal waters are reacting largely with the neutral ion pair (Fe(OH)$_3$).

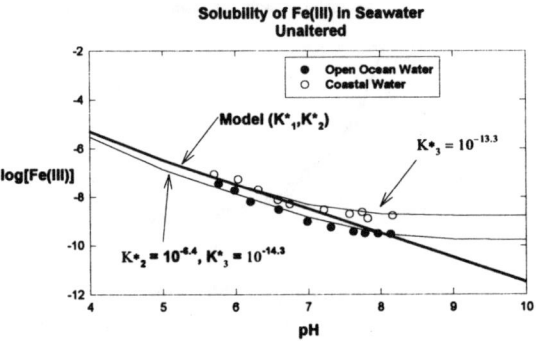

Figure 5. The solubility of Fe(III) in coastal and open ocean waters unaltered as a function of pH [24].

Since the reactive portion of the organic ligand may be depend on pH, this may also be due to changes in the ligand strength. The results obtained by Kuma et al. [24] for UV irradiated coastal and open ocean waters are shown in **Figure 6**. The UV irradiation lowers the solubility in the coastal waters at the higher pH (1.7 nM to 0.6 nM at pH = 8.16); but, has a smaller effect on the open ocean values (0.28 nM to 0.19 nM at pH = 8.16). The UV irradiated coastal waters still have a higher solubility than the open ocean values at the higher pH. This could be due to the incomplete destruction of the coastal organics by UV irradiation.

Recently a number of workers [25, 26, 27, 28] have used voltammetry to examine the concentration and strength of natural organic ligands to complex Fe(III) in seawater. They have found ligand concentrations of [L] = 0.4 - 13 nM and apparent stability constants of $K'_{FeL} = 10^{19}$ to 10^{23} defined by

$$K'_{FeL} = [FeL]/[Fe^{3+}]_F[L] \tag{52}$$

where [L] is the total ligand not complexed by Fe^{3+}. The effect of an organic ligand on the speciation of Fe(III) can be determined from by adding K'_{FeL} [L] terms to the

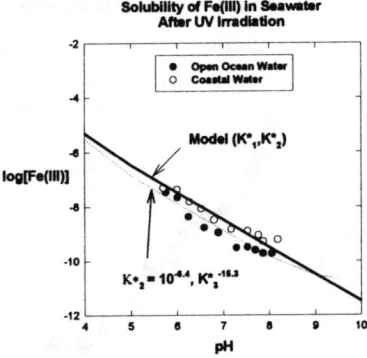

Figure 6. The solubility of Fe(III) in UV irradiated coastal and open ocean waters as a function of pH [24].

inorganic terms given in equation (51) depending upon the number of organic ligands present

$$\alpha_{Fe(III)} = 1/(1 + \sum K^*_{FeX} [X^{n-}]^m_F + \sum K'_{FeL} [L]) \quad (53)$$

and its affect on the solubility can be determined using equation (46) with and without the organic ligand by a series of iterations. The fraction of total Fe(III) complexed with the organic ligand in the solution can be determined from

$$\alpha_{FeL} = (\sum K'_{FeL} [L]) \, \alpha_{Fe(III)} \quad (54)$$

The recent measurements of [L] and K'_{FeL} give increases in the solubility of 10 to 80% (when [L] = [Fe] = 0.2 to 2 nM $K'_{FeL} = 10^{20}$] which is comparable to the decrease found after UV irradiation of the open ocean (32%) and coastal waters (65%).

The new measurements of Kuma et al. [24] for UV irradiated coastal waters are in good agreement (see **Figure 7**) with the earlier measurements of Byrne and Kester [22] and can adequately represented by our speciation model by including the formation of $FeOH^{2+}$ and $Fe(OH)_2^+$ (K^*_1 and K^*_2). The UV irradiated and non UV irradiated open ocean waters solubility measurements (0.2 to 0.3 nM) of Kuma et al. [24] are lower than the model as well as the earlier filtration measurements (0.6 nM) on Sargasso seawater by Byrne and Kester [22]. At present it is not possible to be sure of the cause of these differences (incomplete oxidation of the organic ligands by the UV technique or differences in stabilization of the iron oxides or colloids in the two solutions). Although further measurements are needed to clear up these apparent discrepancies, the new measurements of Kuma et al. [24] and the methods developed to examine the speciation of Fe(III) with natural organic ligands opens up a way to determined the solubility of Fe

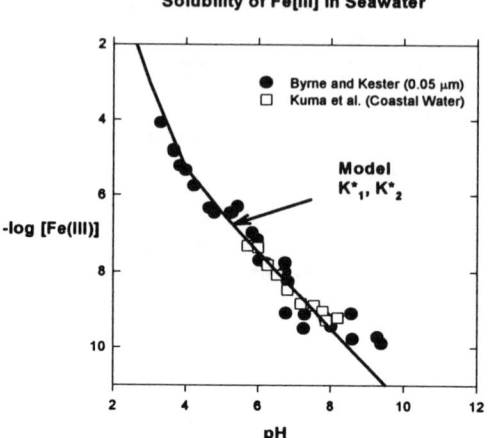

Figure 7. The solubility of Fe(III) as a function of pH measured by filtration in seawater [22,23] compared to model calculations considering the formation of FeOH^{2+} and Fe(OH)$_2^+$ [19]

in seawater and other natural waters such as marine aerosols. Future measurements of ocean waters (coastal to open ocean) over a wide range of pH (2-9) are needed along with voltammetric measurements of the formation of Fe organic complexes.

The model calculations for the solubility at pH = 8.0 (free scale) considering only the formation of FeOH^{2+} and Fe(OH)$_2^+$ [19] yield a solubility of Fe(III) of 0.3 nM in seawater at 25°C. These results are in good agreement with measurements in open ocean surface waters (0.2 nM) and deep waters (0.6 nM) determined by various workers [17, 24-31].

9. Reduction of Fe(III) with Sulfite in Natural Waters

Iron is essential for the growth of phytoplankton in the oceans [17,18,30-34]. In several regions of the ocean, the deposition of atmospheric mineral aerosols may be an important source of iron to surface waters where phytoplankton growth may be limited by the availability of iron [33]. The availability of iron from aerosols may be a function of the solubility and oxidation state. The iron in aerosols and water droplets (clouds, fog, or rain waters) is also involved in a number of chain and redox reactions which influence the chemistry of other important atmospheric species [35,36]. In most atmospheric studies of the solubility of iron, it has been assumed [21] that the element was present as Fe(III). Fe(III) as an oxide, the most common mineral phase, is not very soluble in aqueous solutions. Fe(II) is thermodynamically and kinetically unstable in aerosols and water droplets with O$_2$ and H$_2$O$_2$ [4,37]. Thus, one would expect that any Fe(II) produced in these waters would be rapidly converted to Fe(III). Recent work suggests that Fe(II) could in fact be quite high (20-90% of total iron) in fog water [38] and marine aerosols [39]. The recent work by Zhu et al. [40], however, indicates that the concentrations of Fe(II) are low (7.5% of total iron) in marine aerosols.

The production of Fe(II) in aerosols and water droplets is thought to be due to the photoreduction of inorganic and organic complexes of Fe(III) [40]. This photoreduction is a function of light, pH and the concentration of inorganic and organic ligands in the aqueous solutions. Recent workers [39] have suggested that the reduction of Fe(III) with sulfite, present in most marine aerosols and water droplets, could be important in the production of Fe(II). Reddy et al. [41] also suggested that the sulfite induced autoxidation of Fe(II) to Fe(III) is an important step in the cycling of iron in water droplets.

A number of studies have been made on the effect of Fe(III) on the autoxidation of S(IV) in natural waters due to the interest in the formation of acid rain [see, 42]. The process in acidic solutions is thought to be initiated by the formation of iron sulfite complexes. Recent work [43] indicates that the complexes formed are:

$$Fe^{3+} + HSO_3^- \Leftrightarrow Fe\text{-}SO_3H^{2+} \tag{55}$$

$$FeOH^{2+} + HSO_3^- \Leftrightarrow HOFe\text{-}SO_3H^+ \tag{56}$$

The conditional stability constants for the formation of these iron sulfite complexes (K_{app}) have recently been determined using spectrophotometric techniques. Values of K_{FeSO3H} = 55 M^{-1} and $K_{FeOHSO3}$ = 850 M^{-1} were found by evaluating the pH dependence of the spectra at 400 nm. The reduction of Fe(III) and oxidation of sulfite occurs as an electron transfer reaction that results in the formation of Fe(II) and the sulfite free radical

$$FeOH^{2+} + HSO_3^- \Leftrightarrow HOFe\text{-}SO_3H^{2+} \tag{57}$$

$$HOFe\text{-}SO_3H^+ \xrightarrow{k} Fe(OH)^+ + \cdot SO_3H \tag{58}$$

where $HOFe\text{-}SO_3H^+$ is an inner sphere complex that may undergo a rearrangement to form $HOFe\text{-}OSO_2H$.

The proposed free radical mechanism for the further oxidation of sulfite to sulfate with and without oxygen has been discussed in detail by many workers [see, 42]. The rates of the reduction of Fe(III) and oxidation of S(IV) without oxygen have been studied by a number of workers [44]. Most of these studies have emphasized the oxidation of S(IV) and were made at high concentrations of reactants in a perchlorate media. The back oxidation of Fe(II) by sulfur species to form Fe(III) has also been studied [41]. A limitation of most of these past studies is that the concentrations of Fe(III) were at levels above the solubility of iron at values of pH over 3.0.

Below we present our results [42] for the reduction of nanomolar levels of Fe(III) with S(IV) without oxygen as a function of ionic strength (0 to 6 m), temperature (0 to 40°C), pH (2 to 6.8), and composition (Mg^{2+}, Ca^{2+}, F^-, Br^-, and SO_4^{2-}). By making these measurements at concentrations of Fe(III), which are more representative of the levels in natural waters, we have been able to stay below the solubility limits to pH 6.0. The rates of reduction of Fe(III) with S(IV) can be represented by

$$Fe(III) + S(IV) \xrightarrow{k} Products \tag{59}$$

with the overall rate equation was found to be given by

$$d[Fe(III)]/dt = -k\,[Fe(III)]\,[S(IV)] \tag{60}$$

which is first order with respect to Fe(III) and S(IV). The rate constant was determined by measuring the appearance of Fe(II) using a stopped flow chemiluminescence technique [45, 46]. The disappearance of total Fe(III) ($[Fe(III)]_T = [Fe(III)]_0 - [Fe(II)]$), where the subscript zero denotes the initial concentration) was determined as a function of time under pseudo first order conditions (an excess of S(IV)). The first order rate constant for the disappearance of Fe(III) under these conditions is given by

$$d[Fe(III)]/dt = -k'\,[Fe(III)] \tag{61}$$

where $k' = k\,[S(IV)]$. The effect of ionic strength on the rate of reduction of Fe(III) by S(IV) has been determined in NaCl and seawater at 25°C and pH = 3.5. The results [42] for the overall rate constant $k = k'/[S(IV)]$ are shown plotted versus ionic strength in **Figure 8**. The results in NaCl and seawater have been fitted to the equation

$$\log k = \log k^0 + A\,I^{1/2}/(1 + I^{1/2}) \tag{62}$$

where the value in pure water is $\log k^0 = 4.18 \pm 0.09$ and where $A = -1.1 \pm 0.2$ and -2.2 ± 0.3, in NaCl and seawater, respectively.

The lower rates for the reduction of Fe(III) in seawater were examined by measuring the rates in NaCl solutions with various amounts of seasalts equivalent to their concentrations in seawater ($Na^+ = 0.56M$, $Mg^{2+} = 0.053M$, $Ca^{2+} = 0.01M$, $K^+ = 9$ mM, $Cl^- = 0.6M$, $SO_4^{2-} = 0.028$ M, $Br^- = 800$ μM, $HCO_3^- = 2.2$ mM, and $F^- = 70$ μM). The results [42] of $\Delta \ln k = \ln k(\text{media}) - \ln k(\text{NaCl})$ are shown in **Figure 9**. The addition of F^-, SO_4^{2-}, Mg^{2+}, and Ca^{2+} cause the rate to decrease; while, the addition of K^+ and Br^- cause the rate to increase slightly. The addition of the major seasalts (Na^+, Mg^{2+}, Cl^-, and SO_4^{2-}) yields a relative rate (-0.60) that is higher than the value in seawater (-1.05) or artificial

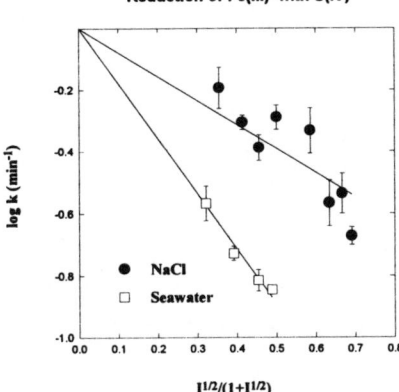

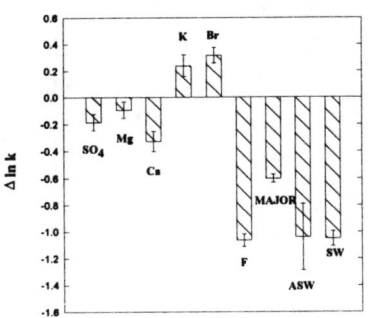

Figure 8. The effect of ionic strength (I) on the rates of reduction of Fe(III) by S(IV) in NaCl and seawater at 25 °C and pH = 3.5 [42].

Figure 9. The effect of composition ($\Delta \ln k = \ln k(\text{soln}) - \ln k(\text{NaCl})$) on the rates of reduction of Fe(III) by S(IV) in NaCl at pH = 3.5 with various sea salts at 25 °C [42].

seawater (-1.04) with all the major components (Na^+, K^+, Mg^{2+}, Ca^{2+}, F^-, Cl^-, Br^-, HCO_3^-, and SO_4^{2-}). The addition of boric acid or deletion of HCO_3^- (not shown) had no effect on the rate. The addition of F^- to NaCl yielded a relative rate (-1.06) that was in good agreement with the seawater results. These experiments show that F^- is the most important anion in seawater that causes the rate to be lower than in NaCl at the same ionic strength. The decrease due to the addtion of SO_4^{2-} Mg^{2+}, Ca^{2+} (-0.62) is nearly balanced by the increase due to the addiltion of K^+ and Br^- (0.53).

The effects of pH on the values of ln k in seawater were also determined and the results [42] are shown in **Figure 10**. The values increase from a pH = 2.0 to a maximum near pH = 4.0 and decrease at higher pH. Our rate constants for the reduction of Fe(III) by S(IV) are strong functions of pH and the solution composition. To analyze the effects of composition on the reaction rate, it is necessary to be able to determine the speciation of the reacting species. This can be done by using the Fe speciation discussed above [19,42] that considers the interactions of Fe(III) with all the components of seawater. Since the formation of $FeSO_3H$ and $HOFeSO_3H$ are important in interpreting the rates, we have reevaluated [42] the effect of pH on the spectroscopically determined values of K_{app} for the formation of Fe(III)-S(IV) [43] using the equation

$$K_{app} = K_{FeSO3H}\, \alpha_{Fe}\, \alpha_{HSO3} + K_{HOFeSO3H}\, \alpha_{FeOH}\, \alpha_{HSO3} \tag{63}$$

The evaluation of these results (**Figure 11**) indicates that only the formation of $HOFeSO_3H$ (638 ± 25 M^{-1}) is needed to fit the data (the smooth curve) over the pH range of 1 to 3.

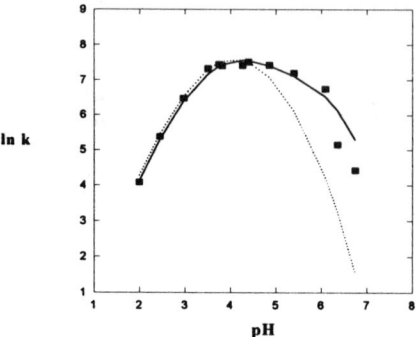

Figure 10. The effect of pH on the rates of reduction of Fe(III) by S(IV) in seawater at 25 °C. The dotted lines are calculated using only k_1 and the solid line is calculated using k_1 and k_2 [42].

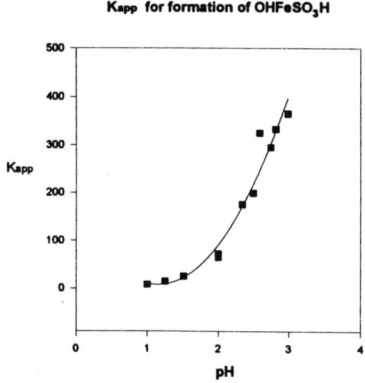

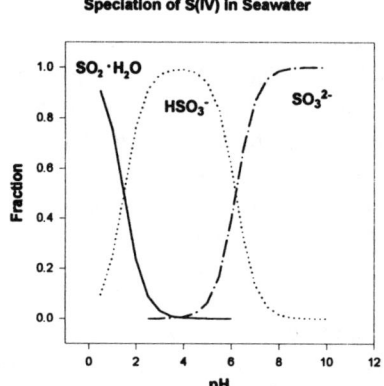

Figure 11. Values of the conditional formation constant, K_{app} [43] as a function of pH. The smooth curve is calculated using $K_{FeSO3H} = 0$ and $K_{HOFeSO3H} = 638 \, M^{-1}$.

Figure 12. Speciation of S(IV) in seawater as a function of pH [42].

The speciation of Fe(III) [19] (**Figure 3**) and S(IV) [16] (**Figure 12**) in seawater as a function of pH show that both $Fe(OH)_2^+$ and $FeOH^{2+}$ as well as HSO_3^- go through a maximum between a pH of 3 to 5, the same region where the ln k goes through a maximum. The addition of S(IV) at the levels used in our experiments (100 µM) had the rate. The addition of F⁻ to NaCl yielded a relative rate (-1.06) that was in good agreement with the seawater results. These experiments show that F⁻ is the most important anion in seawater that causes the rate to be lower than in NaCl at the same ionic strength. The decrease due to the addition of SO_4^{2-} Mg^{2+}, Ca^{2+} (-0.62) is nearly balanced by the increase due to the addition of K^+ and Br^- (0.53). little effect on the speciation of Fe(III) in seawater which is dominated by the formation of FeF^{2+}, FeF_2^+ and $FeSO_4^+$ at low pH and $FeOH^{2+}$ and $Fe(OH)_2^+$ at high pH. The HSO_3^- species is dominant over the pH range of our studies (2 to 7).

These speciation diagrams indicate that the effect of pH on the rate constant for the reduction of Fe(III) can be attributed to

$$k [Fe(III)] [S(IV)] = k_0 [Fe^{3+}][HSO_3^-] + k_1 [FeOH^{2+}] [HSO_3^-] + k_2 [Fe(OH)_2^+] [HSO_3^-] \tag{64}$$

This equation can be simplified to

$$k = k_0 \alpha_{Fe} \alpha_{HSO3} + k_1 \alpha_{FeOH} \alpha_{HSO3} + k_2 \alpha_{Fe(OH)2} \alpha_{HSO3} \tag{65}$$

where α_i are the molar ratios of species i and the individual rate constants are for

$$Fe^{3+} + HSO_3^- \xrightarrow{k_0} Fe^{2+} + \cdot SO_3H \tag{66}$$

$$FeOH^{2+} + HSO_3^- \xrightarrow{k_1} FeOH^+ + \cdot SO_3H \tag{67}$$

$$Fe(OH)_2^+ + HSO_3^- \xrightarrow{k_2} Fe(OH)_2 + \cdot SO_3H \tag{68}$$

The values of k_0, k_1 and k_2 were determined from the calculated values of α_i determined from our speciation model and experimental values of k. The value of k_0 was not needed to represent the results. From a pH of 2 to 4 the results could be adequately represented by using only $k_1 = 6160 \pm 300$ M min⁻¹ (see **Figure 10**). The measurements above a pH = 4 required values of $k_1 = 4650 \pm 500$ M min⁻¹ and $k_2 = 1213 \pm 210$ M min⁻¹ to represent the results (see **Figure 10**). These results indicate that the rate constant for $FeOH^{2+}$ with HSO_3^- is two times larger than for $Fe(OH)_2^+$. The results between pH = 2 to 4 are similar to the earlier results at low pH [43.44] for the formation of the complex $OHFeSO_3H$ as described above. Our finding that the reaction of $Fe(OH)_2^+$ and HSO_3^- is important at a higher pH (between 4 and 6) is possible because we have been able to make our measurements below the solubility limits of Fe(III) (≈ 10 nM at pH = 6).

The effect of composition on the rates of reduction of Fe(III) with S(IV) in NaCl media with added cations and anions can qualitatively be attributed to changes in the interaction of Fe(III) with various anions and S(IV) with various cations. A small part of the decrease in the rates due to the addition of Mg^{2+} and Ca^{2+} can be attributed to a decrease in the concentration of free HSO_3^-. Both of these cations are known to form complexes with sulfite. The increase in the rate with the addition of K^+ may be related to it having weaker interactions with HSO_3^- than Na^+. The decrease in the rate with the addition of SO_4^{2+} and F^- can be attributed to the formation of $FeSO_4^+$ and FeF^{2+} ion pairs that are non reactive to reduction with S(IV). The increase in the rate with the addition of Br^- may be related to it having weaker interactions with Fe(III) than Cl^-.

The effect of F^- is quite large and demonstrates the influence the formation of strong complexes can have on the reaction rates of metals at low concentrations. These effects would not be seen if the rates of reduction of Fe(III) by S(IV) were made at high concentrations of Fe(III). The total effect of the ions in seawater is smaller than the effect of the individual ions due to the interactions of the seasalt ions with each other. For example, Mg^{2+} and Ca^{2+} can form complexes with F^- and SO_4^{2-} and lower the concentration of free F^- and SO_4^{2-} that can interact with Fe^{3+}.

The effect of composition and ionic strength on the rates at pH = 3.5 can be examined quantitatively by considering reaction (67) to be dominant. The lower rates in seawater relative to NaCl at the same ionic strength can be related to the changes in the speciation of Fe(III). From a simplistic view one would expect that the decrease in the rates due to the addition of seasalts would be related to changes in the fraction of $FeOH^{2+}$ in the solutions (the fraction of HSO_3^- is not strongly dependent on the composition). The values of $\Delta \ln k = \ln k(soln) - \ln k(NaCl)$ versus the $\ln (\alpha_{FeOH} \alpha_{HSO3})$ are shown in **Figure 13**. The results show a rough correlation with the fraction of $FeOH^{2+}$ in agreement with the rates being directly related to the levels of $HOFeSO_3H$ in the solutions.

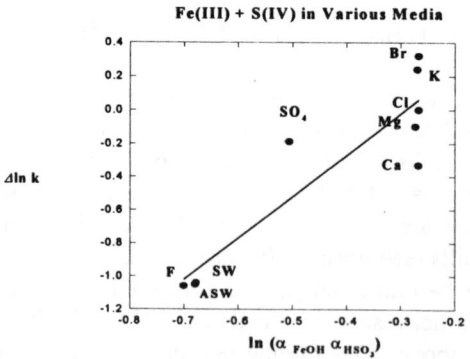

Figure 13. Changes in the rate constant upon the addition of seasalts to NaCl ($\Delta \ln k = \ln k(soln) - \ln k(NaCl)$) versus the natural log of the molar fraction of $FeOH^{2+}$ and HSO_3^- at pH = 3.5 and 25 °C [42].

Our results support the mechanism suggested by Conklin and Hoffmann [44] for the autoxidation of S(IV) with a few minor modifications due to the work of Betterton [43] and our results at high pH. The proposed mechanism is

$$SO_2 + H_2O \overset{K_{1S}}{\Leftrightarrow} H^+ + HSO_3^- \quad (69)$$

$$HSO_3^- \overset{K_{2S}}{\Leftrightarrow} H^+ + SO_3^{2-} \quad (70)$$

$$Fe^{3+} + H_2O \overset{K_1}{\Leftrightarrow} FeOH^{2+} + H^+ \quad (71)$$

$$Fe^{3+} + 2H_2O \overset{K_2}{\Leftrightarrow} Fe(OH)_2^+ + 2H^+ \quad (72)$$

$$FeOH^{2+} + HSO_3^- \overset{K_A}{\Leftrightarrow} HOFe\text{-}SO_3H^+ \quad (73)$$

$$Fe(OH)_2^+ + HSO_3^- \overset{K_B}{\Leftrightarrow} (HO)_2Fe\text{-}SO_3H \quad (74)$$

$$HOFe\text{-}SO_3H^+ \overset{K_I}{\Leftrightarrow} HOFe\text{-}OSO_2H^+ \quad (75)$$

$$(HO)_2Fe\text{-}SO_3H \overset{K_{II}}{\Leftrightarrow} (HO)_2Fe\text{-}OSO_2H \quad (76)$$

$$HOFeO\text{-}SO_2H \overset{k_a}{\rightarrow} FeOH^+ + \cdot SO_3H \quad (77)$$

$$(HO)_2Fe\text{-}SO_2H \overset{k_b}{\rightarrow} Fe(OH)_2 + \cdot SO_3H \quad (78)$$

where K_{1S} and K_{2S} are the first and second dissociation constants for the ionization of $SO_2 \cdot H_2O$, K_1 and K_2 are the stepwise hydrolysis constants for Fe(III); K_A and K_B are the formation constants of Fe(III)-S(IV) complexes and K_I and K_{II} are related to changes in the bonding of Fe-S to Fe-O that may occur in the complex [44]. The overall rate constant is related to reactions (77) and (78) by

$$-d[Fe(III)]/dt = k\,[Fe(III)]\,[S(IV)]$$
$$= k_a\,[HOFe\text{-}OSO_2H] + k_b\,[(HO)_2Fe\text{-}OSO_2H] \qquad (79)$$

The substitution of the equilibrium concentrations gives

$$[HOFe\text{-}OSO_2H] = K_A\,K_I\,\alpha_{FeOH}\,\alpha_{HSO3}\,[Fe(III)]\,[S(IV)] \qquad (80)$$

$$[(HO)_2Fe\text{-}OSO_2H] = K_B\,K_{II}\,\alpha_{Fe(OH)2}\,\alpha_{HSO3}\,[Fe(III)]\,[S(IV)] \qquad (81)$$

where the fractions of Fe^{3+}, $FeOH^{2+}$, $Fe(OH)_2^+$, and HSO_3^- are given by

$$\alpha_{Fe} = 1/\{1 + K_1/[H^+] + K_2/[H^+]^2 + K_3/[H^+]^3 + \Sigma K_{FeXn}\,[X]^n\} \qquad (82)$$

$$\alpha_{FeOH} = \{K_1/[H^+]\}\,\alpha_{Fe} \qquad (83)$$

$$\alpha_{Fe(OH)2} = \{K_2/[H^+]^2\}\,\alpha_{Fe} \qquad (84)$$

$$\alpha_{HSO3} = K_{1S}\,[H^+]/\{[H^+]^2 + K_{1S}\,[H^+] + K_{1S}\,K_{2S}\} \qquad (85)$$

where X is a ligand such as Cl^-, F^-, etc.

The substitution of equations (80) and (81) into equation (79) gives

$$-d[Fe(III)]/dt = k_a\,K_A\,K_I\,\alpha_{FeOH}\,\alpha_{HSO3} + k_b\,K_B\,K_{II}\,\alpha_{Fe(OH)2}\,\alpha_{HSO3} \qquad (86)$$

which is consistent with the rate equation (68) where $k_1 = k_a\,K_A\,K_I$ and $k_2 = k_b\,K_B\,K_{II}$. This reaction mechanism without oxygen is consistent with the mechanism with oxygen with the exception of the possible formation of $(HO)_2Fe\text{-}SO_3H$ at high pH. Further measurements of the reduction of Fe(III) with S(IV) in the presence of oxygen at high pH are needed to access the importance of this species.

The reduction of Fe(III) by S(IV) in controlling the state of iron in aerosols and water droplets was found to be important at the levels of S(IV). In acidic fog and cloud waters [47] concentrations can be as high as 3000 μM; while the values in marine aerosol are normally less than 1 μM. These calculations indicate that the higher levels of Fe(II) found in water droplets [48] compared to marine aerosols [40] may be due to the reduction of Fe(III) by S(IV). The rates of reduction given in this paper should prove useful in modeling the changes in the redox state of iron in acidic aerosols and water droplets. Studies on these rates under the presence of O_2 are presently underway. Further measurements of Fe(II) as a function of S(IV) are needed in water droplets and marine aerosols to prove that they are kinetically linked in natural waters.

Acknowledgments

The author wishes to acknowledge the support of the Oceanographic section of the National Science Foundation and the Office of Naval Research for supporting his Fe studies.

References

1. Millero, F. J. (1982) Use of models to determine ionic interactions in the natural waters, *Thalassia Jugoslavica*, **18**, 253-291.
2. Millero, F. J. (1990) The effect of ionic interactions on the rates of oxidation of metals in natural waters, in: *Chemical modeling in aqueous systems II* ,D. Melchior and R. Bassett (eds.), ACS Books, Washington D.C., Chapter 34, pp. 447-460.
3. Sharma, V.K. and Millero, F.J. (1988) The oxidation of Cu(I) in electrolyte solutions, *J. Solution Chem.* **17**, 581-599.
4. Millero, F. J., Sotolongo, S., and Izaguirre, M. (1987) The kinetics of oxidation of Fe(II) in seawater, *Geochim. Cosmochim. Acta*, **51**, 793-801.
5. Vazquez, F., Zhang, J.Z. and Millero, F.J. (1989) Effect of trace metals on the oxidation rates of H_2S in seawater, *Geophys. Res. Lett.* **16**, 1363-1366.
6. Zika, R.G., Saltzman, E.S., and Cooper, W.J. (1985) Hydrogen peroxide concentrations in the Peru upwelling area., *Mar. Chem.* **17**, 265-275.
7. Millero, F. J. and Schreiber, D. R. (1982) Use of the ion pairing model to estimate activity coefficients of the ionic components of natural waters. *Amer. J. Sci.* **282**, 1508-1540.
8. Pitzer, K. S. (1991) Ion interaction approach: theory and data collection, in K. S. Pitzer (ed.), *Activity Coefficients in Electrolyte Solutions*, CRS, Boca Raton, FL, pp. 75-153.
9. Harvie, C. E., Møller-Weare, N. and Weare, J. H. (1984) The prediction of mineral solubilities in natural waters: The $Na-K-Mg-Ca-H-Cl-SO_4-OH-HCO_3-CO_3-CO_3-H_2O$ system to high ionic strengths at 25°C, *Geochim. Cosmochim. Acta* **48**, 723-752.
10. Møller, N. (1988) The prediction of mineral solubilities in natural waters: A chemical equilibrium model for the $Na-Ca-Cl-SO_4-H_2O$ system, to high temperature and concentration, *Geochim. Cosmochim. Acta* **52**, 821-837.
11. Greenberg, J. P. and Møller, N. (1989) The prediction of mineral solubilities in natural waters: A chemical equilibrium model for the $Na-K-Ca-Cl-SO_4-H_2O$ system to high concentration from 0 to 250°C, *Geochim. Cosmochim. Acta* **53**, 2503-2518.
12. Millero, F. J., and Roy, R. (1996) A chemical model for the carbonate system in natural waters, *Croatia Chemica Acta*, in press.
13. Millero, F. J. and Hawke, D.J. (1992) Ionic interactions of divalent metals in natural waters, *Mar. Chem.* **40**, 19-48.
14. Millero, F. J. (1992) Stability constants for the formation of rare earth inorganic complexes as a function of ionic strength, *Geochim. Cosmochim. Acta* **56**, 3123-3132.
15. Whitfield, M., 1975. The extension of chemical models for seawater to include trace components, *Geochim. Cosmochim. Acta* **39**, 1545-1557.
16. Millero, F. J., Hershey, J. P., Johnson, G. and Zhang, J. (1989) The solubility of SO_2 and the dissociation of H_2SO_3 in NaCl solutions, *J. Atmos. Chem.* **8**, 377-389.
17. Martin, J.H. and R.M. Gordon (1988) Northeast Pacific iron distributions in relation to phytoplankton productivity, *Deep-Sea Res.* **35**, 177-196.
18. Bruland, K.W. and Wells, M. (1995) The chemistry of iron in seawater and its interactions with phytoplankton, Special Issue of *Mar. Chem.* **50**, 1-241.
19. Millero, F. J, Yao, W., and Aicher, J. (1995) The speciation of Fe(II) and Fe(III) in natural waters, *Mar. Chem.* **50**, 21-39.

20. Campbell, D. M., Millero, F. J., Roy, R., Roy, L., Lawson, M, Vogel, K. M. and Moore, C. P. (1993) The standard potential for the hydrogen - silver, silver chloride electrode in synthetic seawater, *Mar. Chem.* **44**, 221-233.
21. Zhu, X., Prospero, J. M., Millero, F. J., Savoie, D. L. and Brass, G. W. (1992) The solubility of ferric ion in marine mineral aerosol solutions at ambient relative humidities. *Mar. Chem.* **38**, 91-107.
22. Byrne, R. H. and Kester, D. R. (1976) Solubility of hydrous ferric oxide and iron speciation in sea water, *Mar. Chem.* **4**, 255-274.
23. Kuma, K., Nakabayashi, S., Suzuki, Y., and Matsunaga, K. (1992) Dissolution rate and solubility of colloidal hydrous ferric oxide in seawater, Mar. Chem. **38**, 133-143.
24. Kuma, K., J. Nishioka, and K. Matsunaga (1996) Controls on iron (III) hydroxide solubility in seawater: The influence of pH and natural organic chelators, *Limnol. Oceanogr.* **41**, 396-407.
25. Gledhill, M. and van den Berg, C. M. G. (1994) Determination of complexation of iron (III) with natural organic complexing ligands in seawater using cathodic stripping voltammetry. *Mar. Chem.* **47**, 41-54.
26. Wu, J. and Luther, G.W. (1994) Complexation of Fe(III) by natural organic ligands in the North west Atlantic Ocean by a competitive ligand equilibration method and a kinetic approach, *Limnol. Oceanogr.* **50**, 1119-177.
27. Rue, E.L. and Bruland, K.W. (1995) Complexation of iron (III) by natural organic ligands in the Central North Pacific as determined by a new competitive ligand equilibration/adsorptive cathodic stripping voltammetric method, *Mar. Chem.* **50**, 116-138.
28. van den Berg, C.M.G. (1995) Evidence for organic complexation of iron in seawater, *Mar. Chem.* **50**, 139-157.
29. Landing W.M. and K.W. Bruland (1987) The contrasting biogeochemistry of iron and manganese in the Pacific Ocean, *Geochim. Cosmochim. Acta* **51**, 29-43.
30. Martin, J.H., S.E. Fitzwater, and R.M. Gordon (1990) Iron deficiency limits phytoplankton growth in Antarctic waters, *Global Biogeochem. Cycles* **4**, 5-12.
31. Martin, J.H., S.E. Fitzwater, R.M. Gordon, C.N. Hunter, and S.J. Tanner (1993) Iron, primary production, and carbon-nitrogen flux studies during the JGOFS North Atlantic bloom experiment, *Deep-Sea Res.* **40**, 115-134.
32. Anderson, M. A. and Morel, F. M. (1982) The influence of aqueous iron chemistry on the uptake of iron by the coastal diatom Thalassiosira weissflogii, *Limnol. Oceanogr.* **27** 789-813.
33. Martin, J.H. and S.E. Fitzwater (1988) Iron deficiency limits phytoplankton growth in the north-east Pacific subarctic, *Nature* **331**, 341-343.
34. Martin, J.H. et al. (1994) Testing the iron hypothesis in ecosystems of the equatorial Pacific Ocean, *Nature* **371**, 123-129.
35. Zuo, Y. and J. Hoigné (1993) Evidence for photochemical formation of H_2O_2 and oxidation of SO_2 in authentic fog water, *Science*, **260**, 71-73.
36. Faust, B. C., C. Anastasio, J. M. Allen, and T. Arakaki (1993) Aqueous-phase photochemical formation of peroxides in authentic cloud and fog waters, *Science* **260**, 73-75.
37. Millero, F. J., and Sotolongo, S. (1989) The oxidation of Fe(II) with H_2O_2 in seawater, *Geochim. Cosmochim. Acta* **53**, 1867-1873.
38. Erel, Y., S. O., Pehkonen, and M. R. Hoffmann (1993) Redox chemistry of iron in fog and stratus clouds, *J. Geophys. Res.* **98**, 18423-18434.
39. Zhuang, G. S., Duce, R. A. and Kester, D. R. (1990) The solubility of atmospheric iron in surface seawater of the open ocean, J. Geophys. Res. **95**, 16207-16216.
40. Zhu X., Prospero, J.M., Savoie, D.L., Millero, F.J., Zika, R.G., and Saltzman, E.S. (1993) Photoreduction of iron (III) in marine mineral aerosol solutions, *J. Geophys. Res.* **98**, 9039-9046.
41. Reddy, K. B. and van Eldik, R. (1992) Kinetics and mechanism of the sulfite-induced autoxidation of Fe(II) in acidic aqueous solution, *Atm. Environ.* **26A**, 661-665.
42. Millero, F. J, Gonzalez-Davila, M., and Santana-Casino, J.M. (1995) Reduction of Fe(III) with sulfite in natural waters, *J. Geophys. Res.* **100**, 7235-7244.
43. Betterton, E.A. (1993) On the pH-dependent *formation -constants of iron (III) sulfur(IV) transient complexes*, J. Atm. Chem. **17**, 307-324.

44. Conklin, M. H., and Hoffmann, M.R. (1988) Metal-ion sulfur(IV) chemistry. 3.Thermodynamics and kinetics of transient iron(III)-sulfur(IV) complexes, *Environ. Sci. Technol.* **22**, 899-907.
45. O'Sullivan, D. W., Hanson, A. K., and Kester, D. R. (1995) Stopped flow luminol chemiluminescence determination of Fe(II) in seawater at subnanomolar levels, *Mar. Chem.* **49**, 65-77.
46. King, D. W., H. A. Lounsbury, and F. J. Millero (1995) Rates and mechanism of Fe(II) oxidation at nanomolar concentrations, *Envir. Sci. Technol.* **29**, 818-824.
47. Warneck, P. (1989) Sulfur dioxide in rain clouds: gas-liquid scavenging efficiencies and wet deposition rates in the presence of formaldehyde, *J. Atm. Chem.* **6**, 99-117.
48. Behra, P., and Sigg, L. (1990) Evidence for redox cycling of iron in atmospheric water droplets, *Nature* **344**, 419-421.

LIGHT AND CHEMICALLY DRIVEN REACTIONS AND EQUILIBRIA IN THE PRESENCE OF ORGANIC AND INORGANIC COLLOIDS

Claudio MINERO
Dept. of Analytical Chemistry, University of Torino, Italy

1. Introduction

Suspended particles in natural waters comprise both inorganic solids and dead or live organic matter, the proportion of each varying widely in time and place. In the euphotic zone of oceans, bacteria, microalgae, zooplankton and mainly their dead remains contribute a large portion (>90%) of the total mass of suspended particles. Various types of colloids and particles are present, from fulvic and humic acids, proteins, polysaccarides, to clays, carbonates, metal oxides and cellular debris, respectively. The concentration range of suspended particles in natural waters spans over several orders of magnitude, from 0.01 mg L^{-1} in the deep ocean to 50,000 mg L^{-1} in turbid estuaries. Eutrophic lakes and closed seas may contain up to a few hundred milligrams per liter of organic matter. Only 20% of it is in the form of dissolved species.

The particle size distribution is usually very large, from 10^{-9} m for colloids to 10^{-2} m for large zooplankton aggregates, following a power law function of the form $n(r)=dN/dr=A\ r^{-p}$, where the exponent p ranges from 2 to 5. This size distribution is observed at all depths in the ocean, even though the total particle concentration decreases sharply with depth in the top mixed layer. The exponent p=4 well approximates data in the size range 1-100 µm. For colloids (size range 1 nm to 1 µm) p value is close to 3 [1]. This implies that the small particles dominate numerically and contribute to most of the area available for sorption, although the total particle volume and thus the total mass is preferentially distributed over the larger size classes.

Humic substances at sea water/ air interface may influence the surface tension and other processes such as exchange of gases, foam formation and concentration of particles and other elements at the surface [2]. The so called microlayer, which is perhaps 100-500 µm thick, depending on the method of collection, may contain up to 10 times the concentration of organic matter compared to the bulk water concentration, and usually has elevated levels of trace metals and toxic organics. Colloids and particles in the microlayer have increased residence times caused by their tendency to adhere to the air /water interface. Moreover, colloids have a large surface area to volume ratios, like particulate phase, and non-existent settling rates, like the dissolved phase.

2. Colloidal Solutions and Interfaces

The presence of colloidal or particulate interphases affects (i) the sorption and distribution of dissolved elements and organic chemical species and their availability; (ii) the acid/base, redox, and complexation equilibria that these species are subjected to, as well as (iii) their reactivity, i.e. the rates at which the complex chemical system attains the equilibrium value. These topics are treated in a variety of books [3-7].

After (ad)sorption the surface is modified. Adsorption of organic substances at solid particles has a complex influence on the particle-particle interactions and the colloidal stability in aqueous systems [8]. Adsorption of ionic species leads to developing of an electric surface charge and surface potential, which can prevent further adsorption, and to long range effects that can modify colloidal stability and aggregation [1,9]. Changes in particle charge and surface characteristics, as well as attenuation of (repulsive) electrostatic interaction by ionic strength, may give rise to coagulation phenomena, alteration in particle size distribution and sedimentation features of particulate matter. Such changes in sedimentation rates are observed in estuaries, that act as efficient trap for particulate load carried in by rivers. Settling particles are important carriers for sorbed metal ions, organic pollutants and anions.

A substantial difference with the usual chemistry in solution is observed in the presence of interfaces. The change in the chemical equilibria due to the presence of colloids was recognized since thirties. The determination of pH with indicators has been faulty in the presence of proteins or their degradation products. The same effect was observed in the presence of surfactants. The variation of pK_a of indicators was in opposite directions depending on the charge of the surfactant polar head. Since then, literature reported pK shifts for many substrates and surfactants [10], some of them of the order of ±4 pK units, transforming weak acids in stronger ones and viceversa. Thus, the evaluation of solution equilibria and related species concentration, equivalence points, buffer effects, or the deduction of true pK_a values, must be performed taking into account the eventual presence of colloids.

Surface catalyzed reactions and surface reactions, eventually with assistance of solar light, are also of primary importance, particularly in the microlayer where colloidal concentration is high. In addition to enzymatic reactions, the extent of which is dependent on the total biomass, and to dissolution and precipitation of solid phases, many abiotic chemical transformations may be mediated by inorganic or organic surfaces. Hydrolysis rates are often controlled by the activities of H^+ and OH^- or transition metal species [11,12], which in turn may be changed in the proximity of an interface. The oxidation and reduction of organic species depend on the concentration of oxygen and hydrogen peroxide, but also on the photoinduced steady-state concentration of aqueous singlet oxygen ($^1O_2$), •OH radicals, peroxyradicals (ROO•) or solvated electrons (e^-_{aq}), which is dependent (directly or inversely) on the surface concentration of Dissolved Organic Matter (DOM), mainly present in the form of colloids. Also light-induced band-gap excitation of semiconductor metal oxide's particles, like Fe_2O_3, ZnO, TiO_2 and some metal sulfides, produces electron/hole pairs able to catalyze the oxidation of organics (by O_2) and the reductive dechlorination.

The chemical interactions for partition of solutes at the interphases are those of the solute with the surface and those of the solute with the solvent. Among these are of primary importance (i) long range interactions, such as the electrostatic (strong coulombic forces between charged species and weaker electrostatic higher-order interactions arising from charge-dipole, dipole-dipole and so on) and hydrophobic ones, caused by the negative interactions with the solvent and leading to solute expulsion from the aqueous phase; and (ii) interactions that require an intimate contact between sorbent and sorbate, such as those arising from chemical reactions at the surface (hydrolysis, complexation, ligand exchange).

Despite the importance of understanding colloid-solute interactions, few studies have been conducted with natural colloids, because of the difficulties associated with isolating colloids from natural dissolved and particulate phase. Methods of analysis and characterization of marine particles and colloids are reported [13]. The field is currently an active area of research [14].

The interaction of the solvent with the surface is usually not explicitly accounted for, since water is the only solvent considered and its effect may be assumed constant. However, the replacement of the surface water with solute molecules or ions may force the species to dehydrate more than in a similar complex in solution. Owing to this and other steric effects, and to the influence of neighboring atoms, chemically similar groups in different geometrical configurations may exhibit very different affinities for solutes. In spite of this, there are good correlations between surface-extracted affinity constants and the corresponding values in solution. This supported in the last twenty years the surface complexation model [15], which describes the surface adsorption reactions as consisting of specific surface interactions (chemical bonding to surface atoms) and electrostatic interactions at the surface. Eventually, hydrophobic expulsion can be accounted for properly. The model allows, for example, to well describe all available data on adsorption of metal-EDTA complexes onto different metal oxides [16], and adsorption of phosphate on manganese dioxide in sea water [17].

For most natural surfaces the electrostatic state of the interface, at least when adsorption of protons does not induce the dissolution (for some metal oxides), is primarily determined by adsorption of protons and the pH of the solution. This is accounted for in Point of Zero Charge (PZC) concept [18]. Organic coating of particles can also modify the natural charge of inorganic colloids [19,20]. At the pH of sea water, most interfaces are negatively charged, favoring preferential adsorption of (metal) cations with respect to anions.

The structure of the particle-solution interface is fundamental for the effects induced by the surface itself and for its physico-chemical modeling. To explain linear sorption isotherms with concentration and the proportionality of sorption to the total organic content of the particulate and to the octanol-water partition coefficient, the models based on mixing of two different types of particulate (inorganic and organic) are alternative to those assuming humic material adsorbed as thin film onto the surface of mineral particles [21]. For electrostatic models the structure of the particle-solution interface fixes the location of the counterions. If the sorbing phase is three-dimensional, such as a gel layer at an oxide surface or a non polar permeable or porous organic coating, and the counterions can be contained within this sorbing phase, the electric field will not extend significantly beyond the particle. By contrast, if sorption occurs at a two-dimensional surface, electrokinetic and electric double layer effects will be observed, depending on the ionic strength. The roughness and the porosity of the surface also exert profound effects on the diffusion and availability of chemical species reacting either with foreign species or light.

The adsorption will be examined in some detail. Only the relevant features of partition in the presence of colloids will be considered for simple chemical equilibria and thermally driven reactions. Light induced reactions will be also examined under a general kinetic scheme in which the complexity caused by partition of reactants will be briefly outlined.

3. (Ad)Sorption

The expression for the chemical potential of a species i under an external electric field is

$$\mu_i = \mu°_i + \mu^{spec}_i + RT \ln x_i + z_i e (\phi - \phi_o) \tag{1}$$

where ϕ_o is the potential in the standard state and μ^{spec}_i is the free energy per mole due to specific (chemical) interactions. The standard chemical potential $\mu°_i$ contains all the interactions of the species i with the medium (and external fields) that are not explicitly considered in eq.(1). For example, it accounts for the free energy changes due to the modification of the solvent structure and electrostatic effects of higher-order than coulombic ones. Entropic and enthalpic contributions arising from interactions in concentrated solution can be conveniently accounted for by Flory-Huggins modeling [22] or regular solution theory. Since these introduce some complexity, we avoid to deal with.

The condition of uniformity of the molar free energy in an equilibrium system imposes that μ_i is constant through the system for every species *i*. In the presence of interphases or surfaces, it is convenient to model the system with few spatial regions where does exist in the time scale of the phenomena of interest a (satisfactory) homogeneity of chemical and physical properties. These spatial regions are called pseudophases and provide a useful separation of the whole system in subsystems. The concepts valid for macroscopic phases can equally apply to them, with the exceptions that electroneutrality and mass conservation principles are requested only for the whole chemical (macroscopic) system. In contrast to a macroscopic phase, a pseudophase can be charged, as will be a (solid) surface. To obey to mass and neutrality conservation principles, the pseudophases are at least two, each other in equilibrium through exchange reactions for the species, regulated by proper transfer constants.

Applying eq.(1) to two pseudophases (1,2) in equilibrium, the transfer constant or partition coefficient K^p_i, is given by:

$$\frac{[I_2]}{[I_1]} = K^o_i K^c_i K^e_i w_i = K^p_i \qquad (2)$$

where concentrations are given in moles per unit volume of the whole system. This has advantages over other definitions, mainly from operative point of view. A bidimensional pseudophase may be treated as three-dimensional, with a volume given by $V_s = S_{tot} v_i / a_i$, where S_{tot} is the total surface area and v_i and a_i are the molar volume and area of the adsorbed molecules. Surface species concentrations can be expressed in mole per liter, per gram of solid, per square meter of solid surface or per mole of solid. However, the molarity scale provides a consistent set of units for all aqueous and adsorbed species. The parameter w_i accounts for the relative dimensionality and size of the pseudophases. With two pseudophases $w_i = f^v_i f_c / (1 - f_c)$, where f_c is the volume fraction of colloidal (or particulate) phase and f^v_i is the fraction of colloidal (or particulate) phase volume that the adsorbed molecules occupy. These quantities can be calculated from geometric considerations or from complementary information (for example the amount of (organic) material adsorbed as thin film onto the surface of mineral particles).

The quantity $K^o_i = \exp(-\Delta\mu^o_i / RT)$ is suitable for further factorization and may mainly account for hydrophobic interactions. It is useful for example to phenomenologically account for salting out effects present in the sea water also on small hydrophylic organic molecules. The quantities $K^e_i = \exp(-z_i e \Delta\phi / RT)$ and $K^c_i = \exp(-\Delta\mu^{spec}_i / RT)$ describe the coulombic work for transferring an ion along a potential gradient and the chemical (specific) work for changing the species coordination, respectively. Remembering the homogeneity required by the definition of the pseudophase, in the case that different chemical interactions are possible, for example on a surface, different sets of transfer constants must be considered or even different pseudophases may be postulated. This approach is equivalent to that based on homologous compounds for complexation equilibria, in which the constant of complexation K (here K^c_i) is considered a (continuous) variable parameter [23]. The pseudophase approach stresses more the effect of mean fields, i.e. those generated by the medium, than those experienced by a molecule in a two-bodies' interaction.

Eq.(2) provides a linear increment of the adsorbed amount with respect to the solution concentration and is suitable for sorption in bulk sorbents or in macrophases, where site occupancy can be assumed limitless. When the number of sites is finite, for example on a surface, an exchange reaction between the species and a reference compound (the solvent) is more suitable, together with the mole balance for surface sites. For a chemical equilibrium with 1:1 stoichiometry in two pseudophases,

$$I_1 + S_2 \overset{K_{ads}}{=} I_2 + S_1 \qquad (3)$$

assuming that the species have the same volume fractions ($v_i = v_S$) and compete for the same fraction of pseudophase volume ($f^v_I = f^v_S$), under the previous notation, from eq.(2) the equilibrium constant is

$$K_{ads} = K°_i \, K^c_i \, K^e_i = \frac{[I_2][S_1]}{[I_1][S_2]} \tag{4}$$

where $K°_i = K°_i/K°_S = \exp(-(\Delta\mu°_i - \Delta\mu°_S)/RT)$, $K^e_i = K^e_i/K^e_S = \exp(-(z_i - z_S)e\Delta\phi/RT)$ and $K^c_i = K^c_i/K^c_S = \exp(-(\Delta\mu^{spec}_i - \Delta\mu^{spec}_S)/RT)$. This last term formally accounts for the free energy change resulting from the removal of solvent from the surface site (ligand) and from the coordination sphere of the species I (acceptor), as it is for solution chemistry. In this respect it is not surprising that there exists a strong correlation of K_{ads} with the complexation constant in solution [24,25].

Given the moles of possible sites in the pseudophase 2 (a surface) per unit of whole system volume C_s, and considering only the species I and S, the solvent related quantities are $[S_2] = C_s - [I_2]$ and $[S_1] = 1/v_S - [I_1]$, respectively. In this case the Langmuir isotherm is obtained:

$$[I_2] = \frac{K_{ads}[I_1]C_s}{1/v_S + [I_1](K_{ads} - 1)} = K^L_i [I_1] \tag{5}$$

where the subscript 1 refers to pseudophase 1, assumed as the solution, and $K_{ads} = K°_i K^c_i K^e_i$ from eq.(4). Eq.(5) provides a saturating behavior of $[I_2]$ versus $[I_1]$. It is interesting to compare the limit of eq.(5) for $[I_1] \to 0$, i.e. when the surface is far from saturation. In this case $K^L_i = K_{ads} C_s v_S \neq K^P_i$, only for the definitions of constants, but still depends on the size of the pseudophase.

Eq.(2) and (5) can be written with the general notation

$$[I_2] = K^r_i [I_1] \tag{6}$$

where the transfer coefficient K^r_i is the partition coefficient K^P_i in the case of eq.(2). Under the condition of site limitation (Langmuir equation) it is given by

$$K^r_i = 1/2(K_{ads}v_s(C_s - C_i) + C_i v_s - 1 + ((1 - C_i v_s)^2 + 2K_{ads}v_s(C_i v_s(C_s - C_i) + C_i + C_s) + K_{ads}^2 v_s^2(C_s - C_i)^2)^{1/2}) \tag{7}$$

where C_i is the analytical concentration of the species i.

When two (or more) adsorbates compete for the same sites or one adsorbent has two (or more) sites of different affinities, one can formulate eq.(8) and (9) respectively.

$$[I_2] = \frac{K^i_{ads}[I_1]C_s}{1/v_S + [I_1](K^i_{ads} - 1) + \sum[J_1](K^j_{ads} - 1)} \tag{8}$$

$$[I_2] = \sum_j \frac{K_{j,ads}\,[I_1]\,C_{s,j}}{1/v_S + [I_1]\,(K_{j,ads} - 1)} \tag{9}$$

The approximations involved in the Langmuir equation are clearly shown in its derivation. It was assumed that molecules are all of the same size and that the adsorbed ones do not laterally interact each other. Effects on activity coefficients other than electrostatic and hydrophobic ones are not accounted for. Owing to the different sizes of possible adsorbable species and solvent, the eq.(5) may oversimplify the adsorption model. In this case other possible adsorption isotherms have been postulated, either on a theoretical [26] or empirical basis [27]. Consider for example the more general case in which the adsorbable species removes several solvent molecules because of its different molar volume or area:

$$I_1 + v_i\,S_2 \;\overset{K_{ads}}{=}\; I_2 + v_i\,S_1 \tag{10}$$

where $v_i = v_i/v_S$. Noting as before that $[S_1] = 1/v_S - v_i\,[I_1]$ and $[S_2] = C_s - v_i\,[I_1]$ from the volume (area) balance at the surface,

$$[I_2] = \frac{C_s\,(K_{ads}\,[I_1])^{1/v_i}\,w_i^{(1-v_i)/v_i}}{v_i\,(K_{ads}\,[I_1])^{1/v_i}\,w_i^{(1-v_i)/v_i} + [I_2]^{1/(v_i-1)}\,(1/v_S - v_i\,[I_1])} \tag{11}$$

where $K_{ads} = K^\circ_i\,K^c_i\,K^e_i\,/(K^\circ_S\,K^c_S\,K^e_S)^{v_i}$. When $v_i = 1$ the Langmuir equation (5) is obtained. Assuming in eq.(11) that the second term at denominator is almost constant and at fixed w_i (i.e. f_c), a generalized Langmuir equation, very similar to that of Hill [28] widely used in biochemistry, is obtained

$$[I_2] = \frac{a\,C_s\,(K_{ads}\,[I_1])^{1/v_i}}{b\,(K_{ads}\,[I_1])^{1/v_i} + c} = \frac{C_s\,K'_{ads}\,[I_1]^{1/v_i}}{b'\,K'_{ads}\,[I_1]^{1/v_i} + 1} \tag{12}$$

Furthermore, when $b'\,K'_{ads}\,[I_1]^{1/v_i} \ll 1$, i.e. when $[I_1] \to 0$, eq.(12) reduces to the empirical isotherm proposed by Freundlich $[I_2] = K''_{ads}\,[I_1]^a$, in which $a = 1/v_i < 1$ (i.e. $v_i > v_S$) and K''_{ads} are adjustable empirical parameters. Interestingly, the Freundlich isotherm can also be derived following Sposito [29] by generalizing eq.(9) to an integral and assuming a log-normal distribution of $K_{j,ads}$. There are many other isotherms specifically developed to take (non-electrostatic) lateral interactions into account. The reader can refer to many textbooks available [5,30]. In addition to consistency and compactness, the present derivation shows also that fit of experimental data with Langmuir or other isotherms normally does not constitute evidence that model hypotheses are supported.

The present modeling view is a good example of how lateral interaction can be accounted for. For all these equations, if some species are charged, during the exchange process the surface acquires charge and K^e_i will depend on $[I_2]$. The actual calculation of K_{ads} needs an iteration procedure. The electrostatic transfer constant takes into account the work for transferring an ion from the solution phase to the second pseudophase. Although sophisticated approaches or the use of measured activity coefficients are necessary for accurately modeling the interactions of seawater ions, in solution can the potential sensed by ions be approximated through the classical Debye-Huckel solution to the Poisson-Boltzmann equation, eventually modified according to Davies [31,32]. At the surface, even in the presence of very salted solutions, the work is simply $z_i e \phi_2$. Thus

$$-\ln K^e_i = \frac{z_i e \phi_2}{k_B T} - \frac{z_i^2 B_l B_u}{2} \left(\frac{\sqrt{\mu}}{(1+ r_i B_u \sqrt{\mu})} - D \mu \right) \tag{13}$$

where B_l=7.19 Å, B_u=0.329 Å^{-1} at 25°C and μ is the ionic strength of the solution, D is the Davies coefficient (0.2 or 0.3, originally), r_i is the radius of the ion (in Å) and ϕ_2 is the second pseudophase (surface) potential. The electrostatic transfer constant can thus be calculated after evaluation of the ionic strength and surface potential [33]. Unless the activity coefficients in the free solution are accounted for, the supposition that $\phi_1 = 0$ leads to erroneous modeling.

The potential sensed by ions in the second pseudophase (or surface) can be calculated by taking into account its charge density and its geometry. The second pseudophase can be a humic acid, a gel or a polyelettrolyte totally permeable to charges, an impenetrable sphere or a flat surface. Since in general the exact solution to the charge density/potential relationship is not possible, approximate analytical solutions depending on the geometry can be used [34], as well as numerical solutions of the Poisson-Boltzmann equation. Several models of the interface give different charge/potential relationships [35].

In the simple case of a spherical colloid of radius R_{coll} and when the surface potential is low, the linearization of the Poisson-Boltzmann equation gives the linear dependence of the potential on the surface charge density.

$$\phi_2 = \frac{Q B_l k_B T}{e R_{coll}} \frac{1}{1+\kappa R_{coll}} \tag{14}$$

where $\kappa = B_u \sqrt{\mu}$ and
(i) the total net charge Q of the particle of radius R_{coll} (in Å), is given by

$$Q = \Sigma z_i [I_2]/N_p + z_s q \tag{15}$$

> in which N_p is the moles of particles, $z_s q$ are the total fixed charges on the particle surface and $[I_2]$ is given by some of the equations for the partition (see eq.2,5,8-9,11-12). Eq.(15) assumes a semipermeable model for the interface.

(ii) the ionic strength of solution is given by

$$\mu = \tfrac{1}{2} \Sigma z_i^2 [I_1]/(1-f_c) \tag{16}$$

The sea water at average salinity of 35 g/kg is mainly composed by Na^+ (10.77 g kg^{-1}), Mg^{2+}(1.29 g kg^{-1}), Ca^{2+}(0.41 g kg^{-1}), K^+(0.40 g kg^{-1}), Cl^-(19.35 g kg^{-1}), SO_4^{2-} (2.71 g kg^{-1}), HCO_3^-(0.14 g kg^{-1}) [36]. For a flat surface the surface charge density ($\sigma_2 = Q/S_{tot}$) is more appropriate than eq.(14) and a proper relationship $\sigma_2 = f(\phi_2)$ has to be used instead of eq.(15). The classical equation of Gouy-Chapman for a diffuse double layer is given in eq.(17) [37].

$$\sigma_2 = 2 \kappa \varepsilon \varepsilon_0 RT/(z_i F) \sinh(z_i F \phi_2 / 2RT) \qquad (17)$$

The potential of the charged pseudophase is thus dependent on the partition of the species, their hydrophobic transfer constants, the possibility of chemical interactions, the nature of the pseudophase (surface) and the solution ionic strength. Given the mass balance ($C_i = [I_1] + [I_2] + ...$) and K°_i for every chemical species in the system, one can calculate the equilibrium concentrations in the pseudophases using one of the eq.(2,5,8,9,11,12), eq.(14) or similar, together with eqs.(15-16).

Figure 1 reports the calculation of the electrostatic transfer coefficient for a spherical colloid bearing 10 µC cm^{-2} of fixed charges in the presence of increasing concentrations of salt MX (the ionic strength can be calculated from eq.(16)). The electrostatic transfer coefficients decrease as the ionic strength is increased, decrease with particle radius for the model without site limitation, but are almost constant with radius in the case both M and X compete for the surface sites.

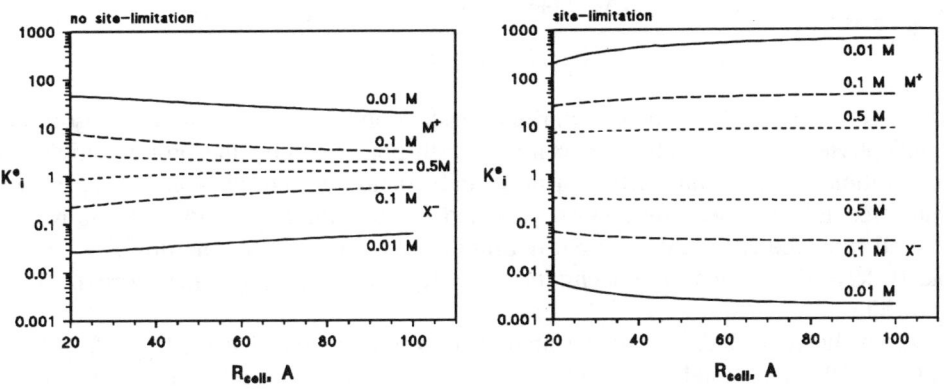

Figure 1. Variation of K^e_i for monovalent species as a function of the particle radius at different added salt concentrations for two models of adsorption. Negative spherical colloids (1×10^{-4} M) having a density of fixed surface charges $\sigma = 10$ µC cm^{-2}. All ion radii 2 Å. $K^\circ_i = 1$ for H^+, OH^-, M^+, and X^-.

The behavior is different when the total charge is proportional to the particle volume. In this case [38] the coulombic factor increases with particle radius. Note that in river or lake waters the electrostatic transfer coefficient and the surface potential is higher than in marine waters, leading to big changes in particle stability, shape and partition of charged species in regions where the two types of water mix, such as in estuaries.

4. Chemical Equilibria

Using the generalized definition of the partition coefficient K^r_i (eq.(6)) it is possible to write and calculate all the possible equilibria in the presence of two (or more) interfaces or pseudophases. For the partitioning of species not subject to equilibria, from the mass balance in the system of two pseudophases

$$[I_1] = C_i / (1+K^r_i)$$
$$[I_2] = C_i K^r_i / (1+K^r_i) \qquad (18)$$

where C_i are the analytical concentrations. The generalization to more pseudophases is obvious.

Consider for example the simple acid/base equilibria $HA = A^- + H^+$. By changing the symbols, this equilibrium stands for a site complexation by a metal or a ligand adsorption on a cationic site at the surface. The equilibrium constant in the second pseudophase is related to that in the first by $K_a^2 = K_a^1 K^r_A K^r_H / K^r_{HA}$. The equilibrium may be also described by a measurable equilibrium constant in the composite system

$$K_a^{meas} = \frac{C_A [H^+_1]}{C_{HA}} = \frac{[A_1](1+K^r_A)[H^+_1]}{[HA_1](1+K^r_{HA})} = K_a^1 \frac{(1+K^r_A)}{(1+K^r_{HA})} \qquad (19)$$

where K_a^1 and $[H^+_1]$ are the equilibrium constant and the measurable quantity in the pseudophase 1, respectively. For example, with potentiometry the proton activity in the solution in equilibrium with a surface can be measured. For pH concepts in the sea water see the excellent discussion provided by Stumm and Morgan [39]. K_a^1 is derived from the chemical potential definition eq.(1) under the condition $\Sigma \nu_i \mu_i = 0$, and is defined in terms of concentration, i.e. $K_a^1 = K_a^* \gamma_{HA}/\gamma_H \gamma_A$, where $\gamma_i = \exp((\mu°_i + z_i e\phi)/RT)$.

From the mass balance on the analytical concentration of acid, $C_{acid} = C_{HA} + C_A$, and eq.(19) the total concentrations of HA and A are easily found as $C_{HA} = C_{acid}[H^+_1]/([H^+_1]+K_a^{meas})$ and $C_A = C_{acid} K_a^{meas}/([H^+_1]+K_a^{meas})$ and, as a consequence,

$$[HA_2] = C_{acid} K^r_{HA} [H^+_1] /B \qquad [A_2] = C_{acid} K^r_A K_a^1 /B$$
$$[HA_1] = C_{acid} [H^+_1] /B \qquad [A_1] = C_{acid} K_a^1 /B \qquad (20)$$

where $B = [H^+_1](1+K^r_{HA}) + K_a^1(1+K^r_A)$. It is evident that the ratio $[HA_2]/[A_2] = K^r_{HA} [H^+_1]/K^r_A K_a^1$, i.e. the distribution of the species in the second pseudophase (for example the colloid surface), is different by a factor K^r_{HA}/K^r_A from that in the first pseudophase (for example the solution), where $[HA_1]/[A_1] = [H^+_1]/K_a^1$ as in the usual solution chemistry.

Under the hypothesis of partition without site limitation, the factor $K^r_{HA}/K^r_A = K^o_{HA} K^c_{HA} K^e_{HA}/ K^o_A K^c_A K^e_A$ depends mainly on the inverse of electrostatic partition coefficient for the species A and, as a consequence, on the surface. When partitioning is subject to site limitation, the proper form for K^r_i has to be used. Eq.(7) is valid for one species in competition with the solvent. In the simple equilibrium case treated here three species may compete (HA, A and the solvent). Thus eq.(8) applies and a different generalized partition coefficient K^r_i has to be used. In this case it is still possible to obtain a cumbersome analytical solution in term of analytical concentrations. However, for more than one species competing for the solvent, numerical solutions are more suitable.

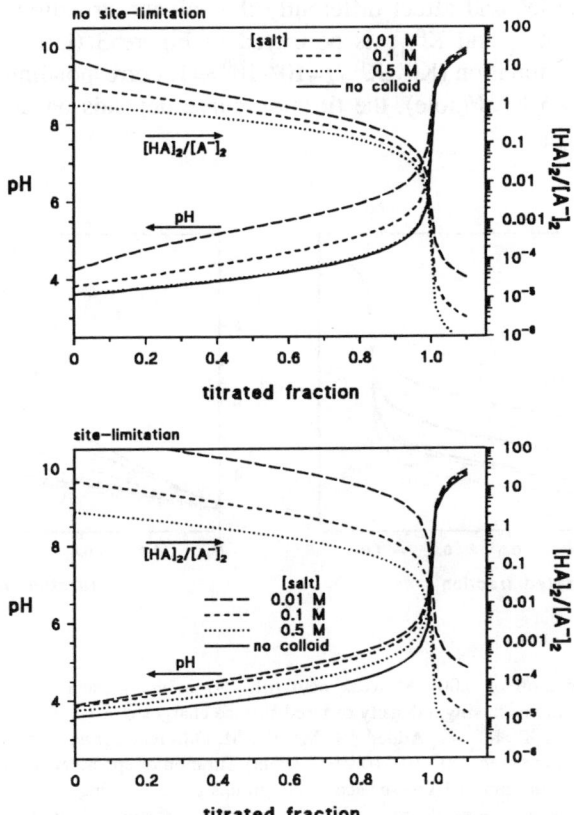

Figure 2. Variation of pH and ratio $[HA_2]/[A_2]$ during the titration of 0.001 M weak acid (pK$_a$=4) in the presence of spherical colloids (1x10^{-4} M) of 30 Å radius having a density of fixed surface charges $\sigma = 10$ μC cm^{-2}. All ion radii 2 Å. For H$^+$, OH$^-$, M$^+$, X$^-$ K^o_i =1. K$^o_{HA}$, K^o_A =10^4. Different curves correspond to different salt concentrations.

The variation of the concentration ratio $[HA_2]/[A_2]$ along the pH variation is illustrated in Figure 2 for the titration of a weak acid (1×10^{-3}M) for two models of adsorption. Note that under the site-limitation condition, the titration curves and the species distribution at the colloid interface are remarkably different from those in the other case.

The total amount of acid in the second pseudophase, and thus the extent of partitioning, depends on the pH of solution, since $C_{2,acid} = [HA_2] + [A_2] = C_{acid}/(1+f_2)$ where

$$f_2 = \frac{[H^+_1] + K_a^1}{[H^+_1] K^r_{HA} + K_a^1 K^r_A} \tag{21}$$

The values of K^r_A and K^r_{HA} are different for the different electrostatic contributions, but may be different also because HA and A^- can have different hydrophobic properties and affect differently the solvent structuration. The effect of different values of K^o_A and K^o_{HA} is portrayed in Figure 3 (left). Note that for the values used in the simulation ($K^o_A, K^o_{HA} = 10^4\text{-}10^6 \gg 1$, corresponding to free energy of adsorption of 6.8-9.5 Kcal/mole), the titration curve depends on the ratio K^o_{HA}/K^o_A and that $C_{2,acid} = C_{acid}$.

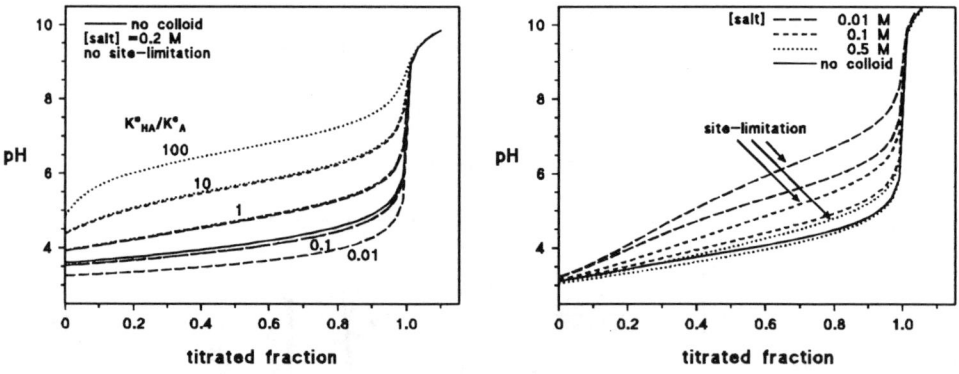

Figure 3. (left) Titration of 0.001 M weak acid ($pK_a=4$) in the presence of spherical colloids (1×10^{-4}M) of 30 Å radius having a density of fixed surface charges $\sigma = 10$ μC cm^{-2}. All ion radii 2 Å. For H^+, OH^-, M^+, X^-, $K^o_i =1$. Added salt MX 0.2 M. Different curves correspond to different ratios K^o_{HA}/K^o_A in the range $10^4/10^6\text{-}10^6/10^4$. (right) Titration of spherical colloids (1×10^{-4}M) of 30 Å radius having a density of fixed surface acidic groups corresponding to $\sigma = 10$ μC cm^{-2}. The effect of different concentrations of added salt under hypothesis of site-limitation or not is shown.

The quantity K_a^{meas} defined in eq.(19) can be considered a conditional acidity constant. However, it varies during the titration and can not be extracted directly from titration curves (see for example Figure 3, right). For other examples, the possible simplifications in the presence of excess salt (as is in the sea water), and the treatment of more complex equilibria, including those of autoaggregation, see the proper literature [40,41].

5. Chemical Reactivity

The rate observed in the presence of colloids is generally different from that observed in the corresponding solution. This has important environmental role [42], since the rates at which several compounds transform, may be different from that measured or reported in colloid-free solutions. For an excellent review of types of reactions and kinds of environmental catalysis see ref.[11].

Three causes contribute to the change observed for the chemical reactivity in the presence of colloids: (i) the partitioning of reactants modifies the free energy between reagents and products; (ii) the reagents can be confined in more restricted spatial regions and their local concentration changes; (iii) in different pseudophases there are medium effects, that is the local physico-chemical characteristics can be different from those in the reference pseudophase. As a consequence, the term catalysis has to be used with caution, and applies only to case (iii). Some reactions are really catalyzed when the catalysis is accomplished by partitioned species. Solvolytic reactions and acid/base catalyzed hydrolysis are effectively catalyzed by colloids that are able to partition these species. Since these species do not change the free energy of the reaction of interest, they do not change it even when adsorbed on the colloid. However, the reaction in the colloidal system always implies partitioning of reactants and the resultant rate is the sum of true-catalytic and compartimentalization effects.

The reaction rate in the presence of two or more pseudophases can be treated using the preceding formalism under the assumption that the rate of equilibrium attainment is higher than the reaction rate. This means that the rate of diffusion of reactants (and products) from one pseudophase to another is higher than the rate of reactant depletion by the local chemical reaction.

For a bimolecular reaction A + B -> products in the presence of several pseudophases, the second order kinetic constant observed in the presence of colloids or interfaces is expressed by a linear combination of the rates in the pseudophases weighed by their volume fraction φ_f.

$$k_{obs} = \Sigma \, \varphi_f \, k_f \, [A_f][B_f]/C_A C_B \tag{22}$$

where $[I_f]$s are the local concentrations in the pseudophases and the rate constant k_f $[M^{-1}s^{-1}]$ is typical of the composition and the (catalytic) properties of the pseudophase. k_f can be independently measured, measured in systems with equivalent composition or evaluated through proper models. The ratios $[I_f]/C_I$ are evaluated through the isotherms (2,5,8-9,11-12) and eq.(18).

The general kinetic relationship (22) provided by the pseudophase model reduces for two pseudophases to eq.(23).

$$k_{obs} = \frac{k_2 \varphi_2 K^r_A K^r_B + k_1(1-\varphi_2)}{(1+K^r_A)(1+K^r_B)} \qquad (23)$$

In systems with two pseudophases, like water with humic acids, surfactant micelles, oxide particle surfaces, biological membranes, and polyelectrolytes, the rate change observed in the presence of spherical colloids is shown in Figure 4 for different values of K^r_A at fixed K^r_B. When colloids have radius of 25-30 Å, a particle concentration 1×10^{-6} M corresponds to 10-20 mg L^{-1} of organic colloidal material.

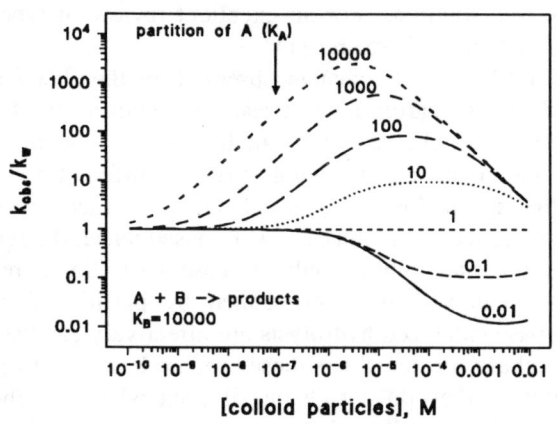

Figure 4. Variation of the observed rate constant normalised for that in solution as a function of the molar concentration of colloid particles having radius of 25 Å. It was assumed $k_{coll} = k_{water}$. K_A and K_B are the global transfer coefficients K^r_i for the species A and B defined in eq.(6-7).

The rate can increase or decrease of several orders of magnitude, depending on the partition properties of the reactants. When both reactants are strongly retained in the colloidal domain, the rate increases. By contrast, when one reagent is partitioned on the colloidal pseudophase (B) and the other (A) is forced in solution (for example by electrostatic repulsion), the rate is reduced, since the probability of reactant's encounter is diminished. On this topic the literature is plenty of reviews [43,44].

The pseudophase model is able to quantitatively predict the effects on the rates of (i) ionic medium for the ionic strength variation and the type of counterions [33], (ii) the surface charge of the colloid, as is the case of mixed micelles of ionic and non-ionic surfactants [45]; (iii) the simultaneous presence of more than two pseudophases, as it is the case of microemulsions [46]. Moreover, from test reactions, thermodynamic and physical properties of the colloidal solution, like the partition of reagents, the charge of the interface and the micro-structuration of the solution [47] can be inferred.

The kinetic model involving site saturation is more complex. Assuming that species A competes for the surface sites and that species B is regulated by eq.(2), the relationship is still analytical. However, when two reactants compete for the surface sites, the problems discussed for the weak acid equilibrium are present and the numerical treatment is required.

Many species can be non-competing for surface sites (for example H^+ or OH^- can often be treated in this way) or the second species is a surface site (in the case of metal oxide dissolution), giving products that fast leave the surface. Following these observations, the bimolecular reaction can be treated as first order [48], or it can be assumed that only one species competes for the surface sites, allowing an analytical treatment based on the use of eqs.(7) and (23). When the reactant B is a surface site, in eq.(22) $C_B=C_s$ and $[B_f]/C_B=1$, and the solution term is null. The resulting relationship for the rate shows a linear behavior versus the surface concentration of the species [15]

$$r = k_{obs}C_A C_s \text{ (experimental)}$$
$$r = k_s f_c [A_s] = k_s f_c C_A C_s K^r_A / (1+K^r_A) \qquad (24)$$

where $K^r_A = f(C_A, C_s, K_{ads})$ is given by eq.(7) and k_s is the reaction rate constant at the surface. When $[A_s]$ is expressed as a function of $[A_{free}]$ from eq.(5), eq.(24) is equivalent to the Langmuir-Rideal equation. Since $f_c = C_s/(1000\ \beta d)$, where β is the number of surface sites per gram of colloids or powder and d is their density (g cm^{-3}), the rate has a parabolic shape as a function of C_s. Eq.(24) can easily applied to the study of the dissolution kinetics mechanism, which has important implications for several geochemical and technological fields, such as weathering and soil transformation, metal dissolution and corrosion, or passivation of metal surfaces.

6. Light-Driven Reactions

Photochemical reactions are transformations initiated by absorption of light and involve either the chromophore or other molecules through secondary reactions. The absorption of light into the sea is due to water itself, dissolved organic matter (DOM) and living and dead (inorganic) suspended particles. Inorganic solutes contribute little, but some transition metals and organic micropollutants may play a significant role. For example, dissolved Fe(III) in low pH waters can be reduced and the concentration of Fe(II) correlates well with the light intensity [49]. Such photoreduction may also take place in surface waters at higher pH where the oxidation of ferrous ion is fast. The net effect of iron photoreduction is to accelerate the dissolution of its minerals and chelates. Other transition metals undergo the same transformation. For example, the depth profile of Cu(I) in sea water has a surface maximum (up to a 15% of total Cu as Cu(I)) consistent with a photochemical reduction mechanism, with concomitant oxidation of ligands and inhibition of Cu(I) oxidation by chloride ions [50].

Inorganic particulate, mainly metal oxides like MnO_2 and Fe_2O_3, are subject to important photochemical reactions, leading to their dissolution, oxidation of metal chelates [51], decomposition of organic compounds and adsorbed DOM, and production of reactive species like OH and H_2O_2 [52]. These oxides are semiconductors. Light absorption promotes an electron from the valence band to the conduction band leaving an electron/hole pair, which, after migration at the surface, may oxidize and reduce electron donor and acceptor molecules adsorbed at the surface, respectively [53].

The organic particulate and colloids, mainly DOM, also absorb solar light. Normally, the excited state can transfer its excess energy to another ground-state chemical species, acting as a sensitizer, and producing reactive transients. Approximately, 0.1-60 nM of reactive radicals are produced in seawater per minute of full sun illumination, with the highest production rates in organic-rich estuarine waters. Morel and Hering [54,56] gave a detailed review of sources and sinks of photoreactants in natural waters.

The general reaction for all these photochemical processes may be schematized in the following Scheme

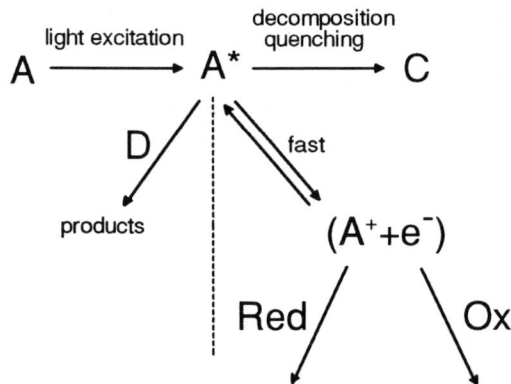

After light excitation, the chromophore (either a molecule, DOM, or a metal oxide particle) can undergo several processes. The reaction scheme takes into account either the reaction of the excited chromophore with the external species D (with rate constant k_D, left) or the charge separation ($A^+ + e^-$, right), and the reaction of these primary reactive species with electron donor (Red, with rate constant k_{red}) and acceptors (Ox, with rate constant k_{ox}). The excess energy can be relaxed to the lattice, the solvent or quenchers present in solution (k_q), or through molecular decomposition (k_{dec}). When A is a molecule, in the absence of further reactions, the sequence A → C represents the direct photolysis. When A* reacts with D, the reaction is called sensibilization or indirect photolysis. When A is DOM and D is $^3O_2$, or even easily oxidable organic matter R, the products are singlet oxygen $^1O_2$, and $A^\bullet + R^\bullet$, respectively. The radicals can further add O_2, giving peroxyradicals. When A is NO_3^-, NO_2^-, Fe(III) complexes, and D is water, the products are $^\bullet OH$ and $^\bullet NO_2$, and reduced iron, respectively. The dimerization of •OH produces H_2O_2 [55]. All these processes are important for the production of reactive species that acts as secondary environmental factors [56].

The case of charge separation applies to semiconductor metal oxides. Also DOM was suggested to produce aquated electrons [57], although only indirect evidence through reaction with easily reducible molecules was found. The kinetic analysis of the left part of the scheme is facile, giving for the rate $r_D = k_D[D] k_f \phi /(k_2+k_D[D])$ (see below for $k_f \phi$ and k_2).

The reactions depicted in the right part of the scheme lead to the following kinetic equations for the primary events

$$d[e]/dt = k_f \phi - (k_q+k_{dec})[e]_p [A^+]_p - k_{red}[e]_p [Ox]_p \quad (25a)$$
$$d[A^+]/dt = k_f \phi - (k_q+k_{dec})[e]_p [A^+]_p - k_{ox}[A^+]_p [Red]_p \quad (25b)$$

where $k_f \phi$ is the net absorbed photon flux leading to A*, $[e]_p$ and $[A^+]_p$ are electron and primary oxidant concentrations available at the surface traps of the semiconductor (hole) or at accessible sites of DOM, respectively. Since the reaction with Red and Ox is at the surface of the particle or inside it, the concentrations of Red and Ox (and D) are local concentrations of (ad)sorbed species (subscript p). In the seawater, at least one concentration of the species Red and Ox can be assumed constant. For example in surface waters, the concentration of dissolved oxygen (Ox), which is a good scavenger of electrons, may be constant.

Under constant illumination, steady state conditions can be assumed and $d[e]_p/dt = d[A^+]_p/dt = 0$. No assumptions are made on the steady state concentrations of $[e]_p$ and $[A^+]_p$. Eventual charging of the particle or DOM is allowed. The system (25) can be analytically solved (see for example ref.[58]) giving for the rate of Red disappearance $d[Red]_p/dt = - k_{ox}[A^+]_p[Red]_p$

$$\text{rate} = k_{ox}k_{red}[Ox]_p[Red]_p \; (-2 \; k_f(k_q+k_{dec})\phi - k_{ox}k_{red}[Red]_p[Ox]_p + (k_{ox}k_{red}[Red]_p[Ox]_p \; (k_{ox}k_{red}[Red]_p[Ox]_p + 4k_f(k_q+k_{dec})\phi))^{1/2} /((k_q+k_{dec}) (k_{ox}k_{red}[Red]_p[Ox]_p - (k_{ox}k_{red}[Red]_p[Ox]_p \; (k_{ox}k_{red}[Red]_p[Ox]_p + 4 \; k_f\phi (k_q+k_{dec})))^{1/2}) \quad (26)$$

Although depending on the values of parameters, this function shows that the rate increases as a function of $[Red]_p$ (or $[Ox]_p$) concentration or light intensity toward a limiting value, with a Langmuirian shape. Yet, in the case of sensibilization (right part of the Scheme) the rate depends linearly on the absorbed light. Under the hypothesis that the rate of A^+/e^- recombination and thermal or quenched decays are slower than that of electron scavenging by oxygen, the rate of Red oxidation is given by a simpler equation [59] that shows an analogous behavior:

$$\text{rate} = \frac{k_f\phi k_{ox}k_{red}[Ox]_p[Red]_p}{k_f\phi(k_q+k_{dec}) + k_{ox}k_{red}[Ox]_p[Red]_p} \quad (27)$$

It is worth noting that a saturating behavior is obtained independently from any adsorption hypothesis. The relationship (24), or similar equations reminiscent of classical theories on surface kinetics [60], like the Langmuir-Rideal or -Hinshelwood equations, have been often applied to the present kinetic problems, assuming that the surface site is reactive after photon absorption. However, those treatments hide the photochemistry behind the surface and are unable to correctly predict the dependences on all the experimental variables and concentrations.

Noting as before that $[Ox]_p$ and $[Red]_p$ are local concentration in the colloidal or particle domain, for sorption $[I]_p=[I_2]/f_c$ and for adsorption on C_s surface sites $[I]_p=[I_2]/C_s$, where $[I_2]$ is given by the moles of the species in the colloidal or surface pseudophase over the system volume. In the case of adsorption, recalling the definitions of the partition coefficient (6) and (18), eq.(27) can be transformed in a form which shows the dependence on the amount of surface sites. However, since also the amount of absorbed light depends on the particle concentrations, at low absorbed photon fluence, one can hypothesize that $k_f\phi = k'\phi\, C_s$. When the light is totally absorbed there is no such a dependence. Usually, the rate increases with a Langmuirian shape also as a function of the absorber concentration. All these dependences have been experimentally observed in oxidative degradation experiments with powdered and colloidal semiconductors [61]. Possibly, also the site saturation constraints may be taken into account.

7. References

1. Buffle J. and Leppard G.G. (1995) *Environ. Sci.Technol.* **29**, 2169.
2. Hunter, K.A., (1980) *Mar. Chem.* **9**, 49
3. Stumm W. (ed.) (1990) *Aquatic Chemical Kinetics. Reaction Rates of Processes in Natural Waters,*, Wiley, N.Y.
4. Stumm W. (ed.) (1987), *Aquatic Surface Chemistry. Chemical Processes at the Particle-Water Interface*, Wiley, N.Y.
5. Morel, F.M.M. and Hering J.G. (1993), *Principles and Applications of Aquatic Chemistry*, Wiley, N.Y.
6. Stumm W. and Morgan J.J. (1981), *Aquatic Chemistry. An Introduction Emphasizing Equilibria in Natural Waters*, Wiley, N.Y.
7. Stumm W. (1992), *Chemistry of the Solid-Water Interface. Processes at the Mineral-Water and Particle-Water Interface in Natural Systems*, with contributions of Sigg L. (chp 11) and Sulzberger (chp 10), Wiley, N.Y.
8. O'Melia C.R. (1990), *Kinetics of Colloid Chemical Processes in Aquatic Systems*, in ref.3, chp 13, pp 447-474
9. O Melia C.R. (1987), *Particle-Particle Interactions*, in ref.4, chp 14, pp 385-403.
10. El Seoud, O.A. (1989) *Adv. Coll. Interface Sci.* **30**, 1
11. Hoffmann M.R. (1990), *Catalysis in Aquatic Environments*, in ref.3, chp 3, pp 71-111.
12. Schwarzenbach R.P. and Gschwend P.M. (1990), *Chemical Transformations of Organic Pollutants in the Aquatic Environment*, in ref.3, chp 7, pp 199-233.
13. Hurd D.C. and Spencer D.W. (eds.) (1991), *Marine Particle: Analysis and Characterization*, Geophysical Monograph 63, American Geophysical Union, Washington DC.
14. Burgess R.M., MeKinney R.A., Brown W.A and Quinn J.G. (1996) *Environ. Sci Technol.* **30**, 1923.

15. Stumm W. and Wieland E. (1990), *Dissolution of Oxide and Silicate Minerals: Rates Depend on Surface Speciation*, in ref.3, chp13, pp 367-400.
16. Nowack B., Lutzenkirken J., Behra P. and Sigg L. (1996) *Environ. Sci. Technol.* **30**, 2397.
17. Yao W. and Millero F.J. (1996) *Environ.Sci Technol.* **30**, 536.
18. Stumm W. (1992), in ref 7, chp 3, p 43.
19. Hunter K.A., Liss. P.S. (1979) *Nature* **282**, 823.
20. Beckett R., Le N.P. (1990) *Colloids and Surfaces* **44**, 35
21. Karickhoff S.W., Brown B.J. and Scott T.A. (1979) *Water Res.* **206**, 831
22. Chin Y.-P., Weber W.J. and Eadie, B. (1990) *Environ. Sci. Technol.* **20**, 161.
23. Buffle J. (1988), *Complexation Reactions in Aquatic Systems. An Analytical Approach*, Ellis Horwood Lim. Publ., Chichester, chp 5.
24. Schindler P.W. and Stumm W. (1987), *The Surface Chemistry of Oxides, Hydroxides and Oxide Minerals*, in ref.4, chp 4, p 98.
25. Schlindler P.W., Walti E. and Furst B. (1976) *Chimia* **30**, 107; Schlindler P.W., Furst B., Dick R. and Wolf P.U. (1976) *J. Coll. Interface Sci.* **55**, 469.
26. Adamson A.W. (1990), *Physical Chemistry of Surfaces*, 5th Ed., Wiley, N.Y.
27. Buffle J. (1988), in ref.23, p.258.
28. Buffle J. (1988), in ref 23, p 262.
29. Sposito, G. (1984), *The Surface Chemistry of Soils*, Oxford University Press, N.Y.
30. Hunter R.J. (1989), *Foundation of Colloid Science*, Clarendon Press, Oxford, vol.2, chp 12.
31. Morel, F.M.M. and Hering J.G. (1993), in ref.5, p 76.
32. Buffle, J. (1988), in ref. 23, chp 5, p 203
33. Minero C., Pramauro E. and Pelizzetti E. (1988) *J.Phys.Chem.* **92**, 4670.
34. Ohshima H., Healy T.W. and White L.R (1982), *J.Coll. Interface Sci.* **90**, 17.
35. Stumm W. and Morgan J.J. (1981), in ref 6, p.610; Buffle J. (1988) in ref.23 p.240
36. Stumm W. and Morgan J.J. (1981), in ref 6, pp 4, 392, 411.
37. Pankow J.F. (1981), *Aquatic Chemistry Concepts*, Lewis Publ., Chelsea, chp 26. pp 603-641.
38. Bartshat B.M., Cabaniss S.E. and Morel F.M.M. (1992) *Environ. Sci Technol.* **26**, 284.
39. Stumm W. and Morgan J.J. (1981), in ref 6, pp 414-416.
40. Minero C., Pelizzetti E. (1992) *Adv. Colloid Interface Sci.* **37**, 319
41. Minero C., Pelizzetti E. (1993) *Pure Appl. Chem.* **65**, 2573.
42. Pelizzetti E., Minero C. and Maurino V. (1990) *Adv.Coll. Interface Sci.* **32**, 271
43. see for example: Burgess J and Pelizzetti E. (1992) *Progress Reaction Kinetics* **17**, 1.
44. Fendler J.H. (1982), *Membrane Mimetic Chemistry*, Wiley, N.Y.
45. Minero C. and Pelizzetti E. (1992), in S.E. Friberg and B. Lindman (eds.), *Organized Solutions*, Dekker, N.Y., chp 21.
46. Minero C., Pramauro E. and Pelizzetti E. (1989) *Coll. Surf.* **35**, 237
47. Minero C. and Pelizzetti E. (1995) *J. Disp. Sci. Technol.* **16**, 1
48. Zepp, R.G. and Wolfe N.Lee (1987), in ref.4, chp 16, p 423-455.
49. McKnight D.M. Kimball B.A. and Bencala K.E. (1988) *Science* **240**, 637.
50. Moffett J.W. and Zika R.J. (1987), in R.G. Zika and W.J. Cooper (eds.), *Photochemistry of Environmental Aquatic Systems*, ACS Symp.Ser. 327, ACS, Washington DC, pp117-129.
51. see for example the oxidation of Co(II)-EDTA in aqueous goethite suspensions, Xue Y. and Traina J. (1996) *Environ. Sci. Technol.* **30**, 1975.
52. Siffert C. and Sulzberger B. (1991) *Langmuir* **7**, 1627.

53. Serpone N. and Pelizzetti E. (eds.) (1989), *Photocatalysis. Fundamentals and Applications*, Wiley, N.Y.
54. Morel, F.M.M. and Hering J.G. (1993), in ref.5, p.489
55. Cooper W.J., Zika R.G., Petasne R.G., Plane J.M.C. (1988) *Environ. Sci.Technol.* **22**, 1156.
56. Hoigné J. (1990), in ref.3, chp 2, pp 43-70.
57. Zepp R.G, Braun A.M., Hoigné J. and Leenher J.A. (1987) *Environ. Sci. Technol.* **21**, 485
58. Minero C., Pelizzetti E., Malato S. and Blanco J. (1996) *Solar Energy* **56**, 421.
59. Minero C. (1995) *Sol En.Mat.Sol.Cells* **38**, 421.
60. Clark A. (1970), *The Theory of Adsorption and Catalysis*, Academic Press, N.Y., chp 11, pp 239-270.
61. Al-Ekabi H., Serpone, N., Pelizzetti E., Minero C., Fox M.A. and Draper R.B. (1989) *Langmuir* **5**, 250.

CHEMICAL SPECIATION OF SOME CLASSES OF LOW MOLECULAR WEIGHT LIGANDS IN SEAWATER

Alessandro DE ROBERTIS, Claudia FOTI, Silvio SAMMARTANO
Dipartimento di Chimica Inorganica, Chimica Analitica e Chimica Fisica, Università di Messina
Salita Sperone 31, I-98166 Messina (Vill. S. Agata), Italy

Antonio GIANGUZZA
Dipartimento di Chimica Inorganica, Università di Palermo
Via Archirafi 26, I-90123 Palermo, Italy

Abstract

The possibility of using simple relationships for the study of the complexing ability of some classes of ligands (carboxylates, amines, aminoacids, phenols) is discussed. Consistent speciation models (which take into account the dependence on ionic strength of protonation and formation constants) together with ligand-class characteristics, are able to facilitate speciation studies. Some examples of complexation studies in seawater are reported.

1. Introduction

The general term *"organic matter in sea water"* (D.O.M. and P.O.M.) refers to very complex molecular structures, such as polypeptides, polysaccharides, lipids, fatty acids, nucleic acids, polyamines, etc., and to a more general and undefined class of organic compounds termed *humic substances*. All these organic compounds are involved in the general biochemical processes of synthesis (*anabolism*) and degradation (*catabolism*), showing chemical behaviour based on similarities in structure and on the presence of certain functional groups, *i.e.*, R-COOH, R-OH, R_2-NH, R-NH_2, R-SH, with R alkylic or arylic organic structure. During catabolism, metabolites of each original macromolecular structure, which keep the primary functional groups, can be formed with lower molecular weight[*]. Dipeptides and single aminoacids can be formed from

---

[*] Owing to the great variability of organic sea water composition, data analyses of the individual low molecular weight organic compounds (rarely exceeding 1 μM) are not clearly defined and are found in a wide range of concentrations. A list of suggested readings on the topic is contained in refs. [1-4,10]

the degradation of proteins and polypeptides; saccharide units, mainly fructose and glucose, derive from the degradation of cellulose; phenolic compounds are present in sea water coming from the degradation of soluble lignine; low molecular weight carboxylic acids, such as citric, succinic, malic, oxalacetic, α-ketoglucaric, are produced during the Krebs cycle; other carboxylic acids are formed from the degradation of fatty acids during the bacterial decomposition of organic detritus; low molecular weight amines, such as diamines and tetramines, are formed during the decomposition of animal tissues, etc. Humic substances, contain all the most important functional groups, including amino and thiolo groups, in different percentages depending on the source of the original organic matter.

All the above functional groups interact, to different extents, with the macro and micro components of sea water, by affecting its chemical speciation. Their coordination capacity depends, in general, on several factors: i) the type and the number of functional groups in the same ligand molecule; ii) the pH value; iii) the qualitative (ionic medium) and quantitative composition (ionic strength) of the solution. Since both the pH value and the macro composition of sea water are constant, attention must be paid to the structure of the ligand molecule, by considering the number of functional groups and the total charge of ligand. Because the *conservative* components (Na^+, K^+, Ca^{2+}, Mg^{2+}, Cl^-, SO_4^{2-}, HCO_3^-, Br^-, F^-, Sr^{2+} and $B(OH)_3$) are typically present in seawater at least 100 to 1000-fold higher concentrations than each individual organic compound or trace metal, all possible interactions of these macro components with each class of functional groups must be considered, in order to build up chemical base models for the speciation of each class of low molecular weight ligands naturally occurring in sea water.

For this reason, a systematic study of the chemical behaviour of different carboxylates, aminoacids, amines, phenolic compounds in different ionic media, including a synthetic seawater containing the six major components (Na^+, K^+, Ca^{2+}, Mg^{2+}, Cl^-, SO_4^{2-}) has been carried out, in our laboratories, in a wide range of ionic strengths and temperatures. Examples for the speciation of some ligand classes are reported on. As concerns trace metals, attention has been focused on hydrolysis, which is the most consistent process in an aqueous solution such as sea water.

2. Speciation Models

Different speciation models have been proposed for both major and minor (or trace) components of seawater [5,6]. Since the goal of speciation is the distribution of a component among various species, one must search for interactions (ion pair or complex formation), for their equilibrium constants, and for activity coefficients, in order to refer to the real situation in the aquatic environment. When doing this one must choose the baseline for significant interactions. Specific interaction models attribute all weak (*e.g.* Na^+-SO_4^{2-}, Ca^{2+}-Cl^-, etc.) or very weak (*e.g.* Na^+-Cl^-) and in some cases also quite strong (*e.g.* Ca^{2+}-SO_4^{2-}) interactions to specific factors influencing the activity coefficients in different media. Ion pair (or complexation) models also take weak interactions into account, to a pre-established extent, and consider total and free activity coefficients, γ_T and γ, which are related by the equation (for a generic X component)

$$\gamma / \gamma_T = [X]_{free} / [X]_{total} \qquad (1)$$

2.1 SPECIFIC INTERACTION MODELS

According to the Specific Interaction Theory [7], mean activity coefficients can be expressed by the equation

$$\log \gamma_{\pm(MX)} = -A |z_M z_X| I_m^{1/2} (1 + I_m^{1/2})^{-1} + 2 v^+ v^- / v \, B_{MX} \, m_{MX} \qquad (2)$$

where A is the constant of the Debye-Hückel theory; z the charge of ion; I_m the molal ionic strength; $v = v^+ + v^-$; m the molality of the salt $M_{v+}X_{v-}$; B_{MX} the interaction coefficient. By assuming that the interaction coefficients are independent of the composition of the solution, eq. (2) can be written, for a mixture of electrolytes, as:

$$\log \gamma_{\pm(MX)} = -A |z_M z_X| I_m^{1/2} (1 + I_m^{1/2})^{-1} + v^+ / v \sum_X B_{MX'} m_{X'} + v^- / v \sum_M B_{M'X} m_{M'} \qquad (3)$$

On the basis of Specific Interaction Theory, other equations, such as those of Pitzer [8] and Bromley [9], have been developed for the determination of activity coefficients in multicomponent solutions. The Pitzer equation differs from eq. (2) in the addition of a second virial coefficient:

$$\ln \gamma_{\pm(MX)} = z_M z_X \, f^\gamma + 2 v^+ v^- / v \, B_{MX} \, m_{MX} + 2 (v^+ v^-)^{3/2} / v \, C_{MX} \, m^2_{MX} \qquad (4)$$

where:

$$f^\gamma = -0.392 \, [I_m^{1/2} (1 + 1.2 \, I_m^{1/2})^{-1} + 1.667 \ln (1 + 1.2 \, I_m^{1/2})]$$

$$B_{MX} = 2\beta^{(0)}_{MX} + [\beta^{(1)}_{MX} (2 I_m)^{-1}] [1 - (1 + 2 I_m^{1/2} - 2 I_m) \exp(-2 I_m^{1/2})]$$

$$C_{MX} = 1.5 \, C^{(\phi)}_{MX}$$

(for 1-1, 1-2, 1-3, 1-4 and 1-5 electrolytes). Bromley proposed a one-parameter equation for the dependence on ionic strength of activity coefficients:

$$\log \gamma_{\pm(MX)} = -A |z_M z_X| I_m^{1/2} (1 + I_m^{1/2})^{-1} + g(B_{MX'} I_m) + B_{MX} I_m \qquad (5)$$

$$g(B, I_m) = [(0.06 + 0.6\, B_{MX})\, |z_M\, z_X|\, I_m]\, [(1 + 1.5\, I_m / |z_M\, z_X|)^2]^{-1}$$

2.2 ION PAIR (OR COMPLEXATION) MODELS

For a metal M in the presence of a ligand L interacting with M, we have

$$[M]/[M]_T = (1 + K_{ML}\, [L])^{-1} = \gamma/(\gamma_M)_T \tag{6}$$

where K_{ML} is the stoichiometric formation constant [note that eqs. (1) and (6) are equivalent]. A series of different complex formation models have been reported, based on different assumptions [5,11]. Equations similar to (4) have been developed for systems in which more than one weak species is formed [12]. The main problems when using complex formation models are: (i) the determination of the explicit function $\log K = f(I)$; (ii) the definition of the baseline for the extent of complexation to be considered; and (iii) the definition of a convention for the properties of single ions and complexes in solution. In a series of studies dealing with dependence on ionic strength [13-17], we used the equation

$$\log K = \log {}^T K - z^*\, I^{1/2}\, (2 + 3\, I^{1/2})^{-1} + C\, I + D\, I^{3/2} + E\, I^2 \tag{7}$$

$$z^* = \Sigma\, (\text{charge})^2_{\text{reactants}} - \Sigma\, (\text{charge})^2_{\text{products}}$$

When all the interactions, with $K > 0.1\, \text{mol}^{-1}\text{dm}^3$, are taken into account, in the range $0 \leq I \leq 1\, \text{mol dm}^{-3}$, we found that the empirical parameters of eq. (7) depend only on the stoichiometry of complex formation reaction: $C = 0.1\, p^* + 0.23\, z^*$; $D = -0.1\, z^*$; $E = 0$; $p^* = \Sigma\, (\text{moles})_{\text{reactants}} - \Sigma\, (\text{moles})_{\text{products}}$ (the parameter E becomes relevant for $I > 1\, \text{mol dm}^{-3}$).

2.3 CONSTANT IONIC MEDIUM METHOD

The constant ionic medium method (*i.e.*, the determination of thermodynamic parameters for complex formation in a large excess of a salt considered as not interacting with reactants) has been widely used in coordination chemistry. However, when dealing with multielectrolyte solutions, such as seawater, it shows some weakness. In particular, most of the studies have been performed in $NaClO_4$, whilst NaCl prevails in natural fluids. In order to overcome this problem, as an example, Millero [18] estimated the ion pair parameters using the linear correlation between the Pitzer $\beta^{(0)}_{MX}$ parameters in NaCl and $NaClO_4$ for monovalent, divalent, and trivalent cations $[\beta^{(0)}_{MX}\, (NaClO_4) = a + b\, \beta^{(0)}_{MX}\, (NaCl)]$. As regards marine chemistry, several studies have been performed (mainly on the protonation of low molecular weight ligands) using synthetic seawater (SSW) as the constant ionic medium, according to different recipes [19,20].

3. Carboxylic ligands

Carboxylic ligands show an appreciable tendency to form complexes with alkali and alkaline earth cations. As reported in a recent review [21] on a systematic investigation from these laboratories, a carboxylic anion L^{z-} can, in principle, form all the species M_iH_jL [with M^{n+} = Mg^{2+}, Ca^{2+}, Na^+, K^+; j = 0, 1...(z-1); i = 1...(z-1); (i+j)$_{max}$ = 1...z]. For example, 1,2,3-propanetricarboxylic acid forms the species NaL^{2-}, $NaHL^-$, NaH_2L^0, Na_2L^-, Na_2HL^0, CaL^-, $CaHL^0$, CaH_2L^+, Ca_2L^+, and analogous species with potassium and magnesium ions. The stability of these complexes regularly increases with the charge of carboxylic ligand, and the function log K = f(z) can be approximated to a straight line, as shown in Figure 1, where log K of $NaL^{(z-1)-}$ and $CaL^{(z-2)-}$ species is plotted vs. z for $1 \leq z \leq 6$ [13,21,22].

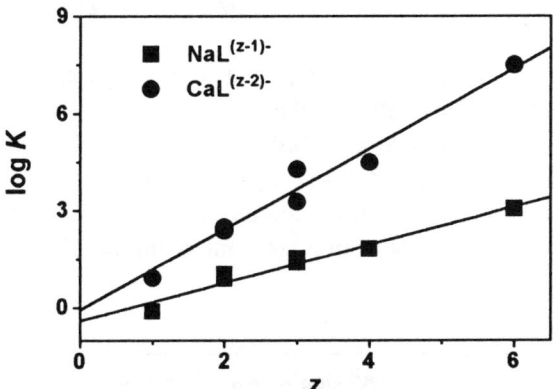

Figure 1. Values of log K of species $NaL^{(z-1)-}$ and $CaL^{(z-2)-}$ for different carboxylic ligands vs. z (z = charge of carboxylic ligand). Ligands reported: z = 1 acetate; z = 2 malonate and phthalate; z = 3 1,2,3-propanetricarboxylate and 1,2,3-benzenetricarboxylate; z = 4 1,2,3,4-butanetetracarboxylate; z = 6 benzenehexacarboxylate.

As regards the dependence on ionic strength of protonation and complex formation constants, all the parameters which define the function log K = f(I) can be expressed as simple relationships of the charges involved in the formation reaction. Recently, the protonation constants of carboxylic acids in different ionic media, have been fitted by some equations (Extended Debye-Hückel, Pitzer and Bromley equations) for dependence on ionic strength. Results obtained from our investigations, showed it was possible to express parameters for the dependence on I by equations depending only on anion charge and independent of the acid considered. As an example, in Figures 2a-b, some interaction parameters of Pitzer [eq. (4)] and Bromley [eq. (5)] equations vs. anion charge, for protonation constants of carboxylic acids (acetic, malonic, succinic, citric, 1,2,3-propanetricarboxylic, 2-methyl-1,2,3-propanetricarboxylic and 1,2,3,4-butanetetracarboxylic acids) in Me_4NCl, have been reported [23].

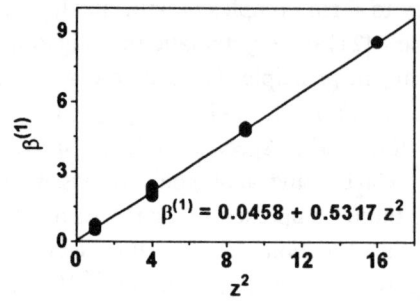

Figure 2a. Interaction parameters of Pitzer equation [eq. (4)] for carboxylic ligands in Me_4NCl vs. z^2 (z = anion charge).

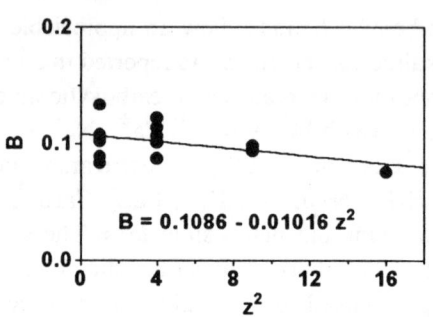

Figure 2b. Interaction parameters of Bromley equation [eq. (5)] for carboxylic ligands in Me_4NCl vs. z^2 (z = anion charge).

Bearing in mind the above considerations, apparent acid-base properties of this class of ligands in seawater should show useful regularities. The preliminary results of a study on the protonation of carboxylic ligands in SSW show that the dependence on salinity is a simple regular function of the negative ligand charge with the same behaviour shown by the stability of Na^+ and Ca^{2+} complexes. As an example, in Figure 3, $\log K_1^H - \log {}^TK_1^H$ ($\log {}^TK^H$ = protonation constant at infinite dilution) vs. $S^{1/2}$ (S = salinity ‰) is shown.

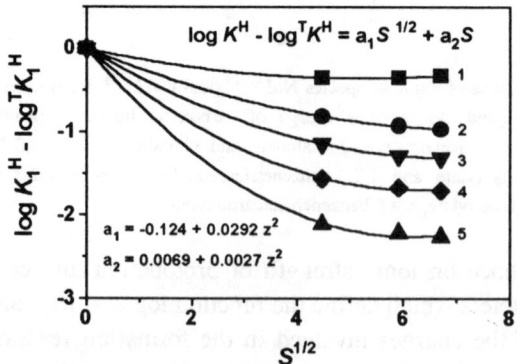

Figure 3. $\log K_1^H - \log {}^TK_1^H$ values of carboxylic acids vs. $S^{1/2}$. Acids considered: 1. acetic, 2. malonic and succinic; 3. citric; 4. 1,2,3-propanetricarboxylic and 2-methyl-1,2,3-propanetricarboxylic; 5. 1,2,3,4-butanetetracarboxylic.

Protonation constants can be expressed as

$$\log K^H - \log {}^T K^H = a_1 S^{1/2} + a_2 S \tag{8}$$

and a_1, a_2 (empirical parameters) are linear function of z^2. Slightly different behaviour has been found for some carboxylic ligands, such as citric and oxydiacetic acids, for which calcium complexes are stronger than those of other carboxylic ligands with the same charge. The higher stability can be attributed to the presence of hydroxide or ethereal groups, which are probably involved in the coordination.

4. Amines

Amines show no tendency to form complexes with alkali metal cations and form weak complexes with alkaline earth cations [24] ($Mg^{2+} \gg Ca^{2+} > Sr^{2+} \geq Ba^{2+}$). The stability of Mg^{2+} and Ca^{2+} complexes is a function of the number of unprotonated amino groups (n_a) according to the rough relationships

$$\log K^{CaA} = -0.4 + 0.44\, n_a \tag{9}$$

$$\log K^{MgA} = -0.9 + 0.45\, n_a \tag{10}$$

Protonated amines form quite stable complexes with organic and inorganic polyanions. In a recent review [25] it has been shown that in these systems a high number of species can be formed, *i.e.* n+m-1 species with m and n maximum values of protonation of amine and polyanion, respectively. The stability of these species is strictly dependent on the charges involved in the formation reactions. In Figure 4 formation constants of proton-amine-chloride complexes *vs.* cation charge are reported. The regular trend towards stability is quite similar to that of other proton-amine-polyanion complexes, and a general equation for the dependence of log K on charges has been found [25]:

$$\log K = -2.3 + 1.42\, z_{cat} + a\, |z_{an}| - 0.45\, \Delta z \tag{11}$$

(z_{cat} = charge of protonated amine, z_{an} = charge of anion, $\Delta z = |z_{cat} + z_{an}|$) with a = 1.44_5 and 1.09_5 for inorganic and organic anions, respectively. Equation (11) was obtained by fitting together all the values of stability constants collected in the review (16 polyamines, 16 polyanions, 481 complexes). Other less important factors affecting stability have been found, such as the distance of charges, the number of aminogroups in the polyamines and the difference of charges in the reactants.

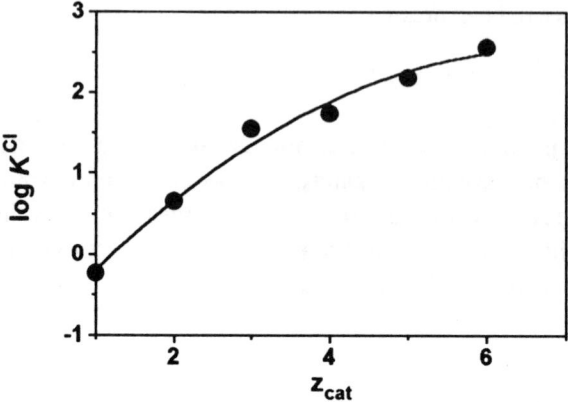

Figure 4. log K values of proton-amine-chloride complexes vs. cation charge (z_{cat}).

When protonated amine-anions of medium are taken into account, the dependence on ionic strength of amine protonation and complex formation constants is the same as for carboxylic ligands. Protonation in SSW can be expressed (for the first protonation constant) by

$$\log K^H - \log {}^T K^H = a_1 S \qquad (12)$$

and for ethylenediamine (en), diethylenetriamine (dien) and tetraethylenepentamine (tetren) we have a linear trend of a_1 as function of z_{cat} (Figure 5).

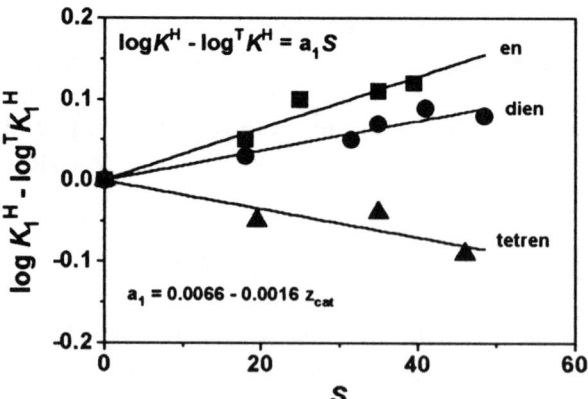

Figure 5. $\log K_1^H - \log {}^T K_1^H$ values of amines vs. S.

5. Aminoacids

As expected, aminoacids show complexation characteristics intermediate between amines and carboxylic ligands. As regards their complexing ability, they can be grouped into three different types [26]. The first group includes glycine, alanine and serine, which form very weak complexes with Cl^-, a weak species with SO_4^{2-} and a fairly stable species with Mg^{2+} and Ca^{2+} The second group includes lysine and histidine, which form a very weak species with Na^+, weak species with Cl^-, and fairly stable species with Mg^{2+}, Ca^{2+} and SO_4^{2-}. The third group includes aspartic, glutamic and iminodiacetic acids, which form a very weak species with Cl^-, two weak species with Na^+, and relatively weak species with Mg^{2+}, Ca^{2+} and SO_4^{2-}. Protonation constants in SSW can be grouped according to carboxylic groups and N-donor groups in the molecule, and the behaviour of $\log K_1^H - \log {}^T K_1^H$ vs. $S^{1/2}$ reflects the complexation characteristics of different types of aminoacids.

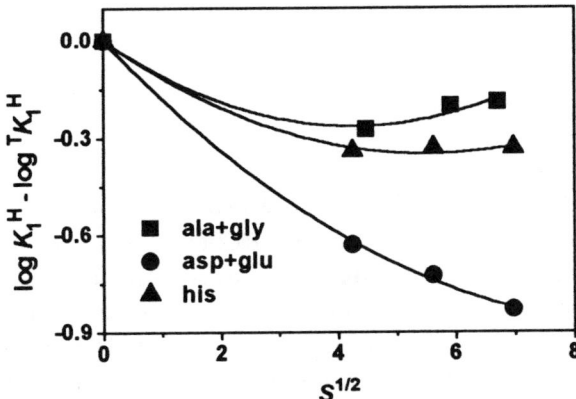

Figure 6. $\log K_1^H - \log {}^T K_1^H$ values of aminoacids vs. $S^{1/2}$.

6. Phenols

All phenolic compounds are O-donor ligands, showing complexing capacities toward metal cations similar to those of carboxylic ligands [27]. Recent studies conducted in these laboratories showed that they form very weak complexes with Na^+ and K^+, with stability constants fairly similar to those found for acetate. Also the formation constants of Mg^{2+} and Ca^{2+} complexes are close to those of acetate. Apparent protonation constants in SSW can be expressed by eq. (8) with values of empirical parameters a and b quite constant for different phenolic compounds (a = 0.106±0.017, b = 0.010±0.004).

7. Hydroxide

Reactions of OH⁻ with metal ions in a marine environment are of great importance. In fact at natural pH values (5 < pH < 9, $pH_{seawater} \approx 8.2$) most cations undergo hydrolysis. Studies have been performed pK_w at different salinities, whilst few investigations can be found on the dependence of hydrolysis constants at different salinities (some studies have been reported at $S = 35$ ‰). Measurements have been taken in these laboratories on the hydrolysis of Cu^{2+}, Ni^{2+}, $(CH_3)_2Sn^{2+}$ and $(CH_3)_3Sn^+$. In general, it has been found that simple ion pair (complexation) models are sufficient to explain the dependence of hydrolysis constants on salinity. As an example, the hydrolysis of trimethyltin, $(CH_3)_3Sn^+ = (CH_3)_3Sn(OH)^0 + H^+$, can be expressed by the equation $\log K = -6.142 - 0.137\ I$ (in $NaNO_3$, not complexing medium). This cation forms weak complexes with Cl⁻ and SO_4^{2-}, and therefore in principle the hydrolysis constant in SSW can be predicted by $\log K = 6.142 - 0.137\ I - \log(1 + K^{Cl}[Cl] + K^{SO_4}[SO_4])$, which is equal to -6.263 (S = 35‰). Measurements performed in this medium give $\log K = -6.265$, in excellent agreement with the predicted value.

8. Conclusions

In this short review we have shown that low molecular weight ligands can be grouped in classes having similar characteristics as regards the formation of not very stable complexes and dependence on ionic strength of activity coefficients and complex formation constants. In other words, parameters of interest in the speciation of natural fluids (and, in particular, seawater) can be easily predicted. The use of appropriate (and consistent) models for the dependence on ionic strength of activity coefficients, together with the knowledge of ligand-class properties make the task of solving speciation problems easier. These data may also be useful for the prediction of intrinsic formation constants for homologous ligands.

9. References

1. Duursma, E.K. (1965) The dissolved organic constituents of Sea Water, in *Chemical Oceanography*, Vol. 1, 1^ ed., Riley, J.P. and Skirrow, G. Eds., Academic Press, New York, pp. 433-475; Duursma, E.T. and K. Mopper, K. (1976) Factors controlling the distribution and early diagenesis of organic material in marine sediments, in *Chemical Oceanography*, Vol. 6, 2^ edt., Riley, J.P. and Chester, R. Eds., Academic Press, New York, pp. 60-113.
2. Williams P.J le B. (1975) Biological and chemical aspects of dissolved organic material in sea water, in *Chemical Oceanography*, Vol. 2, 2^ edt., Riley, J.P. and Chester, R. Eds., Academic Press, New York, pp. 301-363.
3. Lee, C. and Wakeham, S. (1989) Organic Matter in sea water: Biogeochemical Processes, in *Chemical Oceanography*, Vol. 9, 2^ edt., Riley, J.P. Ed., Academic Press, New York, pp. 1-44.
4. Libes, S. (1992) *An Introduction to Marine Biogeochemistry*, John Wiley & Sons, New York.

5. De Robertis, A., De Stefano, C. and Sammartano, S. (1994) Equilibrium studies in natural fluids: a chemical speciation model for the major constituents of sea water, *Chemical Speciation and Bioavailability*, **6**, 65-84.
6. De Robertis, A., De Stefano, C., Foti, C., Gianguzza, A., Sammartano, S. and Signorino, G. (1996) Models of natural fluids: the network of chemical interactions, *Ann. Chim. (Rome)*, **86**, 0000.
7. Biedermann, G. (1986) Introduction to the specific interaction theory with emphasis on chemical equilibria, in *Metal complex in solution*, Jenne, A.E., Rizzarelli, E., Romano, V. and Sammartano, S. Eds., Piccin, Padua, Italy, pp. 303-314.
8. Pitzer, K.S. (1973) Thermodynamics of electrolytes. I. Theoretical basis and general equations, *J. Phys. Chim.*, **77** (2), 268-277.
9. Bromley, L.A. (1973) Thermodynamic properties of strong electrolytes in aqueous solutions, *AlChE J.*, **19** (2), 313-320.
10. Millero, F.J. (1996), *Chemical Oceanography*, CRC Press, Boca Raton, Florida.
11. Millero, F.J. and Schreiber, D.R. (1982) Use of the ion pairing model to estimate activity coefficients of the ionic components of natural waters, *Am. J. Sci.*, **282**, 1508-1540.
12. De Robertis, A., De Stefano, C., Sammartano, S. and Rigano, C. (1987) The determination of formation constants of weak complexes by potentiometric measurements: experimental procedures and calculation methods, *Talanta*, **34**, 933-938.
13. Daniele, P.G., De Robertis, A., De Stefano, C., Sammartano, S. and Rigano, C. (1985) On the possibility of determining the thermodynamic parameters for the formation of weak complexes using a simple model for the dependence on ionic strength of activity coefficients: Na^+, K^+, and Ca^{2+} complexes of low molecular weight ligands in aqueous solution, *J. Chem. Soc. Dalton Trans.*, 2353-2361.
14. Casale, A., Daniele, P.G., De Robertis, A. and Sammartano, S. (1988) Ionic strength dependence of formation constants. Part XI. An analysis of literature data on carboxylate ligand complexes, *Ann. Chim. (Rome)*, **78**, 249-260.
15. Casale, A., Daniele, P.G., De Stefano, C. and Sammartano, S. (1989) Ionic strength dependence of formation constants - XII. A model for the effect of background on the protonation constants of amines and aminoacids, *Talanta*, **36**, 903-907.
16. Daniele, P.G., De Robertis, A., De Stefano, C. and Sammartano, S. (1991) Ionic strength dependence of formation constants. XIII. A critical examination of preceding results, in *Miscellany of scientific papers offered to Enric Casassas*, Alegret, S., Arias, J.J., Barceló, D., Casal, J. and Routers, G. Eds., Bellaterra, Spain, pp. 121-126.
17. De Stefano, C., Foti, C., Gianguzza, A., Martino, M., Pellerito, L. and Sammartano, S. (1996) Hydrolysis of $(CH_3)_2Sn^{2+}$ in different ionic media: salt effects and complex formation, *J. Chem. Eng. Data*, **41**, 511-515.
18. Millero, F.J. (1992) Stability constants for the formation of rare earth inorganic complexes as a function of ionic strength, *Geochim. Cosmochim. Acta*, **56**, 3123-3132.
19. De Stefano, C., Foti, C., Sammartano, S., Gianguzza, A., Rigano, C. (1994) Equilibrium studies in natural fluids. Use of synthetic seawater and other media as background salts, *Ann. Chim. (Rome)*, **84**, 159-175.
20. Whitfield, M. and Clegg, S.L. (1991) Activity coefficients in natural waters in *Activity Coefficients in Electrolyte Solutions*, Pitzer, K.S. Ed., CRC Press, Boca Raton, Florida, pp. 279-434.
21. Daniele, P.G., De Stefano, C., Prenesti, E. and Sammartano, S. (1994) Weak complex formation in aqueous solution, *Current Topics in Sol. Chem.*, **1**, 95-106.
22. De Robertis, A., De Stefano, C. and Foti, C. (1996) Studies on polyfunctional O-ligands. Protonation and alkali and alkaline earth metal complex formation of benzenehexacarboxylate, *Ann. Chim. (Rome)*, **86**, 155-166, and references therein.
23. Foti, C., Gianguzza, A. and Sammartano, S. (submitted) Protonation constants of carboxylic acids in aqueous tetramethylammonium chloride at different ionic strengths and at T = 25 °C, *J. Solution Chem.*
24. Capone, S., De Robertis, A., De Stefano, C. and Scarcella, R. (1985) Thermodynamics of formation of magnesium, calcium, strontium and barium complexes with 2,2'-bipyridyl and 1,10-phenantroline, at different ionic strengths in aqueous solution, *Talanta*, **32**, 675-677. Casale, A., De Robertis, A., Licastro, F. and Rigano, C. (1990) Salt effects on the protonation of ethylenediamine: a complex-formation model, *J. Chem. Res.*, (S) 204-205, (M) 1601-1620.
25. Daniele, P.G., Prenesti, E., De Robertis, A., De Stefano, C., Foti, C., Giuffrè, O. and Sammartano, S. (1997) Binding of inorganic and organic polyanions by open-chain protonated polyamines in aqueous solution, *Ann. Chim. (Rome)*, **87**, 0000.
26. De Stefano, C., Foti, C., Gianguzza, A., Rigano, C. and Sammartano, S. (1994) Chemical speciation of aminoacids in electrolyte solutions containing major components of natural fluids, *Chemical Speciation and Bioavailability*, **7**(1), 1-8.
27. Demaniov, P., De Stefano, C., Gianguzza, A. and Sammartano, S. (1995) Equilibrium studies in natural waters: speciation of phenolic compounds in synthetic seawater at different salinities, *Env. Tox. and Chem.*, **14**(5), 767-773.

COMPUTER TOOLS FOR THE SPECIATION OF NATURAL FLUIDS

Concetta DE STEFANO, Silvio SAMMARTANO
*Dipartimento di Chimica Inorganica, Chimica Analitica
e Chimica Fisica. Università di Messina.
Salita Sperone 31, I-98166 Messina (Vill. S. Agata), Italy*

Placido MINEO, Carmelo RIGANO
*Dipartimento di Scienze Chimiche. Università di Catania.
Viale A. Doria 6, I-95125 Catania, Italy*

Abstract

Some computer tools, useful in speciation studies, are examined in this paper. In particular problems related to the calculation of formation constants and free concentrations in multicomponent systems are described, together with some subsidiary calculations. A new general computer program for fitting linear and nonlinear equations (LIANA) is reported.

1. Introduction

Computers are nowadays among the most important instruments in all scientific fields. The speciation of natural fluids requires computer aids for the solution of a wide series of problems. All these studies, which require the solution of several mass balance equations for calculating the free concentration of all the components present in a multimetal-multiligand system, require the use of appropriate and robust computer programs. In this paper we will examine some computer methods for the determination and use of thermodynamic parameters, and, to a lesser extent, describe other useful and quite general computer programs.
Natural (and biological) fluids may in practice contain all the inorganic cations and anions, several low molecular weight organic compounds, and some (very important) macromolecular compounds, such as humic substances. Two interesting features must be evidenced: (i) the relative constancy of the composition for many natural fluids, and (ii) the abundance of some inorganic ions (Na^+, K^+, Ca^{2+}, Mg^{2+}, Cl^- and SO_4^{2-}) which are always major components. All the ions and substances present in a natural fluid

interact with each other, and behave in different ways according to the different relative compositions of different fluids. Moreover, solution-sediment interactions must be taken into account. As a first approximation, one must primarily estimate the thermodynamic parameters for all the possible interactions in solution (together with temperature and ionic strength dependence), activity coefficients of all the species, and interaction parameters with other phases. Numerical analysis has been extensively applied to chemical problems [1-5] and in some cases chemometrics [6-8] are very useful in environmental studies.

2. Calculation of Free Concentrations

Let us consider the generic reaction among components X_i to give the species S (p_i = stoichiometric coefficients, z_i = charges)

$$p_1 X_1^{z_1} + p_2 X_2^{z_2} + ... + p_i X_i^{z_i} + ... p_n X_n^{z_n} = S^{z_s}$$

with $i = 1...n$ (n = no. of components), and having the equilibrium constant

$$\beta = [S] \{ [X]_1^{p_1} [X]_2^{p_2} ... [X]_i^{p_i} ... [X]_n^{p_n} \}^{-1}$$

By taking into account activity coefficients γ_i and γ_s we can express the equilibrium constant at infinite dilution as

$$^T\beta = \beta \, \gamma_s \, (\gamma_1^{p_1} \gamma_2^{p_2} ... \gamma_i^{p_i} ... \gamma_n^{p_n})^{-1}$$

Calculation problems for equilibria in solution in general need the solution of a set of mass balance equations. This can be written

$$\mathbf{X} = \mathbf{x} + \mathbf{p} \, \mathbf{s}$$

($\mathbf{X}$ = vector of analytical concentrations; $\mathbf{x}$ = vector of free concentrations; $\mathbf{p}$ = matrix of stoichiometric coefficients; $\mathbf{s}$ = vector of species concentrations), or

$$\mathbf{L} = \mathbf{B} + \mathbf{p}^T \mathbf{l} \tag{1a}$$

[$\mathbf{L}$ = vector of log(species concentrations), $\mathbf{B}$ = vector of logβ, $\mathbf{l}$ = vector of log(free concentrations)]. The system of equation can be solved by different methods, to obtain free concentrations, and therefore species concentrations. The most popular numerical methods used for this problem are briefly described below [9-14].
a) The "ratio" method, first proposed by Perrin and Sayce [12], for which the iteration formula is (i = iteration index, q = damping factor):

$$x^{(i+1)} = x^{(i)} R^{(i)}$$

$$R = (X/X_{calcd})^{1/q}$$

b) The Newton-Raphson method, for which the iteration formula is

$$x^{(i+1)} = x^{(i)} + G^{-1}e \qquad (2)$$

(where G is the matrix with generic element $g_{ij} = \partial X_i/\partial x_j$; e is the vector with generic element $e_i = X_{i,exp} - X_{i,calcd}$) has been widely used in the computation of free concentrations, as well as in many nonlinear chemical problems. It is possible to use quite an efficient method which consists in a Newton-Raphson type iteration procedure with a damping routine based on the 'ratio' method. When dealing with experimental conditions and/or species stability constants at different (and/or variable) ionic strengths, one must use activity coefficients in the mass balance equations. Therefore eqn.(1a) becomes

$$L + \Gamma_c = B_T + p^T l + p^T \Gamma_s \qquad (1b)$$

[B_T = vector of thermodynamic (zero ionic strength) overall formation constants (log); Γ_c = vector of component activity coefficients (log); Γ_s = vector of species activity coefficients (log)]. The Newton-Raphson technique allows x_k values to be calculated by the iterative procedure (2). At a certain stage n, the calculated analytical concentration is taken from

$$X_{k,calcd}^{(n)} = X_k^{(n)} + [\Sigma p_{ik} \beta_i \Pi x_j^{p_{ij}}]^{(n)}$$

This technique offers the important advantage that few iterations are generally needed to reach convergence, but there are two main difficulties: (a) when G is singular or near-singular, inversion is impossible; (b) iteration is often very unstable and, at some stage, the correction $G^{-1} e$ may lead to divergence. These difficulties must always be kept in mind when the Newton-Raphson technique is used, but are particularly severe when dealing with highly nonlinear equations and when $k>5$. In order to avoid underflow and/or overflow problems, logarithms are used in calculating species concentrations. To overcome difficulty (a), scaling was applied to matrix G and to vector e. The following procedure proved best at improving the stability of the iterations. If at a certain stage n, the ratio

$$R_k^{(n)} = X_k / X_{k,calcd}^{(n)}$$

for the k-th component lies outside the range $1/\rho < R_k^{(n)} < \rho$ (ρ is a limit chosen in the range $1 < \rho < 10$), then the free concentration of the k-th component is damped by the equation

$$x_{k,damped}^{(n)} = x_k^{(n)} (R_k^q)^{(n)} \qquad [q = 1/(p_{ik})_{max}]$$

(*i.e.*, q is the reciprocal of the largest stoichiometric coefficient of the species containing the k-th component). This procedure is applied to the component for which $\ln|R_k|$ assumes the maximum value, and is repeated until R_k for all components satisfies the condition $1/\rho < R_k^{(n)} < \rho$. Then a new Newton-Raphson iteration step is performed. For the solution of the system [eqn.(2)] the compact Gauss method can be chosen (modified Gauss elimination method, more easily programmable on computers). Dependence on ionic strength (calculated by iteration) of formation constants can be taken into account by adding a linear term to log(formation constant), [eqn.(1b)]. The above described algorithm was used for writing the various versions of the program ES4EC program [9-11]. Many other computer programs have been written to deal with the calculation of equilibrium concentrations (also in the presence of different phases) [12-14].

3. Calculation of Thermodynamic Parameters

3.1 FORMATION CONSTANTS OF COMPLEX SPECIES

For the refinement of formation constants the Gauss-Newton technique can be used [4]

$$\mathbf{s} = (\mathbf{A}^T \mathbf{W} \mathbf{A})^{-1} (\mathbf{A}^T \mathbf{W}) \mathbf{e}$$

[$\mathbf{A}$ = matrix of partial derivatives $\partial y/\partial p$ (y = independent variable); $\mathbf{W}$ = weight matrix; $\mathbf{e}$ = residuals vector; $\mathbf{s}$ = shifts vector]. Different independent variables, y, have been chosen in the refinement of formation constants from potentiometric data, such as E (e.m.f.), v (titrant volume), X (analytical concentrations), $\bar{n}$ (average number of ligands bound to the metal), Z (average number of protons displaced per metal ion). In the new program STACO [15] we devised to minimize (v = titrant volume)

$$U = \Sigma w (v_{exp} - v_{calcd})^2$$

Weights for each point of titration curves are given as (s^2 = variance)

$$w = 1/s^2 \ ; \ s^2 = s_v^2 + (\partial v/\partial E)^2 s_E^2$$

This program can deal with potentiometric titrations at different ionic strengths.

3.2 OTHER THERMODYNAMIC PARAMETERS

Formation enthalpies, ΔH^0, can be obtained from the temperature dependence of equilibrium constants or by direct calorimetry. In this case we calculate ΔH^0 values of the equation

$$-Q_{corr} = \sum \Delta H_i^0 \, \Delta n_i$$

(Q_{corr} = experimental heat, corrected for the dilution; Δn_i difference in mmoles of i-th species). Calculations using linear least squares and free concentration determination allow the problem to be solved [16].

In speciation studies it is very important to determine the dependence of activity coefficients on ionic strength (in different ionic media), *i.e.* to determine parameters, such as specific interaction coefficients, which define this dependence. A useful computer program for such calculations (linear and nonlinear problems) is reported in section 4.

3.3 WEAK COMPLEX FORMATION FROM PROTONATION CONSTANTS IN DIFFERENT MEDIA

Weak complexes are often important in defining rigorously the composition of natural fluids, where the formation of these species is highly probable because of their high concentration of alkali and alkaline metal ions. In the study of the interaction between low molecular weight ligands and these ions three main problems must be taken into account: (i) the choice of a suitable reference background that complexes neither the ligand nor the metal under study; (ii) the choice of a model for the dependence on ionic strength, because high and variable concentrations often make the constant ionic medium method unsuitable; (iii) the choice of a suitable calculation method.

As regards the calculation methods, we will describe the algorithm on which the computer program ES2WC [22] is based. This program allows the determination of thermodynamic parameters in the field of weak interactions. ES2WC can simultaneously treat conditional protonation data obtained at different temperatures and ionic strengths to calculate formation constants and all the parameters for their dependence on temperature and ionic strength. For the general formation reaction

$$pM^{z_M} + L^{z_L} + qH^+ = M_pLH_q^{(pz_M + z_L + q)}$$

the formation constants and the mass balance equations are written as follows (setting the stoichiometric coefficient of the ligand equal to unity)

$$\beta_{pq} = [M_pLH_q] / ([M]^p [L][H]^q)$$

$$[M]_{tot} = [M] + [L] \Sigma p\beta_{pq}[M]^p[H]^q$$

$$[L]_{tot} = [L] (1 + \Sigma\beta_{pq}[M]^p[H]^q)$$

$$[H]_{tot} = [H] + [L] \Sigma q\beta_{pq}[M]^p[H]^q$$

and, when the formation of weak complexes is neglected

$$[L]_{tot} = [L'] (1 + \Sigma\beta'_{0q}[H]^q)$$

$$[H]_{tot} = [H] + [L'] \Sigma q\beta'_{0q}[H]^q$$

where primes indicates conditional quantities. The average number of protons bound to the ligand is defined by

$$\bar{n} = \Sigma q\beta_{pq}[M]^p[H]^q / (1 + \Sigma q\beta_{pq}[M]^p[H]^q)$$

$$\bar{n}' = \Sigma q\beta'_{0q}[H]^q / (1 + \Sigma q\beta'_{0q}[H]^q)$$

As a first step, the conditional protonation constants are simply calculated from primary experimental data by the appropriate computer program, $\bar{n}'$ is then obtained at $2q_{max}-1$ values of [H] corresponding to characteristic points of the titration curve, ($\bar{n}'$ is a quasi-experimental quantity and $\bar{n}$ is a calculated one, $\bar{n}' \cong \bar{n}_{exp}$). The β_{pq} values can be calculated by minimizing the residual squares sum over nk points (n = no. of titrations; k = no. of protonation constants) instead of all experimental data

$$U = \Sigma_i(\bar{n}_i - \bar{n}'_i)^2 \quad (i = 1...nk)$$

3.4 LITERATURE

A large amount of literature about calculation problems concerning speciation studies is available for the reader. In addition to the references cited above, useful information sources on these subjects are refs. [17-21].
The importance of stability constants in all the branches of chemistry is beyond doubt. This is more evident in environmental chemistry, where the species and the free concentrations of a component play a fundamental role in determining the mobility and toxicological impact of that component. A large database of thermodynamic parameters has been accumulated [23-27] for both formation constants and activity coefficients. Since the conditions of natural systems are very different as concerns

temperature and ionic strength, another very important factor in defining rigorously the complex formation between components is a knowledge of the temperature coefficients and the ionic strength dependence of formation constants. ΔH^0 values for complex formation are given in several compilations [24,26]; nevertheless, gaps in knowledge of temperature dependence of logK exist for many systems. The same is true for the dependence on ionic strength of formation constants.

4. A Flexible Computer Program for Linear and Non-linear Fitting

Physical, chemical, biological and in general natural processes are mostly described by non linear equations. In general, the path followed by most scientists is to write new algorithms and computer programs for each type of problem met in their experimental and/or theoretical work, or in some cases, to use fairly general routines (and some further subroutines) with a main program written for the specific problem. This time-consuming approach, often requiring a long set up time, is acceptable only when the problem studied is quite important and a lot of systematic calculations have to be performed; the second approach depends on the flexibility of available routines and on the compatibility of the general algorithm with the problem under study. A further difficulty arises from the availability of high capacity machines in terms of compiling and execution speed. In fact big, fast machines can deal with very complicated algorithms and not very efficient computer programs, in which several subroutines are added to the "main" in order to enable the solution of a specific problem, but this method implies higher running times and higher core requirements of some order of magnitude with respect to specific applied algorithms and computer programs. Therefore, this method is impracticable in most laboratories where calculations have to be performed in a short space of time using personal computers. Nowadays, in general, simple and fast computer programs not requiring large core, are not available in literature, even if general routines have been published to be run in rather big machines. Nevertheless, new efficient small machines, languages, compiler, and operating systems, render the problem easier than in the past, and therefore the matter should be reconsidered in the light of present computer calculation capabilities. Bearing in mind the above considerations, we thought it of some interest to build up a computer program based on a rigorous algorithm, devised to fit any non linear equations (and linear), that can be run in small personal computers.
This program, called LIANA, (LInear And Nonlinear Analysis), was written in Pascal language, using the general Levenberg-Marquardt-Gauss-Newton Method, with the

following relevant features*:
  i) it can deal with the refinement of parameters for any linear and nonlinear equation;
  ii) equation is defined (using a BASIC-like language) by the user;
  iii) in order to make the use of the program easier, the main equation can be subdivided into many partial equations;
  iv) different equations having all or some common parameters can be considered simultaneously;
  v) different weights can be assigned to experimental variables;
  vi) graphical facilities are offered by the program;
  vii) it is also possible to deal with two other (similar) problems, *i.e.* resolution of equations and function optimisation.
In Fig.1 the flow chart of LIANA is reported.

5. Some Utility Programs

5.1 PHYSICAL AND CHEMICAL PARAMETERS FOR SYNTHETIC SEA WATER

A synthetic sea water (SSW), containing different salts was proposed for use as a standard medium in equilibrium studies related to marine chemistry [28] (NaCl 0.42740; Na_2SO_4 0.02919; KCl 0.01112; $CaCl_2$ 0.01121; $MgCl_2$ 0.05552; concentrations in mol/Kg of solvent; S = 35‰). Molalities at different salinities can be calculated by this relationship

---

*LIANA (ver. 2.0; 1996) is a program of general fitting written in Turbo Pascal 5.5 (BORLAND), it works on an Ms-DOS platform (Ver. 5.0 and over).
It requires at least 1.0 MByte of random access memory and it is preferable, if the graphics are going to be used, that a (HIMEM.SYS + EMM386.EXE) memory manager be installed in order to be able to load the overlay module, the driver and the font in high memory.
It warns itself of possible remove programs resident (TSR) in low memory.
Besides fitting procedure, the program includes three other modules that allow: the resolution of system equations, the optimization of a function and the management of a data base of functions.
General characteristics :
– Numeric coprocessor (automatic recognition)
– Software emulator (for PC without coprocessor), or supported chip 80-x87
– Supported graphic card (automatic recognition)
– MDA (no graph), CGA, EGA, VGA, HERCULES

Max no. of equations	15
Max no. of points; couples $x(i)$-y	2100
Max no. of independent variables $x(i)$	30
(Number of points) x (number of ind. var.) must be <	10500
Max no. of equations	15
Max no. of analytical derivatives	15
Max no. of parameters (analytical derivatives used)	30 (15)
Max no. of partial functions	9
Max no. of characters for single equation in explicit form	4000
Max no. of characters for single partial function in explicit form	250

COMPUTER TOOLS FOR THE SPECIATION OF NATURAL FLUIDS

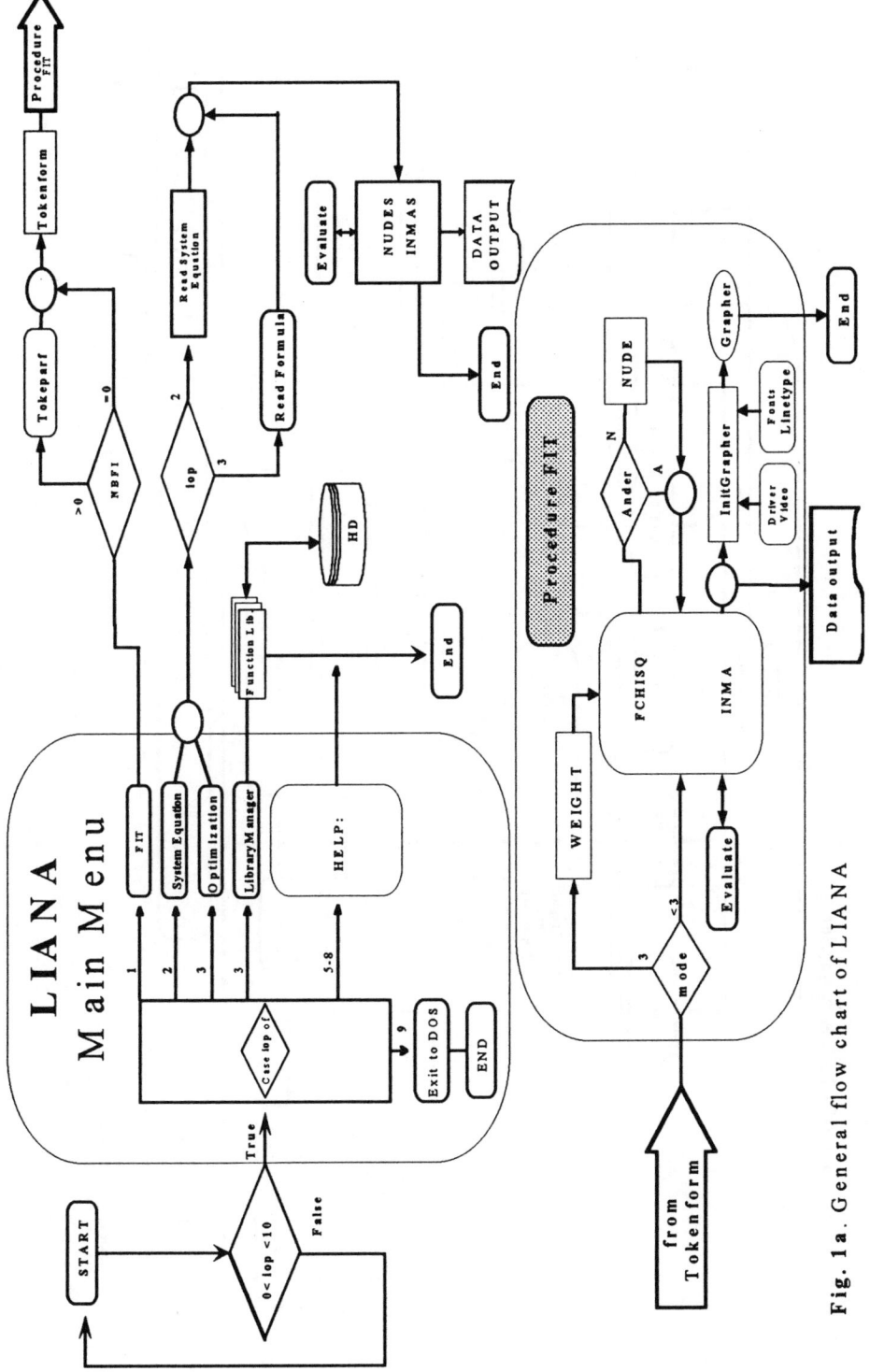

Fig. 1a. General flow chart of LIANA

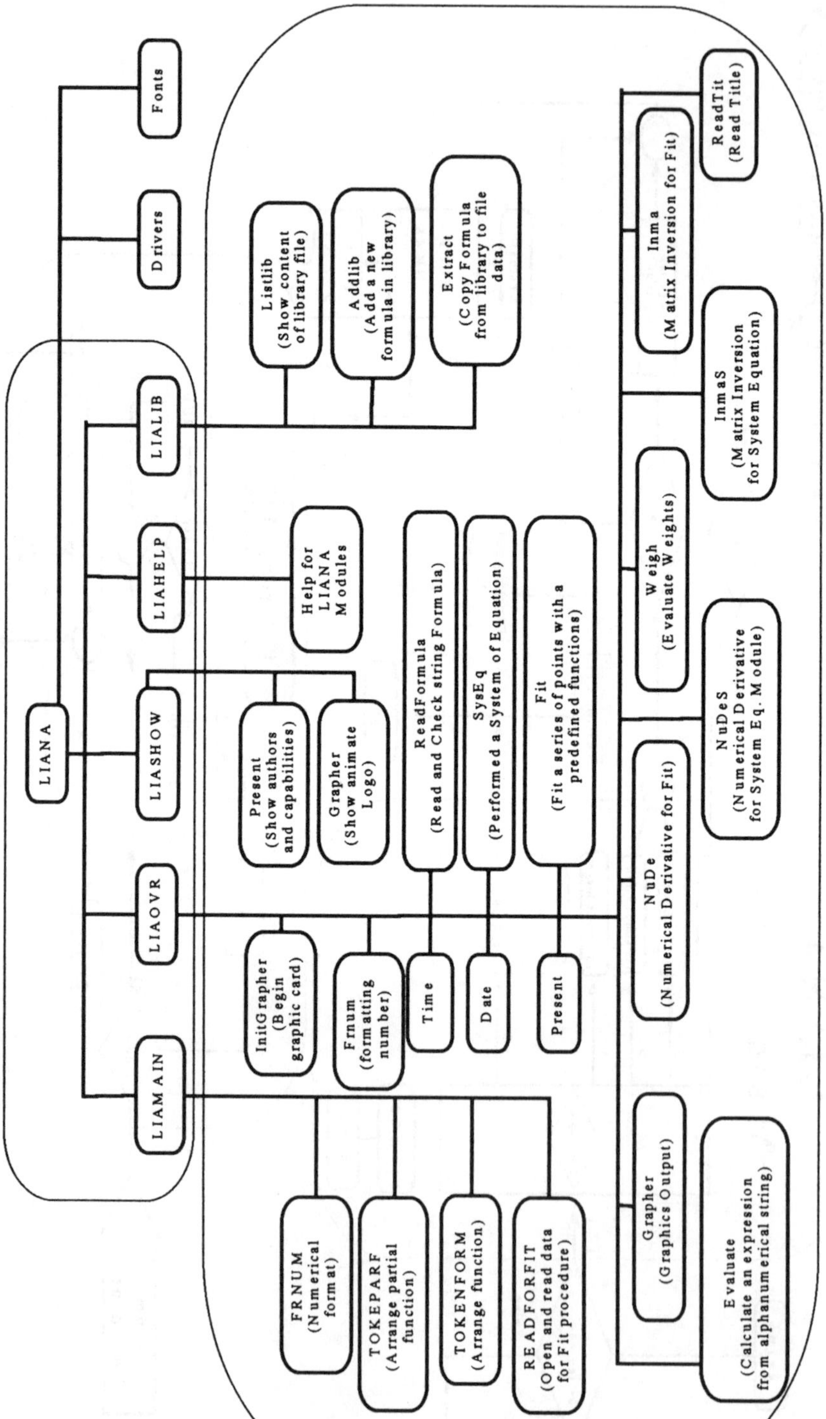

Fig. 1b. Structure of LIANA modules

Table 1. Equations used in the computer program TDA.

$$\log K_T = \log {}^T K_\Theta + \left(\frac{1}{\Theta} - \frac{1}{T}\right)\frac{\Delta H}{R'} + \left(\frac{\Theta}{T} + \ln\frac{T}{\Theta} - 1\right)\frac{\Delta C_p}{R'} +$$

$$+ \frac{1}{2}\frac{9}{R'}\left(\frac{T}{\Theta} - \frac{\Theta}{T} - 2\ln\frac{T}{\Theta}\right)\frac{d(\Delta C_p)}{dT} + G(I)\, A(T) + I\, C(T) + \Phi$$

$$\Delta H^0{}_T = {}^T\Delta H^0{}_\Theta + (T-\Theta)\,\Delta C_p + \frac{1}{2}(T-\Theta)^2 \frac{d(\Delta C_p)}{dT} + 2\,R'\,T^2\,[G(I)A'(T) + I\,C'(T) + \Phi']$$

$$\Delta C_{p,T} = {}^T\Delta C_{p,\Theta} + (T-\Theta)\frac{d(\Delta C_p)}{dT} + 4\,R'\,T\,[G(I)\,A''(T) + I\,C''(T) + \Phi'']$$

R' = 0.019145   T,Θ = reference temperature in °K

$$X(T) = X + (T-\Theta)\frac{dX}{dT} + \frac{1}{2}(T-\Theta)^2 \frac{d^2X}{dT^2} \qquad (X = A; C; D; E; F)$$

$$X'(T) = \frac{dX}{dT} + (T-\Theta)\frac{d^2X}{dT^2}$$

$$X''(T) = \frac{dX}{dT} + \frac{1}{2}(3T-2\Theta)\frac{d^2X}{dT^2}$$

Extended Debye-Hückel equation	Pitzer type equation
$G(I) = -z^* \dfrac{\sqrt{I}}{1+B\sqrt{I}}$	$G(I) = -z^* \dfrac{\sqrt{I}}{1+1.2\sqrt{I}} + 1.667\ln(1+1.2\sqrt{I})$
A = 0.51 at T = 25 °C	A = 0.39 at T = 25 °C
	$f_1 = 1 - (1+2\sqrt{I})\,e^{-2\sqrt{I}}$
	$f_2 = -1 + (1+2\sqrt{I}+2I)\,e^{-2\sqrt{I}}$
$\Phi = I^{3/2} D(T) + I^2 E(T)$	$\Phi = I^2 D(T) + f_1 E(T) + f_2 F(T)$
$\Phi' = I^{3/2} D'(T) + I^2 E'(T)$	$\Phi' = I^2 D'(T) + f_1 E'(T) + f_2 F'(T)$
$\Phi'' = I^{3/2} D''(T) + I^2 E''(T)$	$\Phi'' = I^2 D''(T) + f_1 E''(T) + f_2 F''(T)$

$$m_s = m_{35}\, 27.56572\, S/(1000-1.005714S)$$

where S is the conventional salinity. To make those studies easier, conversion factors for all types of concentration scales generally used have been calculated and given, in order to convert thermodynamic parameters from one scale to another and to compare the data given in different scales by different authors. The scale conversion for equilibrium constants is given by

$$\log K_m = \log K_c + \log (c/m)$$

Analogous conversion equations are given for ΔH^0 and ΔC_p^0. Two simple computer programs have been written to give the numerical values useful in equilibrium studies on natural fluids [28].

5.2 CALCULATION OF APPARENT PROTONATION CONSTANTS

Finding the numerical values of apparent protonation constants is the inverse problem to that described in section 3.3. By considering the equality $\bar{n}' \cong \bar{n}$, if effective β_{pq} values are known, then apparent protonation constants can be calculated. A small computer program (ES6) has been written to deal with this problem (these authors, unpublished results).

5.3 THERMODYNAMIC DATA ANALYSIS (TDA)

Thermodynamic formation data ($\log K$, ΔH^0, ΔC_p^0) obtained both experimentally and from literature sources, can be analyzed altogether to find the parameters for dependence on ionic strength in an appropriate I, T range. The equations used in the computer program TDA are given in Table 1. The dependence on T is taken into account by the Clarke and Glew [29] equation; for the dependence on I extended Debye-Hückel or Pitzer [30] type equations are used.

6. Conclusions

Several different calculation problems are met in speciation studies. In this paper we have shown that appropriate computer programs can solve most of these problems, including searching for and analyzing existing data. In particular we would emphasise the usefulness of flexible general programs, such as LIANA, which are able to solve different problems.

For informations about the calculation computer programs cited in this paper, please contact us at the following E-mail addresses:
- sammartano@chem.unime.it
- gmineo@dipchi.unict.it

7. References

1. Gans, P. (1992) *Data fitting in the chemical science*, Wiley, New York.
2. Moshier, S.L. (1989) *Methods and Programs for Mathematical functions*, Ellis Horwood, Chichester.
3. Griffits, P. and Hill, I.D. (Eds.) (1985) *Applied Statistics Algorithms*, Ellis Horwood, Chichester.
4. Dixon, L.C.W. (1972) *Nonlinear optimisation*, The Englsh Universities Press, London.
5. Ralston, A. (1965) *A First Course in Numerical Analysis*, International Student Edition, Mc Grow-Hill Kogakusha.
6. Kowalski, B.R. (Ed.) (1983) *Chemometrics. Mathematics and Statistics in Chemistry*, NATO ASI Series C Vol. **138**, D. Reidel Publ. Co., Dordrecht.
7. Massart, D.L., Dijkstra, A. and Kauffman, L. (1978) *Evaluation and optimization of laboratory methods and analytical procedures*, Elsevier, New York.
8. Štrouf, O. (1986) *Chemical Pattern Recognition*, Wiley, New York.
9. De Robertis, A., De Stefano, C., Rigano, C. and Sammartano, S. (1986) The Calculation of Equilibrium Concentrations in Large Multimetal/Multiligand Systems, *Analytica Chim. Acta*, **191**, 385-398; and references reported therein.
10. De Stefano, C., Princi, P., Rigano, C. and Sammartano, S. (1989) The Calculation of Equilibrium Concentrations. ES4EC1: A Fortran Program for Computing Distribution Diagrams and Titration Curves, *Computer and Chem.*, **13**, 343- 359.
11. De Stefano, C., Mineo, P., Rigano, C. and Sammartano, S. (1993) Ionic Strength Dependence of Formation Constants. XVII. The Calculation of Equilibrium Concentrations and Formation Constants, *Ann. Chim. (Rome)*, **83**, 243-277.
12. Perrin, D.D. and Sayce, I.G. (1967) Computer Calculation of Equilibrium Concentrations in Mixtures of Metal Ions and Complexing Species, *Talanta*, **14**, 833-842.
13. Ingri, N., Kakolowicz, W., Sillen, L. G. and Warnquist, B. (1967) High Speed Computers as a Supplement to Graphical Methods-V. Haltafall, a General Computer Program for Calculating the Composition of Equilibrium Mixtures, *Talanta*, **14**, 1261-1286.
14. Leung, V.W.-H., Darvell, B.W. and Chan, A.P.-C. (1988) A rapid algorthm for solution of the equations of multiple equilibrium systems-RAMES, *Talanta*, **35**, 713-718 (see also other papers of this series published in *Talanta*; **37**, 413-423; **37**, 425-429; **38**, 875-888; **38**, 1027-1032; **39**, 1057-1059.
15. De Stefano, C., Foti, C., Giuffrè, O., Mineo, P., Rigano, C. and Sammartano, S. (1996) Binding of Tripolyphosphate by Aliphatic Amines: Formation, Stability and Calculation Problems, *Ann. Chim. (Rome)*, **86**, 257- 280.
16. De Robertis, A., De Stefano, C. and Rigano, C. (1989) Computer Analysis of Equilibrium Data in Solution. ES5CM Fortran and Basic Programs for Computing Formation Enthalpies from Calorimetric Measurements, *Thermochim. Acta*, **138**, 141-146.
17. Leggett, D.J. (Ed.) (1985) *Computational Methods for the determination of formation constants*, Plenum, New York.
18. Meloun, M., Havel, J. and Högfeldt, E. (1988) *Computation of Solution Equilibria*, Ellis Horwood, Chichester.
19. Gans, P., Sabatini, A. and Vacca, A. (1996) Investigation of Equilibria in Solution. Determination of Equilibrium Constants with the Hyperquad Suite of Programs, *Talanta*, **43**, 1739-1753.
20. Motekaitis, R.J. and Martell, A.E. (1988) *Computational Methods for the Determination of Formation constants*, VCH, New York.
21. Gaizer, F. (1979) Computer Evaluation of Complex Equilibria, *Coordination Chemistry Reviews*, **27**, 195-222.
22. De Robertis, A., De Stefano, C., Sammartano, S. and Rigano, C. (1987) The Determination of Formation Constants of Weak Complexes by Potentiometric Measurements: Experimental Procedures and Calculation Methods, *Talanta*, **34**, 933-938.
23. National Institute of Standard and Technology, NIST (1994) *Activity and Osmotic Coefficients in Aqueous Solutions*, PC-based Database, Gaithersburg, MD 20899
24. Pettit, D. and Powell, K. (1993) IUPAC *Stability Constants Database*, Academic Software, Otley, UK
25. Murray, K. and May, P.M. (1995) *Joint Expert Speciation System*, Jess Primer, Murdoch, (Western Australia).
26. National Institute of Standards and Technology, NIST (1993) Critical Stability Constants of Metal Complexes, PC-based Database, Gaithersburg, MD 20899
27. Falck ,W.E. and Mestress Ridge, D. (1993) *Chemval2: Thermodynamic Database*, CEC report.
28. De Stefano, C., Foti, C., Sammartano, S., Gianguzza, A. and Rigano, C. (1994) Equilibrium Studies in Natural Fluids. Use of Synthetic Seawater and Other Media as Background Salts, *Ann. Chim. (Rome)*, **84**, 159-175.
29. Clarke, E.C.W. and Glew, D.N. (1965) Evaluation of Thermodynamic Functions from Equilibrium Constants, *Trans. Faraday Soc.*, 539-547.
30. Pitzer, K.S. (1991) Ion Interaction Approach: Theory and Data Correlation, in Pitzer, K.S. *Activity Coefficients in Electrolyte Solutions 2nd Ed.*, CRC Press, Boca Raton (FL), 75-154.

METAL IONS AND ORGANOMETALLIC COMPOUNDS IN SEA WATER AND IN SEDIMENTS: BIOGEOCHEMICAL CYCLES

P. J. CRAIG and D. MILLER
*Department of Chemistry, De Montfort University,
Leicester LE1 9BH, UK*

1. Occurrence of Organometallic Compounds in Sea Water and Sediments.

Organometallic compounds are normally defined as compounds which contain a metal carbon covalent (sigma) bond, eg methyl mercury, CH_3Hg^+. Other types exist, particularly in the area of synthetic chemistry, but the metal carbon single bond is the type which is of environmental interest (the methyl cyclopentadienyl manganese gasoline additive used in North America is the main exception; here a pi bond between the carbon moiety and the metal is involved). In general the metals which are of interest in the organometallic context are Periodic Table Main Group metals rather than Transition metals. With respect to inorganic metal ions (see 2 below) there is environmental interest in both Main Group and Transition types. In the example above, CH_3Hg^+, it can be seen that an ion is quoted. In the environment this positive species would be expected to be coordinated or bonded to one or other of the numerous ligand type atoms which exist in the environment, eg to S, O or N atoms from proteinaceous materials in sediment or biological tissue, or to chloride in sea water. In general these counter ions are not considered to play as important a role in the toxicological properties of the species as the organometallic moiety, and as a result are sometimes neglected in analysis.

The main utilitarian reason for the present considerable interest in organometallic compounds in the environment lies in the fact that they are often more toxic than the inorganic species from which they derive, arsenic species being the main exception to this generality. The properties of such a molecule, in which a part may be soluble in lipid like material and a part may have aqueous solubility, can enhance transport and delivery to sensitive sites within an organism and can also reduce the rate of elimination from the organism.

Organometallic compounds may enter the environment as such, or they may be formed there. Numerous organometallic compounds are synthesised and used as products (eg alkyl leads, butyl tins) and they may be dispersed to the general environment through use (indeed they may be designed to operate via dispersion, eg marine anti fouling paints). Where they are formed in the environment this is nearly always by a process of methylation. There are a few reports of ethyl compounds apparently being formed in the environment, but this is rare. The process of methylation can be considered to be either a genuine enzymatic process or an abiotic transmethylation. Pragmatically this doesn't matter as the end result is the same. The proximate source of the methyl group may a biological methylating agent such as S-adenosylmethionine or, in a polluted environment, it could be another methyl metal species in a transmethylation. The process of environmental methylation is summarized in Scheme 1.

SCHEME 1. - *Environmental Methylation*

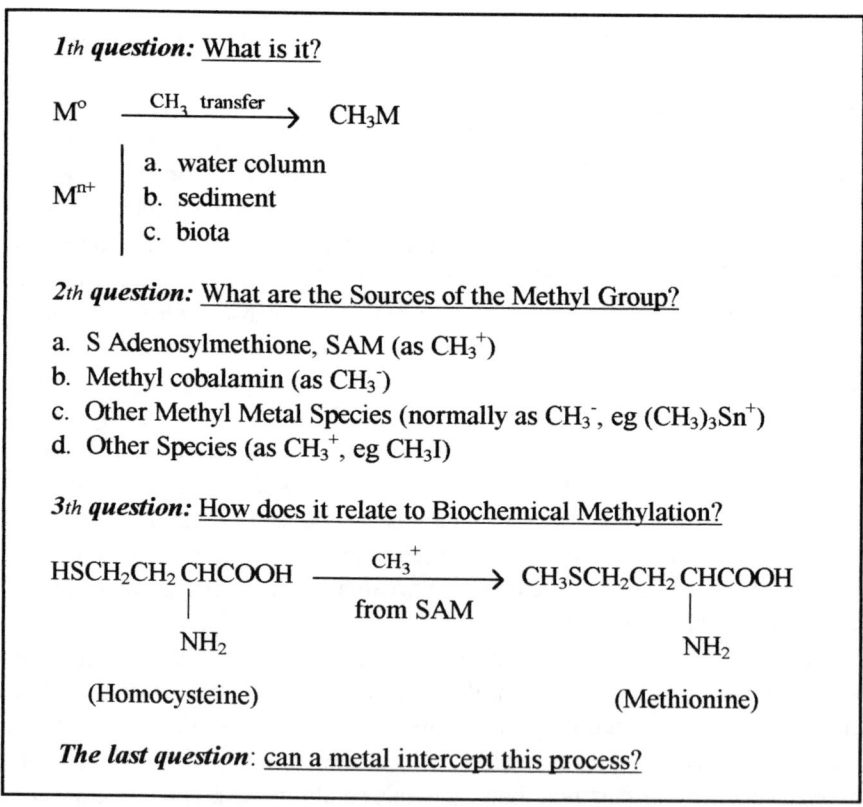

As a result of either product use or environmental methylation, numerous organometallic compounds have now been detected in the natural environment. These are summarized in Schemes 2 and 3. These Schemes demonstrate the existence in the environment of a wide variety of compounds that may be considered to arise from a single entrant, (*ie* tributyltin) by processes of methylation and dismutation.

SCHEME 2 - *Organometallic Components in the Environment*

	Metal Compounds	Used?	Formed?
Mercury	CH_3HgX	No	Yes
	C_2H_5HgX	Yes	Yes
	C_6H_5HgX	Yes	No
	$ROCH_2CH_2HgX$	Yes	No
Tin	$(CH_3)_nSnX_4-n$	Yes (n= 1,2,3)	Yes (inc X = H)
	$(C_4H_9)_nSnX_4-n$	Yes (n= 1,2,3)	No
	$(C_8H_{17})_nSnX_4-n$	Yes (n= 2,3)	No
	$(C_6H_{11})_3SnX$	Yes	No
	$(C_6H_5)_3SnX$	Yes	No
	$(C_4H_9)_2CH_3SnX$	No	Yes
Lead	$(CH_3)_4Pb$	Yes	Perhaps
	$(C_2H_5)_4Pb$	Yes	No
	Methyl-ethyl lead Ions and mixtures	Yes	Perhaps (methyls)
Arsenic	$(CH_3)_3As^+CH_2COO^-$ etc	No	Yes
	$CH_3AsO(OH)_2$	Yes	Yes
	$(CH_3)_2AsO(OH)$	Yes	Yes
	$(CH_3)_3As$	No	Yes
	$(CH_3)_2AsH$	No	Yes
	$PhAsO(OH)_2$	Yes	No
Antimony	$CH_3SbO(OH)_2$	No	Yes
	$(CH_3)_2SbO(OH)$	No	Yes
	$(CH_3)_3Sb$	No	Yes
Selenium	$CH_3Se\,(CH_2)_2CHCOOH$ $\vert$ NH_2	No	Yes
	$CH_3Se^+(R)(CH_2)_2CHCOOH$ $\vert$ NH_2	No	Yes
	CH_3SeH	No	Yes
	$(CH_3)_2Se$	No	Yes
	$(CH_3)_3Se^+$	No	Yes
	$(CH_3)_2Se_2$	No	Yes
	$(CH_3)_2SeO_2$	No	Yes
Various	$(CH_3)_nGeX_4-n$	No	Yes (n=1,2,3)
	$(CH_3)_2Tl^+$	No	Yes
	CH_3CoB_{12}	Yes	Yes
	$AdenCoB_{12}$	No	Yes
	$(CH_3)C_5H_5Mn(CO)_3$	Yes	No

Several of the species in Scheme 3 may exist together at the same location in the aquatic and sediment environment, and they may have a range of different toxicities to a given organism (eg the <u>oyster crassostera gigas</u> or similar species). This illustrates the need for analytical procedures which can identify and quantify all of the species of a given organometallic present at a given location, ie speciation. The compounds in Scheme 3 have variously been detected in marine and estuarine waters, sediments, biota and evolving to atmosphere. Levels range from parts per million (ppm) in some sediments and biota to parts per billion (ppb) and less in water. Areas affected are those in close proximity to harbours and marinas rather than the open ocean, with residual effects still being measured despite legal controls or reductions on tributyl tin by several countries in recent years. The main area of concern is toxic effects to non target organisms, some of which are effected at sub ppb levels. Scheme 3 shows what is, in effect, a pollution situation.

SCHEME 3 - *Butyl and Methyl Tin Species Detected in the Natural Environment*

$(CH_3)_2Sn^{2+}$	$BuSnH_3$	$(CH_3)_3SnH$
$(CH_3)_3Sn^+$	CH_3Sn^{3+}	$CH_3SnH_3^+$
$BuSn^{3+}$	$(CH_3)_4Sn$	$BuSn(CH_3)_2^+$
Bu_2Sn^{2+}	$(CH_3)_2SnH_2$	$BuSn(CH_3)_2^+$
Bu_3Sn^+		

In contrast the biogeochemical cycle shown for arsenic in Scheme 4 is a natural product cycle (the carbon, nitrogen and similar cycling processes are well known; where they occur for metallic species they are known as biogeochemical cycles). In the case of arsenic, which occurs at an average level of 23 nmol kg^{-1} in seawater as arsenate, a series of reductions [As(V) to As(III) followed by oxidative methylations via S-adenosylmethione] leads in the marine environment to the range of compounds shown. The dimethyl arsenic sugar series occurs in algal species with arsenobetaine being found in higher marine animals, often at the ppm level. The driving force for the cycle appears to lie in a substitution by arsenate for phosphate intake in marine areas which are depleted in phosphorus followed by methylation. In the terrestrial environment various moulds and fungi may produce methyl arsine species (Scheme 4)[1]. In addition to the compounds shown here, the tetra methylarsonium species $(CH_3)_4As^+$ and several trimethyl arsenic sugars have been detected recently. In seawater itself by contrast to biota, the only methyl compounds seen are the simple methyl arsenic acids. Despite recent observations from freshwater and terrestrial locations[1,2] methylarsenic species do seem to be more readily generated in the marine environment. The quarternary species arsenobetaine [$(CH_3)_3As^+CH_2COO^-$] seems to

---

[1] Very recently, the full range of compounds noted in Scheme 4 have been detected also in the terrestrial environment as well as the marine environment.

occur in most marine animals[2], including crustacea and fish. Arsenocholine, [$(CH_3)_3As^+CH_2CH_2OH$, which is probably the environmental precursor of arsenobetaine] and also trimethylarsine oxide [$(CH_3)_3AsO$, probably a metabolite of arsenobetaine], have been found in marine and freshwater species[3-6].

SCHEME 4. - *The Biogeochemical Cycle for Arsenic*

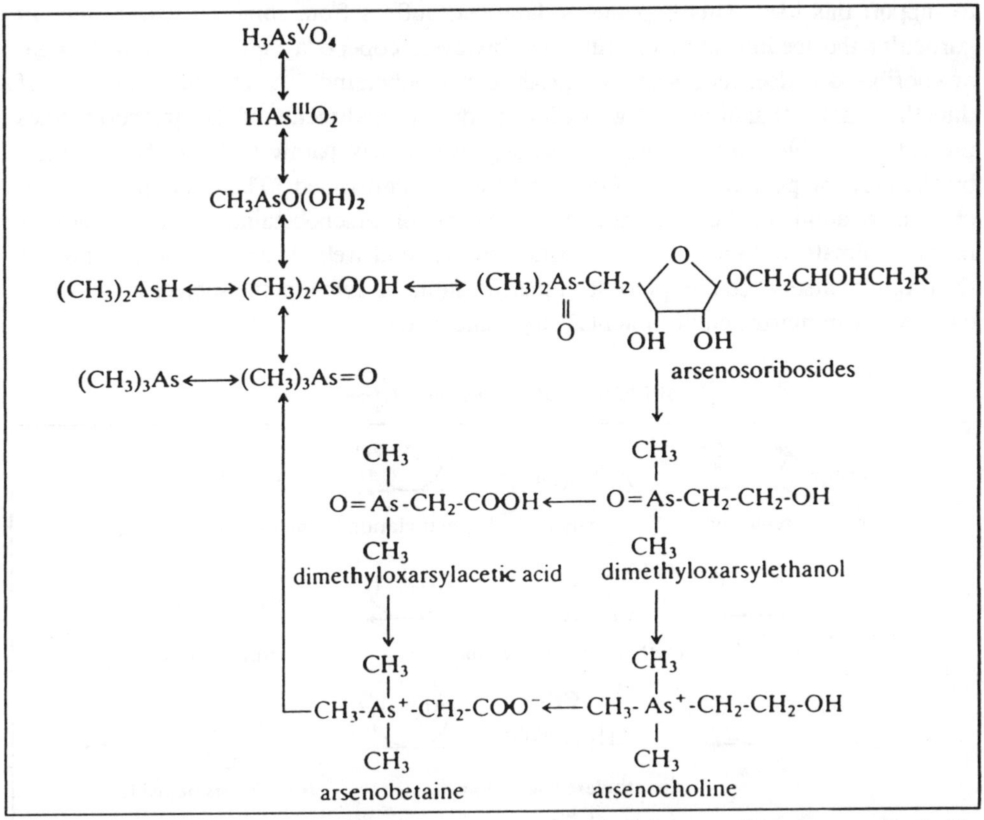

$$H_3As^VO_4$$
$$\downarrow$$
$$HAs^{III}O_2$$
$$\downarrow$$
$$CH_3AsO(OH)_2$$
$$\downarrow$$

$(CH_3)_2AsH \leftrightarrow (CH_3)_2AsOOH \leftrightarrow (CH_3)_2As\text{-}CH_2\text{-[ribose]-}OCH_2CHOHCH_2R$

arsenosoribosides

$(CH_3)_3As \leftrightarrow (CH_3)_3As=O$

Left branch:
$$O=As(CH_3)_2\text{-}CH_2\text{-}COOH \leftarrow O=As(CH_3)_2\text{-}CH_2\text{-}CH_2\text{-}OH$$
dimethyloxarsylacetic acid dimethyloxarsylethanol

$$CH_3\text{-}As^+(CH_3)_2\text{-}CH_2\text{-}COO^- \leftarrow CH_3\text{-}As^+(CH_3)_2\text{-}CH_2\text{-}CH_2\text{-}OH$$
arsenobetaine arsenocholine

(by Andreae, M.O., Organoarsenic compounds in the environment, in Ref. 5 general sources; reproduced with Editor's permission)

Acetylarsenocholine[4,7] and the tetramethylarsonium ion [$(CH_3)_4As^-$] have been measured in marine organisms[8,9]. Although the generality of the mechanism of methyl arsenic synthesis. ie the successive methyl group oxidative transfers, from S-adenosylmethionine, followed by reductions (ie the Challenger mechanism shown in Scheme 5), undoubtedly occurs, the details of the process are not yet all understood. There does not appear to be a biosynthesis of arsenobetaine in seawater. The sugars appear to be synthesised in algae. Arsenobetaine is found in higher marine species, and may arise by organosugar transformation or from food. Fish and crustacea do not seem able to methylate inorganic arsenate directly and the higher marine animals seem also unable to convert the arsenosugars to arsenobetaine. It may be that arsenobetaine and arsenocholine, present at very low concentrations in seawater, may be responsible for the presence of arsenobetaine in mussels and other marine animals[10-12]. The recent identification of dimethylarsinylriboside derivatives (arsenosugars) in shellfish[13] and the suggestion of a theoretical mechanism for the conversion of trimethyl-arsenoribosides into arsenocholine (Bettencourt 1990, Francesconi et al, 1992) seems to support this idea. This hypothesis, however, suffers from some difficulties too. In particular the feeding of fish, crustacea, bivalves, copepoda and bacteria with algal arsenoribosides does not seem to produce arsenobetaine[16-18]. The identification of dimethyloxarsylethanol as a degradation product in sediments of the arsenoribosides present in *Eklonia radiata* tissues also suggests a new pathway. Nevertheless, most biochemical steps have never been clarified or confirmed. There is therefore no obvious relationship between the concentration of arsenobetaine in the tissues of marine animals and their trophic position in the food-web. Therefore the question of the origin of arsenobetaine, present in marine animals, remains the subject of an active controversy in marine environmental biogeochemistry.

SCHEME 5. - *Biomethylation of Arsenic*

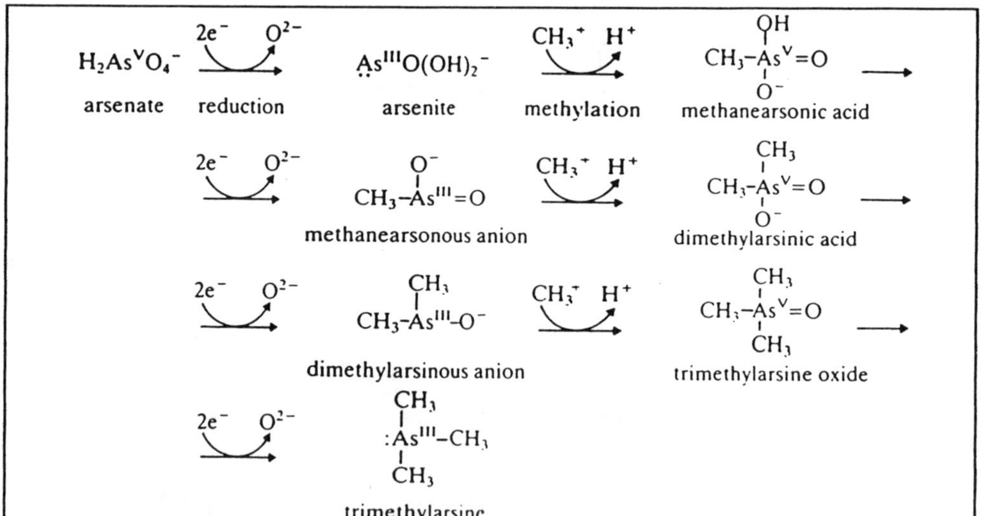

$$H_2As^VO_4^- \xrightarrow{2e^-\ O^{2-}} As^{III}O(OH)_2^- \xrightarrow{CH_3^+\ H^+} CH_3-As^V(OH)(=O)(O^-) \longrightarrow$$
arsenate — reduction — arsenite — methylation — methanearsonic acid

$$\xrightarrow{2e^-\ O^{2-}} CH_3-As^{III}(=O)(O^-) \xrightarrow{CH_3^+\ H^+} (CH_3)_2-As^V(=O)(O^-) \longrightarrow$$
methanearsonous anion — dimethylarsinic acid

$$\xrightarrow{2e^-\ O^{2-}} (CH_3)_2-As^{III}-O^- \xrightarrow{CH_3^+\ H^+} (CH_3)_3-As^V=O \longrightarrow$$
dimethylarsinous anion — trimethylarsine oxide

$$\xrightarrow{2e^-\ O^{2-}} :As^{III}(CH_3)_3$$
trimethylarsine

(by Andreae, M.O., Organoarsenic compounds in the environment, in Ref. 5 general sources; reproduced with Editor's permission)

Although methyl antimony species have been detected in the marine environment, there is no evidence so far for the existence of antimony analogues of arsenobetaine or the arsenosugars. There are now some reports of the volatization of antimony as methyl derivatives by fungi or by mixed cultures. Methyl antimony species have recently been detected in plants[19].

Despite the detection on countless occasions of alkyl lead species in the aquatic, sediment, biological and atmospheric environments, there is, in the opinion of this author, still no conclusive demonstration of a natural environmental or biological methylation process for lead. All the measurements of methyl or ethyl lead seems accountable from gasoline additives, with various processes of transport and/or food chain effects accounting for the observations.

Historically, methyl mercury was the first example of a biomethylation process. There have been many measurements of methyl mercury in the marine and sediment environments, and in nearly all cases the mercury arises from the inorganic form by environment methylation. The mechanism is still not entirely clear; it has been generally assumed for many years that the process is a form of SN_2 substitution by CH_3 from methyl cobalamin, the Vitamin B12 derivative, ie Equation 1.

$$CH_3CoB12 + Hg^{2+} + H_2O \rightarrow CH_3Hg^+ + H_2OCoB12^+ \quad (1)$$

If this is the case then the oxidation state of the mercury remains as Hg(II). However a reduction-oxidation mechanism as for arsenic remains possible, possible CH_3 donors such as methyl iodide[20] can methylate elemental mercury. In this case the situation shown in Equation 2 would exist.

$$Hg(II)L_2 \rightarrow Hg(O) + CH_3^+(SAM) \rightarrow CH_3Hg(II)^+ \quad (2)$$

The environmental properties of methyl mercury are well known and in many ways exemplify the generality of organometallic behaviour in the environment. Methyl mercury is eliminated from organisms much more slowly than in inorganic mercury(II) and is, on this account, more toxic. Within the environment, concentrations of methyl mercury genuinely dissolved in seawater or other waters are tiny, nearly all is present in particulate matter bound to the quasi solid. Concentrations in bottom sediments are much large and in biota higher still, in general proportion to the position of the biota in the food chain.

TABLE 1. - *Mercury concentrations and distributions in the Ottawa River, Canada*

Component	Total mercury concentration (ng g^{-1})	Methylmercury content	Fraction of all Mercury present
Filtered water	0.013	< 1	1.3
Suspended matter	1140	< 1	1.8
Bed sediments	80.6*	3.8	96.7
Higher plants	14.2	20	
Benthic invertebrates	223	40	0.2†
Fish	162	85	

* In polluted sediments levels may be typically 100 times greater than this; but the proportion of CH_3Hg^+ is typical. † But note implication for bioconcentration and toxicity effects. (from ref. 21, reproduced with Author's permission).

Many measurements are available, and the details presented in Table 1 of the classic 1978 Ottawa River study illustrate the above generalities. The overall biogeochemistry cycle for mercury is shown in Scheme 6[22].

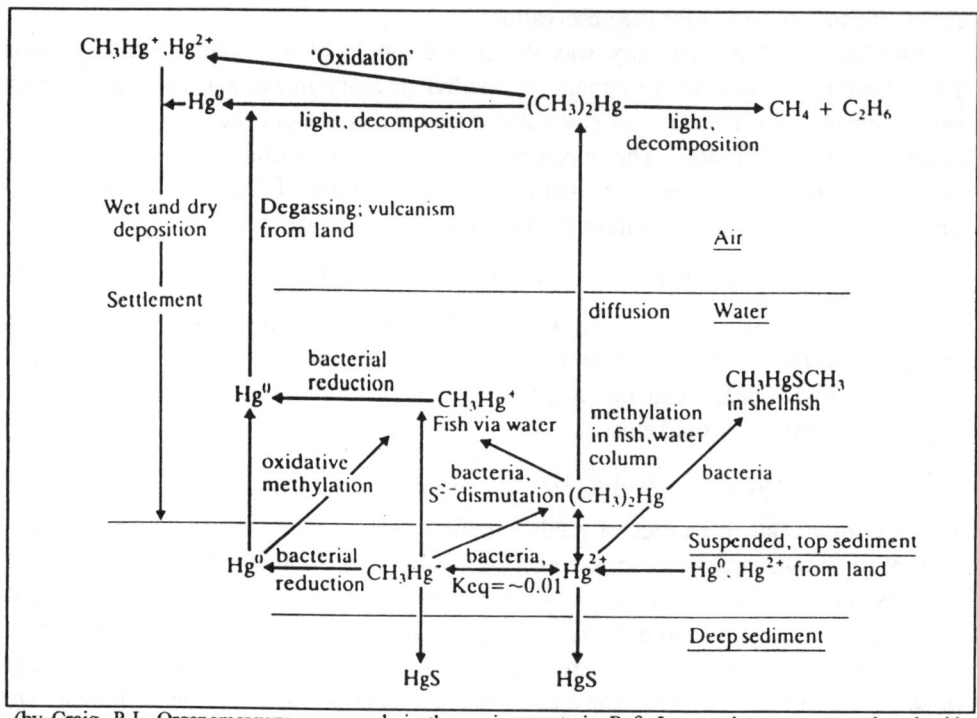

SCHEME 6. - *The Biogeochemical Cycle for Mercury*

(by Craig, P.J., Organomercury compounds in the environment; in Ref. 5 general sources; reproduced with Editor's permission)

Higher marine organisms can contain methyl mercury levels in the hundreds of ppm region without coming to any apparent harm. These levels may be high enough to disallow regular human consumption of course. In addition to the formation of methyl mercury, which is non volatile and is transportable as such only by physical movement of the matrix (water, sediment, biota), there have been numerous observations of the generation and volatization of dimethyl mercury. It is unclear whether or not there is a further biomethylation of methyl mercury or whether the second methyl group arises as a result of a sulphur mediated dismutation (Equation 3)[22].

$$2\ CH_3Hg^+ + S^{2-} \rightarrow (CH_3Hg)_2S \rightarrow (CH_3)_2Hg + HgS \tag{3}$$

This process has been calculated to mobilize up to 12% of the methyl mercury present in a sediment over a period of one year, and may be an explanation of the atmospheric observations of dimethyl mercury[23].

In nature, germanium, like arsenic, is generally found in the form of inorganic compounds associated with igneous rock. Germanium distribution in environmental compartments is caused by natural processes such as weathering of the rocks, geothermal activity and anthropogenic processes such as coal combustion. A comparison with arsenic is useful here. As mentioned above, by processes of organism mediated transformations, the existence of organic forms of arsenic in natural water is well documented. The arsenic transformation is caused by primary producers in both freshwater and seawater. Monomethylarsonic acid and dimethylarsinic acid are the only organic forms of arsenic found in the sea water (as distinct from biota). Their high amounts in sea water, up to 30% of total arsenic amounts, are usually found in highly productive areas and a similar situation seems to be seen for germanium.

As for arsenic, organic forms of germanium, *ie* monomethylgermanium and dimethylgermanium, are also ubiquitous in sea water. They account for more than 70% of the total known germanium compounds present in sea water. However, the process that leads to their formation is still unclear. Attempts to locate natural marine and freshwater sources have been unsuccessful, but it is likely that their formation may be due to microbially mediated processes in anoxic sediments. However, these are not likely to be controlled only by the type of the bacteria itself but also by interaction of the microbial community with other chemical and physical parameters[24,25].

In general, marine organisms exhibit higher selenium concentrations than terrestrial animals[26,27]. The biogeochemical cycling of selenium in the marine environment involves both inorganic and organometallic forms[28]. Recently methyl selenium species have been measured in ocean waters and a similar chemistry compared to sulphur seems to operate.

2. Metal Ions in Seawater and Sediments.

It is useful first to consider the main naturally occurring ions in seawater (Table 2).

TABLE 2. - *Major Ions in Seawater* (S = 35 ‰ ; conc. in g kg^{-1})

Ion	Concentration	Ion	Concentration
Cl$^-$	19.354	Na$^+$	10.77
SO$_4^{2-}$	2.712	Mg^{2+}	1.200
Br$^-$	0.0673	Ca^{2+}	0.4121
F$^-$	0.0013	K$^+$	0.399
B	0.0045	Sr^{2+}	0.0079

(by Chester, R., *Marine Geochemistry*, Unwin Hyman, 1990; Ref. 7, general sources; reproduced with Author's permission.)

The less common ions are shown in Table 3; this is almost a statement about the metals in the Periodic Table (ie, nearly all have been observed) or about the state of the analytical art at the time. Nevertheless, a compilation made in the mid 1980s (Table 2) provides some basic data upon which environmental effects can be superimposed.

TABLE 3. - *Speciation of Some Trace Elements of Interest in Seawater**

Element	Probable Main Species	Concentration Range† (average in brackets)
P	$H_2PO_4^{2-}$, HPO_4^-, $MgHPO_4$	1 - 3.5 umol kg^{-1} (2.3)
V	HVO_4^{3-}, $H_2VO_4^-$, $NaHVO_4^-$	20-35 nmol kg^{-1} (30)
Mn	Mn^{2+}, $MnCl^+$	0.2 - 3 nmol kg^{-1} (0.5)
Fe	$Fe(OH)_3$	0.1 - 2.5 nmol kg^{-1} (1.0)
Cu	$CuCO_3$, $CuOH^+$, Cu^{2+}	0.5 - 6 nmol kg^{-1} (4)
Zn	Zn^{2+}, $ZnOH^+$, $ZnCO_3$, $ZnCl^+$	0.05 - 9 nmol kg^{-1} (6)
Ge	$Ge(OH)_4$, $H_3GeO_4^-$	7 - 115 pmol kg^{-1} (70)
As	$HAsO_4^{2-}$	15-25 nmol kg^{-1} (23)
Se	SeO_4^{2-}, SeO_3^{2-}, $HSeO_3^-$	0.5 - 2.3 nmol kg^{-1} (1.7)
Cd	$CdCl_2$	0.001 - 1.1 nmol kg^{-1} (0.7)
Sn	$SnO(OH)_3^-$	1 - 12 pmol kg^{-1} (4)
Sb	$Sb(OH)_6^{3-}$	1 - 2 nmol kg^{-1} (?)
Hg	$HgCl_4^{2-}$	2 - 10 pmol kg^{-1} (5)
Pb	$PbCO_3$, $Pb(CO_3)_2^{2-}$, $PbCl^+$	5 - 175 pmol kg^{-1} (10)

* Oxygenated seawater. † S = 35‰ ; adapted from Chester R, (1990) - data originally from Ref 29.

The metal ion levels observed are a product of the various types of input and efflux in the marine environment. Input is from atmosphere, from rivers and from upwelling from the ocean floor. The first two may be subject to anthropogenic influences.

TABLE 4 - *Transfer of Metals from Atmosphere to Sea Surface* (ng cm^{-1} yr^{-1})

Element	North Sea	Western Mediterranean	South Atlantic	Tropical North Atlantic	Tropical North Pacific
Mn	30,000	5000	2900	5000	1200
Fe	25,500	5100	5900	3200	560
Cu	1300	96	220	25	8.9
Zn	8950	1080	750	130	67
As	250	54	45	-	-
Cd	43	13	9	5	0.35
Hg	-	5	24	2.1	-
Pb	2650	1050	660	310	7.0

(from Clark, R.B., *Marine Biology*, Oxford University Press, 1989; Ref. 1 general sources; reproduced with Author's permission)

For the open ocean, the levels observed for most elements are little influenced by man, other than in polluted and semi enclosed near shore seas. The comparative effects of input by man into the different types of marine locations can be seen from the atmospheric transfer data given in Tables 4 and 5.

Owing to the very strong sedimentation occurring in estuaries, these locations sequester a large quantity of the metals which are carried down by the rivers. Table 6 shows the relative concentrations of metals in river waters and river particulates compared to ocean waters and bays. Clearly the processes which generate or disperse organometallic compounds in the environment can operate most noticeably in those locations where the substrate metals have higher concentrations. Most organometallic environmental work, whether fieldwork measurements or theoretical studies, has been carried out in river or estuarine locations.

TABLE 5 - *Anthropogenic Compared to Natural Emissions of Selected Metals to Atmosphere (ktonnes yr^{-1})*

Metal	Natural Sources	Anthropogenic
As	7.8	24
Cd	0.96	7.3
Cu	19	56
Ni	26	47
Pb	19	449
Se	0.4	1.1
Zn	4	314

(from Clark R.B. *Marine Biology*, Oxford University Press, 1989; in Ref 1 general sources; reproduced with Author's permission)

TABLE 6 - *A Comparison of the Concentrations of Selected Metal Ions in River and Sea Waters*

Element	River (Dissolved)	River (Particulate)	Ocean Water	Deep Ocean Clays
As	1.7	5	1.5	13
Cd	0.02	(1)	0.01	0.23
Cu	1.5	100	0.1	200
Fe	40	48000	2	60000
Mn	8.2	1050	0.2	6000
P	115	1150	60	1400
Pb	0.1	100	0.003	200
Sb	1	2.5	0.24	0.8
V	1	170	2.5	150

(from Martin and Whitfield, 1984, in Ref. 4 general sources; reproduced with permission of Authors)

3. Some general sources of reference

1. Clark, R.B (1989), *Marine Biology*, Oxford University Press, 1SBN 0-19-85465-8.
2. O'Neil, P (1993), *Environmental Chemistry*, Chapman &Hall. 1SBN 0-412-48490.
3. Manahan, S.E. (1991), *Environmental Chemistry*, Lewis, 1SBN 0-87371-425-3.
4. Wing, C.S., Boyle, E., Bruland, K.W., Burton, J.D and Goldberg, E.D. (eds.), (1983), *Trace Metals in Sea Water*, Plenum Press, 1SBN 0-306-41165-2.
5. Craig, P.J (ed.), (1986), *Organometallic Compounds in the Environment*, Longman, 1SBN 0-582-46341-0.
6. Craig, P.J., (ed.), *Applied Organometallic Chemistry*, (primary journal), John Wiley, 1SBN 0268-2605 (published articles in this field).
7. Chester, R., (1990), *Marine Geochemistry*, Unwin Hyman, 1SBN 0-04-551108-X.
8. Bernhard, M., Brinkman, F.E., and Sadler, P.J., (eds.), (1986), *The Importance of Chemical Speciation in Environmental Processes*, Springer Verlag, 1SBN 0-387-15362-4.
9. Harrison, R.M., and Rapsomanikis, S., (eds.), (1989), *Environmental Analysis Using Chromatography Interfaced with Atomic Spectroscopy* (Ellis Horwood), 1SBN.
10. Riley, J.P. and Chester R., (eds.), (1983), Chemical Oceanography, Academic Press, London.

4. References

1. Kuroiwa, T., Ohki, A., Naka, K., and Maeda, S., (1994), *Applied Organometallic Chemistry*, **8**, 325-33.
2. Edmunds, J. and Francesconi, K., (1988), *Applied Organometallic Chemistry*, **2**, 297-302.
3. Norin, H and Rhyage, A., (1983), *Chemosphere*, **12**, 299-315
4. Norin, H., Christakopoulos, A., Rondhal, L., Hagman. A., and Jacobson, S. (1987), *Biomedical and Environmental Mass Spectrometry*, **14**, 117-125
5. Hanaoska, K., Matsumoto, T., Tagana, S., and Kaise, T., (1988), *Applied Organometallic Chemistry*, **2**, 371-76.
6. Lawrence, J.P., Michalik, P., Tam, G., and Conacher, H.B.S., (1986), *J. Agric. Food Chem*, **34**, 315-19
7. Christakopoulos, A., (1988), *Quaternary Organoarsenic Compounds in Aquatic Organisms*, PhD Thesis, University of Stockholm, Sweden
8. Shiomi, K, Horiguchi, Y. and Kaise, T., (1988), *Applied Organometallic Chemistry*, **2**, 385-90.
9. Cullen, W.R., and Dodd, M., (1989), *Applied Organometallic Chemistry*, **3**, 79-88
10. Shibata, Y and Morita, M (1992), *Applied Organometallic Chemistry*, **6**, 343-49
11. Challenger, F. (1945), *Chemical Reviews*, **36**, 315-361
12. Gailer, J., Francesconi, K.A., Edmonds, J.S., and Irgolic, K.J., (1995), *Applied Organometallic Chemistry*, **9**, 341-349

13. Larsen, G.H., (1995), *Fresenius J Anal Chemistry*, **352**, 582-588
14. Bettencourt, A.M.M and Andreae, M.O., (1991), *Applied Organometallic Chemistry*, **5**, 111-118
15. Francesconi, K.A., Edmonds, J.S., and Stick, R.V., (1992), *Applied Organometallic Chemistry*, **6**, 247-249
16. Cooney, R.V., and Benson A.A. (1980), *Chemosphere*, **9**, 335-341
17. Benson, A.A., (1988), in *Biological Alkylation of Heavy Elements*, Royal Society of Chemistry Special Publication, (Craig, P.J. and Glockling, F. eds.)
18. Francesconi, K.A. and Edmonds, J.S., (1994), in *Arsenic in the Environment*, Part I, Cycling and Characterization (ed. Nriagu, J.R.)
19. Dodd, M., Pergantis, S.A., Cullen, W.R., Li, H., Eigendorf, G.K. and Reimer, K.J., (1996), *Analyst*, **121**, 223-228
20. Craig, P.J. and Rapsomanikis, S., (1985), *Environ Sci Technol*, **19**, 726-729
21. Kudo, A., Miller, D.R., Akai, H., Mortimer, D.C., DeFreitas, A.S., Nagase, H., Townsend, D.R. and Warnock, R.G., (1978), *Progress Water Technology*, **10**, 329-39
22. Craig, P.J. (in Ref 5 general sources above), 92
23. Craig, P.J. and Moreton, P.A., (1984), *Mar Pollution Bull*, **15**, 406-408
24. Lewis, B.L. and Meyer, H.P., (1993), *Metal Ions in Biological Systems*, **29**, 79
25. Lewis, B.L., Froelich, P.N. and Andreae, M.O., (1985), *Nature*, **313**, 303.
26. Siu, M. and Berman (1989), in *Occurrence and Distribution of Selenium* (ed. Inhat, M.), CRC Press
27. Cooke, T.D. and Bruland, K.W. (1987), *Environmental Science Technol.*, **21**, 1214.
28. Chau, Y.K., (1986) in Ref 5, general sources above, 164-267
29. Bruland, K.W., (1983) in *Chemical Oceanography* (eds. Riley J P and Chester, R.) Academic Press, London, **8**, 157-220.
30. Martin, J.M. and Whitfield, M, (1983), in *Trace Metals in Sea Water* (in Ref 4, general sources above), 265-297

NUTRIENTS IN THE SEA

Distribution, circulation and relation to primary production in the marine environment

MAURIZIO RIBERA d'ALCALA'
VINCENZO SAGGIOMO
*Laboratory of Biological Oceanography
Stazione Zoologica "A. Dohrn", Villa Comunale
80121 Napoli, Italy
E-mail: maurizio@alpha.szn.it; saggiomo@alpha.szn.it*

GIUSEPPE CIVITARESE
*Istituto Talassografico "F. Vercelli"
Consiglio Nazionale delle Ricerche
Viale R. Gessi, 34123 Trieste, Italy
E-mail: civitarese@ts.cnr.it*

1. Historical Background

The relationship between nutrients and plankton or, more in general, living organisms in the sea is certainly one the most studied problems since the dawn of modern oceanography. And one of the most debated, too.

In this presentation only a synthetic, conceptual account will be given to familiarize analytical chemists with the relevant processes that determine the variability of nutrient distributions and utilization in the ocean.

Strange enough, the correlation between nutrients (soluble forms of basic elements) and the "fertility" of aquatic environments is a relatively recent acquisition in human knowledge.

During the neolithic phase of human history, *man* learned how to till the soil and to harvest selected plants. Yet, only during the eighteenth century the systematic addition of manure and compost to preserve the *humus* became a widespread practice. The analogous activity in aquatic environment, e.g. shellfish- or fish-farming, was mostly carried out in semi-enclosed systems (e.g. coastal lagoons), based on the natural fertilization of the fluxes at the interfaces.

Agriculture produces a "crop" whereas aquaculture gives a "catch". It is thus understandable why the problem of how to increase the yield of a crop, and the related question of what "substances" can do it in the soil, comes first than the apparently easier problem of what the animals are eating in the water.

In addition, we can assume that no specific fertilization of aquatic environments to stimulate production was carried out before this century, whereas agriculture has always attempted to stimulate plant production on the land.

With the birth of the modern chemistry at the end of the eighteenth century it became clear that living organisms are made up of a few basic elements with the addition of many others, but in trace concentrations.

The generic idea of *humus*, which prompts a complex view of soil dynamics, was replaced at the turn of mid-nineteen century by the suggestion that mineral salts were in fact the factors regulating the primary production in the soil.

This conducted Justus von Liebig [1, 2], who was the first to carry out a systematic study on the relationship between chemistry and agriculture, to formulate his renowned Law of the Minimum according to which ".. a piece of land...becomes barren for any kind of plant when...one only of these constituents [mineral salts]...has been so far removed, that the remaining quantity is no longer sufficient for a crop."[2].

At that time, the fertility of the sea was completely ascribed to the *humus* brought to the sea by rivers [2]. Only a few years later, V. Hensen and, later, his colleague H. Lohman put forward the hypothesis that microscopic organisms (plankton) could be the primary food for sustaining the life in the sea [2]. The following step was understanding what kind and origin had the resources for plankton growth and if these could entirely control their abundance.

The problem was not at all trivial, even at that time.

At a first approximation, two main hypotheses were formulated in the following years. The "agricultural hypothesis" based on the von Liebig's theory of "nutrient limitation" *sensu latu* and the "grazing hypothesis", i.e. the assumption that grazing by animal plankters was the factor controlling the abundance of microscopic algae [2, 3, 4], with the assumption that both factors were somehow affected by water movements.

Those studies are the roots of Biological Oceanography.

We will focus on the first of those hypotheses, synthesizing relevant processes determining nutrient availability for plant growth in the ocean.

2. Nutrients and primary production

Inspired by agricultural studies, the term "nutrients" was bound, since then, to the mineral forms of nitrogen, phosphorus and silica, the last one being clearly involved in the composition of diatom frustules. We all know that oxygen, hydrogen and carbon, plus a long list of ions and trace elements are involved in the biology of a living cell. Yet, extending to the marine environment the von Liebig's results, nitrogen and phosphorus were considered the putative candidates to be the limiting factors. Gran [2] considered the possibility that iron could be the ultimate limiting factor in remote environments, but analytical methods were not accurate enough to test this prophetic hypothesis.

The processes in which nutrients are involved are numerous. The most important for the life on the Earth is by far the assimilatory reduction of carbon and nitrogen by photosynthetic organisms.

Assimilatory reduction of carbon can also take place in chemiosynthetic organisms, but the contribution of this process to the total oceanic carbon fixation is estimated to be less than 0.1% [5].

Global data, global averages etc. are certainly affected by a significant error, sometimes up to a factor of 3 to 5. Nevertheless they are sufficiently representative of the relative contributions of different global processes to be worth considering.

The photosynthesis in the global ocean accounts for 35÷45 % of global carbon assimilation, and the organisms living in the pelagos, contribute to 90÷95 % of it [6].

The main actors of the pelagic realm are small unicellular autotrophic organisms, (phytoplankton); quite recently they have been subdivided in different classes according to their size (e.g. picoplankton, nanoplankton etc.).

The process of reducing carbon and nitrogen to synthesize biomolecules is referred to as primary production, whereas its ratio to the amount of living matter (biomass) responsible for it, is called primary productivity. Both are dominantly carried out by the phytoplankton.

TABLE 1. Annual averages of nutrient and chlorophyll *a* concentrations and primary production for different oceanic regions

Region	Latitude	Longitude	NO_3 (mmol·m^{-3})			Chlorophyll *a* (mg·m^{-2})	Primary Production (gC·m^{-2}·y^{-1})
			0m	100m	150m		
North Sea	59.5° N	1.5° E	6.36	10.7	-	1.6	~ 90
African Upwelling	19.5° N	16.5° W	4.21	19.38	20.37	4.1	> 180
North Atlantic	60.5° N	19.5° W	5.77	12.26	12.99	1.9	~ 90
Tropical Atlantic	27.5° N	59.5° W	0.58	0.65	1.45	0.05	< 36
East Mediterranean	33.5° N	32.5° E	0.22	0.35	0.56	0.22	< 36
West Subeq. Pacific	9.5° N	179.5° W	0.78	3.70	7.50	0.05	< 36
East Subeq. Pacific	24.5° S	89.5° W	0.75	0.57	1.13	0.05	< 36
North Pacific	57.5° N	139.5° W	7.48	13.45	13.80	2.1	~ 180
Peruvian Upwelling	12.5° S	76.5° W	5.63	19.48	20.93	1.8	> 180
South-West Atlantic	50.5° S	59.5° W	12.86	-	-	1.8	>180
South-West Pacific	16.5° S	145° E	0.01	0.01	2.38	< 0.03	< 36
Equatorial Pacific[*]	5° S	90° E	10.0	-	-	0.24	< 36

Sources: Nitrate concentrations [8]; Chlorophyll *a* concentrations [9]; Primary production [10];
[*] for this site nitrate and chlorophyll *a* values are reported from [12]

In Table 1 the annual averages of nitrate and biomass concentrations (measured as amount of chlorophyll *a* contained in the cells per unit volume) as well as estimated ranges of primary production for different oceanic sites are reported (chloropyll *a* and primary production numbers are depth-integrals of concentration values).

The data in Table 1 show a significant covariance of nitrate concentration with biomass and primary production. This result holds true for most of the oceanic realms with the exception of Equatorial Pacific where high nitrate concentrations do not produce high biomass values.

Light and temperature affect photosynthetic activity. But sites reported in the table span over different ranges in light availability or, when located at the same latitude, often exhibit dissimilar values of primary productivity.

On the other hand, annual trends in temperature are, at a first approximation, coupled with solar irradiance and temperature dependence of metabolic activity (e.g. photosynthesis) is weaker than the observed spatial variability [12]. It is then apparent that nutrients, in our case nitrogen, are a prime factor to account for the differences reported in Table 1.

A significant increase with depth is evident in nitrate concentrations in all sites, though with relevant spatial variability.

In conclusion:
- high biomass and production generally characterize at sites where nutrient are available with high concentrations;
- at specific locations, high nutrients values do not produce high biomass values [these areas are named HNLC (High Nutrients Low Chlorophyll) areas];
- below 100-150 m depth, nutrient are always present, but significant differences in their concentration values can be found in different regions.

Therefore we can assume that phytoplankton biomass and production depend on nutrient availability.

Thus, relevant questions for our subject are:
- what controls nutrient distribution;
- how nutrient availability could affect the dynamics of the communities;
- if peculiar oceanic provinces exist.

Since a full lecture [7] is devoted to the role of iron in controlling primary production in the sea, the following discussion will be focused on nitrogen and phosphorus.

3. Sources

A downward increase in nutrient concentrations is a general feature of global ocean [11, 13]. Surface values span over a wide range of concentrations according to latitude, dynamics of the region (see below) and interactions with the biota. The same occur for deeper layers, but with lower variability [13]. The surface layer (0 through 100÷180 m) accounts for less than 5% of the ocean total volume. Therefore deep concentrations determine average values and the overall pool.

Accepted values for mean oceanic concentration of phosphate [14] and nitrate [15] are 2.2 and 35 mmol·m^{-3}, respectively. Ocean total volume being $1.35·10^{18}$ m^3, PO_4 and NO_3 pools amount to $3·10^{15}$ and $4.7·10^{16}$ mol, respectively.

The above quantities are also the internal reservoirs of the system, one of the possible sources of nutrients for the organisms.

Additional sources are atmospheric inputs and land runoff, mostly consisting of riverine contribution.

Atmospheric inputs are relatively hard to quantify and "uncertainties on a global scale are still often a factor of 3 to 5" [16]. Numbers ranging from 0.8 to $43 \cdot 10^{11}$ mol·y^{-1} for nitrogen [16, 17, 18] and $3.2 \cdot 10^{10}$ mol·y^{-1} for phosphorus [16] are reported.

As for land runoff, $50 \div 70 \cdot 10^{10}$ mol·y^{-1} is the range reported by Duce [16] for phosphorus, where approximately 10% is the dissolved fraction, whereas estimates for nitrogen span from 5 to $30 \cdot 10^{11}$ mol·y^{-1} [17, 18].

In the above subdivision of different sources of nitrogen and phosphorus for inorganic dissolved nitrogen and phosphorus in the sea, we did not consider biological transformations.

Internal transformations such denitrification and nitrification do not change the nitrogen internal pool significantly, but they change the available chemical species in solution which, in turn, affects the nitrogen utilization. In particular, denitrification removes reactive nitrogen from the water column, thus acting as a nitrate sink.

Denitrification accounts for the removal of $40 \div 70 \cdot 10^{11}$ mol·y^{-1} of oxidized nitrogen [17].

Nitrogen fixation, which is the reverse process, adds reactive inorganic nitrogen to the system and contributes significantly to the flux of the element into the ocean. The general consensus about this process is that current estimates should be revised upward. McCarthy and Carpenter [17] reported a global value of $1.4 \cdot 10^{12}$ mol·y^{-1}, whereas Laws [18] suggested the higher value of $2 \cdot 10^{12}$ mol·y^{-1}.

Other processes that participate in removing nutrients from the system through burial or long term precipitation are not considered here for clarity.

The above data are synthetized in the following table.

Table 2 - Relevant sources for nitrates and phosphates in the Ocean

	Nitrogen	Phosphorus
Internal pool	$4.7 \cdot 10^{16}$ mol (NO$_3$)	$3.0 \cdot 10^{15}$ mol (PO$_4$)
Atmospheric input	$0.8 \div 43 \cdot 10^{11}$ mol N·y^{-1}	$3.2 \cdot 10^{10}$ mol P·y^{-1}
Land runoff	$5 \div 30 \cdot 10^{11}$ mol N·y^{-1}	$50 \div 70 \cdot 10^{10}$ mol P·y^{-1}
Nitrogen fixation	$14 \div 20 \cdot 10^{11}$ mol N·y^{-1}	-
Denitrification	$40 \div 70 \cdot 10^{11}$ mol N·y^{-1}	-

The uncertainties in the estimates preclude any conclusion on the overall budgets for the two elements. It is normally assumed that both are in steady state, at least on the centennial time scale.

In addition, it is worth to note that:
- the renewal time is larger then 10^4 y for both elements;
- the NO$_3$/PO$_4$ pool ratio is around 15.7, but large errors can affect the reported numbers;

- the internal source is by far the most important. But it is quite remote from the surface layer, whereas the external inputs bring the elements directly to the surface layer which (see above) is the depleted one.

4. Processes

4.1 THE GENERAL CONSTRAINTS OF PHYTOPLANKTON PRODUCTION

In a preceding paragraph, we anticipated that light affects primary production. This is tautological, since primary production in the sea is due to photosynthesis.

It is assumed that all the irradiance within the interval of 400-700 nm, regardless of the wavelength, can be utilized for the photosynthesis and this is named PAR (Photosynthetic Available Radiation). Therefore PAR is measured as number of photons, instead of an energy flux. 1 E (Einstein) is equivalent to 1 mol of photons. Since PAR spans from 10^{-3} to 10^{-6} $E \cdot m^{-2} \cdot s^{-1}$ in the marine environment, the unit generally used for PAR is $\mu E \cdot m^{-2} \cdot s^{-1}$.

Figure 1 shows a typical PAR profile for Mediterranean Sea in summer.

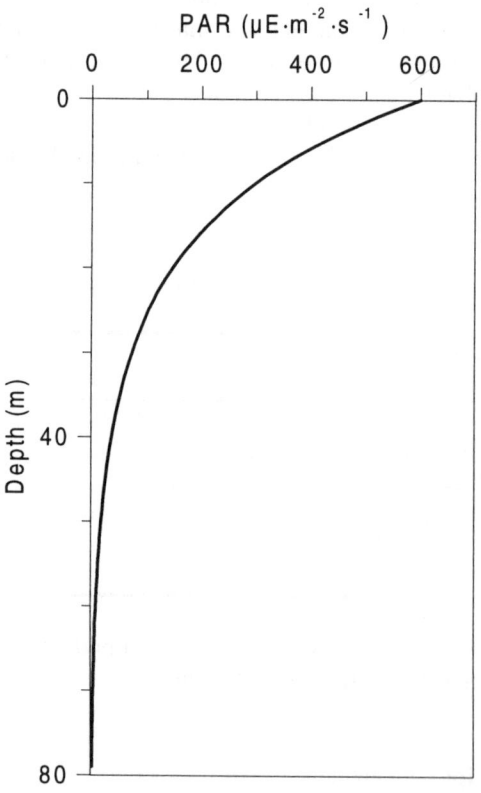

Figure 1 - Typical PAR profile

Phytoplankton cells can actively photosynthesize at very low levels of PAR. Photon fluxes down to 10÷15 $\mu E \cdot m^{-2} \cdot s^{-1}$ can be enough. From the profile in Figure 1 is evident that the thickness of the layer where primary production can occur is generally within the first 100 m, but generally is much thinner. This layer is termed "photic zone"

To photosynthesize, cells must assimilate nutrients which are not so abundant in the surface layer.

Contemporary availability of light and nutrients, that generally exhibit opposite gradients is the general external constraint of marine photosynthesis.

In addition, because of recurrence of similar molecules in the biota, the chemical composition of the cells is more or less constant, at least for major elements. Therefore the molar ratios among the different elements are generally constant and, for marine organisms, are denominated Redfield ratios, by the name of the scientist that first pointed out this aspect [19]. The constancy of Redfield ratios can be regarded as an internal, evolutionary constraint to the growth of photosynthetic organisms in the sea.

Thus, for primary production to take place, phytoplankton cells must be in the surface layer and have to uptake nutrients in an approximately constant ratio. For nitrogen and phosphorus the Redfield ratio (N/P ratio) is 16.

4.2 TRANSPORT MECHANISMS AND EXOSOMATIC ENERGY

Any mechanism capable of transporting nutrients to the photic zone contributes to enhance primary production. The covariance between primary production and nutrients observed for the surface layers in the global ocean (table 1) is thus the result of two interacting processes: physical transport and biological uptake, that eventually results in a partial removal toward the deeper layers through sinking or active movement of organisms. The second mechanism is referred to as "biological pump".

In remote areas of the global ocean (Southern Pacific, Antarctica), atmospheric transport and land runoff are less effective. The only available resources are the internal pool and, for nitrogen, biological fixation.

More in general, the internal reservoir is the only one that is always potentially available, provided that proper physical mechanisms carry out a net vertical transport from underlying layers. In mid to low latitudes, vertical fluxes are further impeded by thermal stratification of the water column, which increases the difference in potential energy among the layers, thus requiring additional energy input to overcome it.

Net vertical transport of solutes and suspended matter in the sea is basically due to three different processes:
− Vertical advection
− Turbulent diffusion
− Convective overturn

The first process moves large water parcels, without affecting too much their characteristics. Vertical advection takes place when divergence or convergence occur in the fluid. Those in turn are generated by the response of the fluid to different forcings (internal pressure gradient, wind forcing, gravity) on a rotating sphere, such as the Earth.

Turbulent diffusion, as opposed to molecular diffusion that relies on the quasi-random motion of molecules, spans from stirring to mixing. It can be regarded as the relative motion of very small water parcels, one respect to the other, in such a way that, ideally, at the end of the process properties of mixing members are homogenized. Turbulence is much more effective and fast in mixing the fluid than molecular diffusion.

A lot more has to be understood about turbulence, because not always can be considered as completely random process. Stirring as opposed to mixing has also to be considered, when turbulence takes place.

Because turbulence is ultimately a dissipative process (kinetic energy transformed to heat through internal friction), it requires energy. Wind stress, breaking of waves, friction on the sea bottom for currents and tides are, among the others, the processes though which turbulent energy is generated in the fluid.

Convective overturn is the third way of producing vertical transport in the water column. Its characteristics are intermediate among the other two. Water parcels can be bigger than in turbulent processes, but mixing generally occurs during convective processes. The forcing for this process is the decrease of potential energy of lighter water overlaying denser water. The property of lighter water that floats on denser water is called "buoyancy". To generate convection, buoyancy has to be extracted from the upper layer, so as to make it denser than underlying water. Heat and water vapor transfer from the liquid to the atmosphere decrease buoyancy of the water.

Buoyancy extraction occurs through energy transfer from the ocean to the atmosphere. Yet, sooner or later equilibrium would be reached among the two, if it were not for the movements of the atmosphere that, besides local heat and vapor exchange, depend on additional energy sources.

Therefore, all the three basic processes to transfer nutrients in the depleted photic zone require an input of external energy. This is different from the chemical energy contained in some of the metabolites or the electromagnetic energy utilized for the assimilatory reduction of carbon and nitrate by the cells. This energy is denominated "exosomatic" or "auxiliary" by Margalef [20, 21]. There are many kinds of auxiliary energy in the terrestrial ecosystem. In our case auxiliary energy is needed to carry nutrients in the zone where cells can perform photosynthesis, i.e. the photic layer. As Frontier [21] pointed out this is the energy that allows two necessary variables, metabolites and light, to be together in the same place. It is worth noting that exosomatic energy does not accumulate in the living matter as metabolic energy does.

The energy of the wind responsible for atmospheric transport of nutrients is another form of auxiliary energy as well as the one involved in the evaporation that generates the rain and, ultimately, the land runoff.

In conclusion, nutrient distributions in the sea depend significantly on the occurrence of different transfers of auxiliary energy.

The processes schematically summarized above are sketched in Figure 2.

MacCarthy & Carpenter [17] and Duce [16], among the others, attempted to give an estimate of the different fluxes of nutrients to the photic zone, as they relate to the different processes. In order to compute the values for turbulent diffusion and vertical advection, the Authors chose typical values indirectly derived from other data. We extrapolated their results to the whole ocean, assuming that the ranges for the coefficient of eddy diffusion and the advective component can be somehow generalized, with the exclusion of upwelling areas, where vertical transport is much stronger than the values used by Duce. The results of this exercise are listed in Table 3 where data from table 2 are also merged for comparison.

As Duce [16] already pointed out, the most relevant source of nutrients for primary production is the internal reservoir which, for phosphate, can account for more than 90% of the total amount transported.

The data for land runoff and atmospheric contribution reported in Table 3 are spatial averages of total inputs to the ocean surface. Their meaning is certainly questionable. Both processes exploit mostly terrestrial reservoirs, particularly land runoff. A strong gradient from the area close to the continental margins toward the open ocean has to be expected. In addition, processes occurring in the coastal regions (sedimentation, enhanced primary production stimulated by terrestrial inputs, semi-isolated coastal

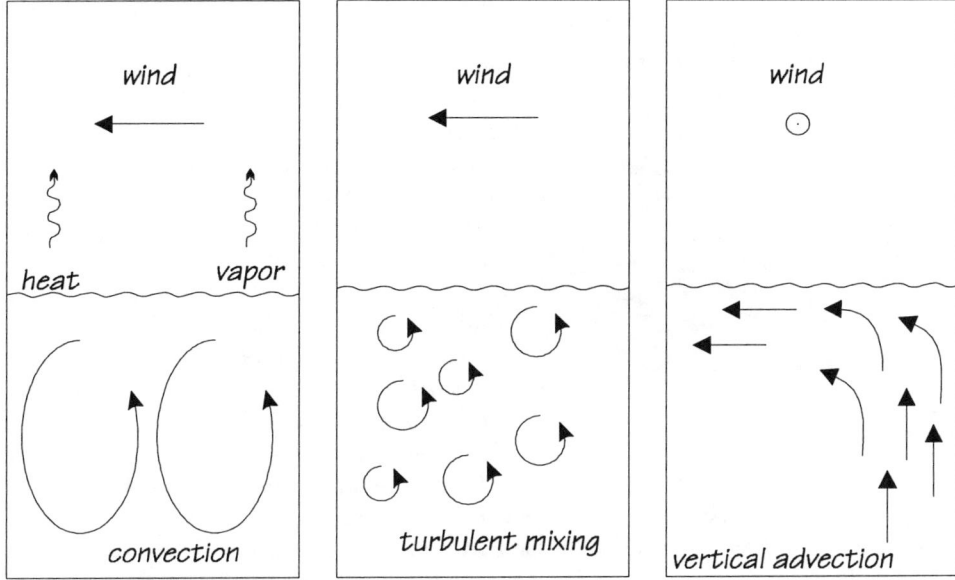

Fig. 2 - Schematics of processes affecting vertical transport of nutrients

dynamics) contribute to the removal of a significant part of terrestrial inputs from the transport to the open ocean. Interestingly, the land runoff was considered at the beginning of oceanographic studies as the only possible source of nutrients for the ocean.

Table 3 - Comparison of typical nutrient fluxes to the photic zone

	Nitrogen	Phosphorus
Atmospehric input	$0.6 \div 32$ μmol N·m^{-2} d^{-1}	0.24 μmol P·m^{-2} d^{-1}
Nitrogen fization	$1 \div 1.5$ μmol N·m^{-2} d^{-1}	
Vertical advection	$0 \div 40$ μmol N·m^{-2} d^{-1}	$0 \div 5$ μmol P·m^{-2} d^{-1}
Turbulent diffusion	$20 \div 800$ μmol N·m^{-2} d^{-1}	$2 \div 50$ μmol P·m^{-2} d^{-1}
Land runoff	$4 \div 22$ μmol N·m^{-2} d^{-1}	$4 \div 5$ μmol P·m^{-2} d^{-1}

Fluxes in the proximity of coasts are generally higher than in open ocean sites. This makes coastal areas very rich in nutrients and, correspondingly, in biomass.

4.2 DYNAMIC STRUCTURES IN THE OCEAN

The processes schematically sketched in Figure 2 occur in the ocean on different space and time scales and are associated with dynamic structures.

As already anticipated, the Earth rotation plays a fundamental role in determining the characteristics of water movements. Because of this rotation, a parcel of water initially

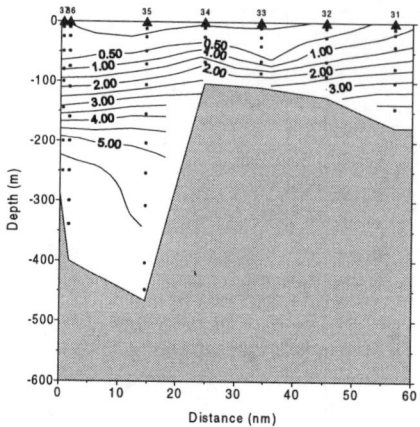

Fig. 4 - Nitrate isolines in the Sicily Channel

pushed in one direction accelerates to the right of the original direction in the northern emisphere (to the left in the southern one). This acceleration, produced by a fictitious force, the Coriolis term, is at the origin of recurrent structures like eddies and gyres. Likewise, Eckman transport, which is at the origin of upwelling events, also depends on the Coriolis acceleration.

As for the atmosphere, the motion of the fluid creates instabilities and the disruption of an uniform flow producing the detachment of small eddies, the formation of vortexes, the mixing of different water types. In addition, the latitudinal variation of the Coriolis term creates extended areas of convergence and divergence that contribute to vertical movements of sea water.

Coastal upwelling systems such as the Peruvian or the African ones (see Table 1) are sustained by the divergence along the coast due to water transported away from it by the winds. This divergence favors an upward movement of deeper water, that fertilizes the photic layer. Upwelling areas are among the most productive regions in the world ocean, because of the active link between internal reservoir and photic layer.

The eddies or the gyres, on the other hand, are vortexes of different size, where vertical transport can take place at the outer boundary (anti-cyclons) or at the center (cyclons).

Figures 3-5 present three of such processes occurring in the Straits of Sicily, the channel that connects Eastern to Western Mediterranean.

The first map (Figure 3) shows the surface contours of dynamic height which can be regarded as a depiction of surface currents. The meandering of the jet will eventually produce small eddies that will last for several weeks.

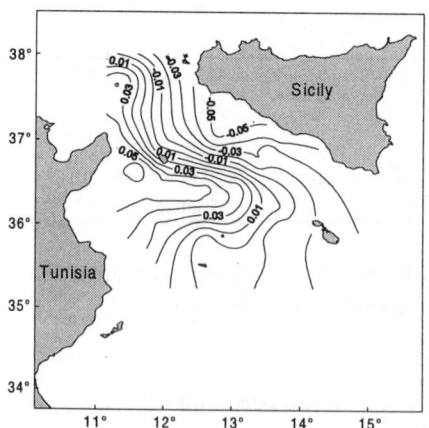

Fig. 3 - Contours of dynamic height (10-150 m) in the Straits of Sicily

In Figure 4 upwelling of water from the intermediate layer toward the surface is clearly detectable from the values of nitrate concentration. Higher nitrate values reach the photic zone, ultimately giving rise to plankton growth.

Figure 5 reports the structure of a gyre, 100 km wide, positioned at the center of the channel. The doming of intermediate water, traced by lower oxygen values, brings near the surface higher nutrient content, making it accessible for diffusion in the surface layer.

The global ocean so far described is a two layer system, with a very thin upper part, photosynthetically active, and a very thick deeper layer which stores most of the nitrogen and phosphorus. Vertical processes, favoring transport from the subsurface reservoir to the surface, connect the two layers. Advection due to upwelling is the most effective among those mechanisms.

One more element is necessary to complete the picture of nutrient transport. It is the "Great Ocean Conveyor" [22], a circulation pattern that encompasses Atlantic, Antarctic, Indian and Pacific Ocean moving around approximately $6.5 \cdot 10^{14}$ $m^3 \cdot y^{-1}$ of sea water. The Great Conveyor not only acts as a link among the different oceans, but is the most effective dynamic pattern of the global ocean in transporting solutes both horizontally and vertically. This because the Conveyor starts as a vertical transport (the sinking of cold water in the North Atlantic) and along its path upward flow compensates for the initial vertical transport.

The Great Conveyor is one of the mechanisms that allow at large scale the nutrients in the deep pool to eventually visit the surface again.

5. The Limiting Factor

At locations where the "external" constraint to plankton growth (i.e. co-occurrence of nutrients and light) could have been temporarily removed, biological uptake might be faster than "physical" supply. Then, more and more nitrogen and phosphorus would be combined in biomolecules within cells and ambient concentration would decrease.

Assuming an equal rate of uptake for all the nutrients and a constant chemical composition of the cells, a point would be reached when one of the nutrients would be no more available and net growth would be no more possible. The least available nutrient in this simplified scheme would be considered the limiting nutrient. This argument is very similar to the one put forward by von Liebig more than a century ago (see above).

To introduce the concept of nutrient limitation we made several assumptions.
- Solar irradiance is not limiting the photosynthetic capacity of the cells;
- nutrient supply is smaller than nutrient assimilation by algae;

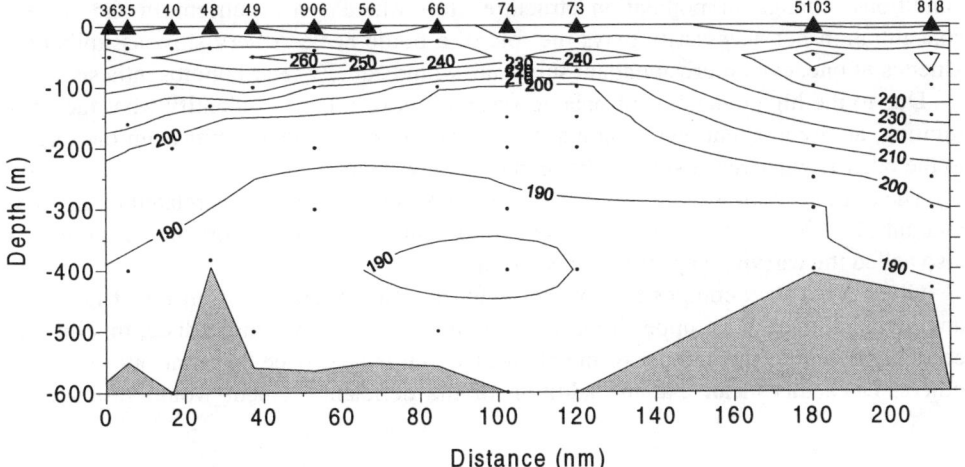

Fig. 5 - Oxygen isolines at the center of Sicily Channel

- assimilation rate is similar for all nutrients;
- grazing, sinking, senescence that would limit cell numbers are not relevant;
- physiological state of cells does not change with time;
- species composition of the community stays stable;
- biochemical composition of cells is basically fixed.

The above conditions very seldom co-occur in the marine environment, which precludes the possibility of easily determining if and what element is limiting plankton growth. In addition, a careful screening of the list also suggests that the concept of limitation has to be analyzed in more detail.

Howarth [23] distinguished between:
- physiological limitation
- net community production
- net ecosystem production.

Because of the relevance of this topic we will devote some time in analyzing the above definitions.

Physiological limitation takes place when the lack of one or more specific substrates impedes the optimal photosynthetic performance of the cells in a given light field. Many laboratory experiments conducted to assess this phenomenon produced very controversial results [24]. No conclusive evidence exists that very low nutrient concentrations (at least for nitrogen and phosphorus) depresses photosynthetic activity *in situ*. The only convincing case of observed physiological limitation *in situ* has been recently described for iron [11, 25], but it has been very hard to prove for other nutrients (but see [26] for a convincing evidence about nitrogen).

The above simply means that cells do probably adapt to depleted environment, regulating their physiology to a slower supply of nutrients but still photosynthesizing at quasi-optimal rate.

This drives us to the second kind of limitation, that affects net community production. The last is simply the amount of carbon that can be "transformed" in living cells. Increase in primary production obtained through enrichment experiments *in vitro* cannot be extended *tout court* to the real environment. Many other processes can contribute to keep the cell number low (grazing, senescence, change in physiological conditions, change in population structure etc.) whereas the nutrient in the lowest concentration not necessarily plays the role of a bottle-neck. A very slow assimilation kinetics of one, even more abundant substrate, would equally retard photosynthesis.

Due to the high dynamics of pelagic system it is very hard to identify what factor is limiting, at any moment, net community production. A continuous shift from one factor to the other is what very likely occurs in natural environment.

The third concept is the most relevant for global cycles, because is related to the total amount of carbon that can be transformed in organic form. On a short time scale this is also called the carrying capacity of the system.

Given a constant composition of the organisms and an efficient transfer of elements among them, there is an upper limit to the biomass the system can produce; this limit is certainly bound to the amount of the element which is below the constant ratio with the others. Does this mean that the addition of the deficient element would produce an increase in total biomass on a short run ? The answer is not as trivial as it appears. Ecosystems are quite complex structures. They evolve and optimize their functions according to the available resources. If resources increase, several homeostatic process could take place so as to damp the fluctuation on a short time scale or, *viceversa* the system could be possibly unprepared to channel the increased resource within the system. This could result in exporting new biomass, without increasing the total biomass within the system. This is the case, for example, of episodic blooms due to occasional inputs of nutrients that do not produce a general increase of carbon at ecosystem level, even if they are determined by a temporary increase in resources.

Furthermore, when analyzing nutrient limitation in semi-enclosed environment such as coastal lagoons, fjords, bays etc., fluxes must be considered instead of concentrations. This subject has been recently reassessed by Söderström [27]. Concentrations are state variables. Assimilation and growth are fluxes. When assimilation equals supply, the concentration can be very low without any implication for growth limitation.

We can now comment on a basic paradigm of chemical and biological oceanography: the recurrence of Redfield ratios in the world ocean.

We introduced Redfield ratios as numbers indicating elemental molar ratio in marine organisms. Interestingly, Redfield started from the opposite perspective. He observed that ratios among inorganic compounds related with living matter in the ocean were constant. From this observation he guessed that this stability could result from a mutual adaptation of marine organisms and the aquatic environment as a product of the evolution, and tested this hypothesis measuring the elemental composition of several organisms.

The elemental ratio in marine plankton, the organisms analyzed by Redfield, had indeed similar ratios of what he had measured as dissolved components in the environment. Then he concluded that this characteristic could not have been occurred by chance and argued that the biota had in fact determined the present composition of seawater.

Redfield was substantially right, because marine bacteria can fix elemental nitrogen or regenerate it through denitrification, and elemental nitrogen abundance even in the ocean is two order of magnitude higher than that of fixed nitrogen. A regulatory mechanism at the scale of the global ocean to cope with available phosphorus is therefore at hand. Is this mechanism so effective and farsighted ?

Some controversy exists about this topic. If biota can regulate nitrogen flux, how nutrient limitation can ever occur ? (here we limit our discussion to nitrogen and phosphorus limitation, a classical dispute in oceanography [15]). Most of work so far conducted on this topic concluded that if any nutrient limitation exists then nitrogen is certainly the putative element to be the limiting one. For nitrogen to be a limiting factor at the level of net ecosystem production, which is the case we are discussing here, nitrogen regulation should be kinetically constrained, originating short term depletion of utilizable nitrogen, or denitrification should be uncoupled with the dynamics of photic layer, or nitrogen regulation should depend on another limiting factor. This last hypothesis is strongly supported by increasing knowledge on iron distribution in the ocean, iron being an essential element for nitrogen fixing bacteria. Within this conceptual framework, iron would not only directly limit marine photosynthesis but would indirectly affect it, limiting nitrogen fixation [28].

On the other hand careful analysis suggests that coastal environment experiences phosphorus limitation, if any.

A way to reconcile those contrasting evidences is to assume that the global ocean has some weak links unable to overcome the existing polarization in concentrations and processes. In other words, coastal areas are not completely connected with open ocean, therefore unbalance of nutrients in coastal area cannot compensate for the opposite unbalance in the open sea.

Likewise, any attempt of balancing nutrient ratio in the photic zone is regularly disrupted in deeper layers, where, through denitrification, inorganic nitrogen is continuously kept out of balance. Apparently biota do not succeed in compensating for environmental unbalances with regard to their biochemical needs. Or, higher order homeostatic mechanisms exist that take advantage of the observed polarization.

Redfield ratios are then a powerful tool to investigate on processes that are quite differentiated and ultimately affect the spatial variability of nutrient distribution.

To put it in a different way, Redfield ratios indicate any process which dismantle the optimal composition obtained via photosynthesis.

6. Conclusions

Nutrients are unevenly distributed in the sea. Large scale circulation, local dynamics and different inputs from the external sources (atmosphere, land) are prime factors in determining the observed variability.

Phytoplankton abundance and primary production are clearly related with nutrient availability. In fact, marine photosynthesis is what prevents nutrient accumulation in the surface layer that would occur because of the internal vertical transport from the richer, deeper layers and because of external inputs that enter the ocean through the surface layer.

Since photosynthesis depends on light availability, net ecosystem production in different areas of the world ocean is controlled by latitudinal gradient in solar irradiance and the average flux of nutrients to the upper photic zone.

A comparison of internal versus external inputs suggests that, at present time, internal resources contribute the most to the above flux, at least for what concerns nitrogen and phosphorus. Therefore all hydrodynamic processes causing vertical transport of nutrients play a determinant role in controlling the productivity of the open ocean. Their extent and characteristics are critical in shaping oceanic provinces.

On the other hand, the strong asymmetry existing between land and ocean, in terms of their properties and functioning, generates strong gradients in external inputs to the sea, which are not dampened by physical and biological processes occurring at the boundary. Likewise, deep ocean biogeochemical processes do not parallel the ones occurring in the surface layer.

The present picture of nutrient cycles and fluxes at a global scale is still too coarse to allow definitive conclusions. While significant knowledge has been accumulated on some relevant processes determining the observed dynamics in the marine environment, the inadequacy of the existing data set, produces budget estimates that are affected by large errors, up to one half order of magnitude.

This prevents a complete understanding of what occurred in the past and the forecasting future scenarios.

A systematic, high quality and better resolved in time and space monitoring activity is needed for the world ocean. This can be accomplished only with the momentum given by new techniques that would allow remote continuous observations.

7. Acknowledgements

The authors wish to thank the Distributed Active Archive Center (Code 902.2) at the Goddard Space Flight Center, Greenbelt, MD 20771, for producing the data in its present format and distributing them. The original data products were produced by the Nimbus Project Office in collaboration with the NASA Goddard Space Flight Center Space Data and Computing Division, the NASA GSFC Laboratory for Oceans, and the University of Miami Rosenstiel School of Marine and Atmospheric Science. Goddard's share in these activities was sponsored by NASA's Mission to Planet Earth program.

8. References

1. Liebig J. von (1855) in L.R. Pomeroy, *Cycles of essential elements (Benchmark papers in Ecology Vol. I)*, Dowden Hutchinson & Ross, Stroudsburg, 1974, 11.
2. Baar, De, H.W.J. (1994), *Progress in Oceanography*, 33, 347.
3. Wyatt T. & Jenkinson I.R. (1993), *Fishery Oceanography*, 2: 231.
4. Mills E. (1989), *Biological Oceanography. An early hstory, 1870-1960*, Cornell University Press, Ithaca, 378pp.
5. Laubier L. (1989) in M.Denis (Ed.), *Océanologie, actualité et prospective*, Centre d'Océalogie de Marseille, Marseille, 61.
6. Margalef R. (1991), *Teoría de los sistemas ecológicos*, Publicacions Universitat de Barcelona, Barcelona, 290pp.
7. Millero F.J. (1997), *The influence of iron on carbon dioxide in surface seawater*, this volume.
8. World Ocean Atlas 1994, U.S. Department of Commerce, National Oceanic and Atmospheric Administration, National Oceanographic Data Center, Washington, CD-ROM no. 4
9. Distributed Active Archive Center, Goddard Space Flight Center, Greenbelt, MD
10. Berger W.H. (1989) in in W.H. Berger, V.S. Smetacek and G. Wefer (eds), *Productivity in the Ocean: Present and Past*, John Wiley and Sons, New York, 429.
11. Martin J.H. et al. (1994), *Nature*, 371, 123.
12. Barber R. T. (1991) in P.G. Falkowski and A.D. Woodhead, *Primary Productivity and Biogeochemical Cycles in the Sea*, Plenum Press, New York, 89.
13. Levitus S. et al. (1993), *Progress in Oceanography*, 1993. 31: 245.
14. Sarmiento J.L., Herbert T.D., Toggweiler J.R. (1988), *Global Biogeochemical Cycles*, 2, 115.
15. Codispoti L.A. (1989) in W.H. Berger, V.S. Smetacek and G. Wefer (eds), *Productivity in the Ocean: Present and Past*, John Wiley and Sons, New York, 377.
16. Duce R.A. (1986) in P. Buat-Ménard (ed.), *The role of Air-Sea Exchange in Geochemical Cycling*, 497.
17. McCarthy J.J. & Carpenter E.J. (1983) in E.J. Carpenter and D.G. Capone, *Nitrogen in the Marine Environment*, Academic Press, London, 487.
18. Laws E.A. (1983) in E.J. Carpenter and D.G. Capone, *Nitrogen in the Marine Environment*, Academic Press, London, 459.
19. Redfield A.C., Ketchum B.H. and Richards F.A. (1963) in M.N. Hill (Ed.), *The Sea (vol 2)*, Interscience, New York, 26
20. Margalef R. (1978), Oceanologica Acta, 1, 493-510.

21. Frontier S. & Pichod-Viale D. (1993), *Écosystèmes, structure, fonctionnnement, évolution*, Masson, Paris, 447pp.
22. Broecker W. (1991), *Oceanography*, **4**, 79.
23. Howarth R.W. (1988), *Annual Revue of Ecology*, **19**, 89.
24. Cullen J., Yang X. and MacIntyre H.L. (1991) in P.G. Falkowski and A.D. Woodhead, *Primary Productivity and Biogeochemical Cycles in the Sea*, Plenum Press, New York, 69.
25. Kolber Z.S. et al. (1994), *Nature*, **371**, 145.
26. Falkowski P.G. et al. (1991), *Nature*, **352**, 55.
27. Söderström J. (1996), *Sarsia*, **81**, 81.
28. Falkowski P.G. submitted to *Nature*.

SAMPLING TECHNIQUES FOR SEA WATER AND SEDIMENTS

Gabriele Capodaglio
Department of Environmental Sciences, University of Venice
Dorsoduro 2137, I-30123 Venezia, Italy

1. General aspects

Sampling is the first step in the analytical procedure; therefore, as with every other part of the procedure, sampling too must be controlled to avoid systematic errors being introduced into the final result. This idea is frequently expressed in the statement: "The analytical results cannot be better than the sample on which the analyses are performed".

The quality of the sampling depends on many parameters, which are defined by the characteristics of the system being studied and of the analytical procedure. The characteristics of the system can be described in terms of physical structure, size and homogeneity as regards both space and time. The analytical procedure can be described in terms of sample characteristics that must be analysed: the size, physical structure, homogeneity and time stability [1].

The sample must keep the properties the object had at the time of sampling or must change its properties in the same way as the object. Thus any change in the sample introduced by, for example, the sampling device, exposure to the air, the sample container or micro organisms should be avoided. The sampling procedure is a succession of steps performed on the object which ensure that the collected part possesses the specified sample quality.

The primary sample quality parameters are representativeness and stability; for changing systems additional parameters are discriminating power and speed. One requirement is that the sampling procedure must fulfil the condition that analysis of the sample collected shows no difference in object quality compared to the object itself (the sample is representative of the studied object). This means that two samples of the same object would not be distinguishable from each other: this condition is difficult to obtain, as a consequence of the bias associated with the complete analytical procedure (including all the analytical steps from sampling to analysis).

Factors contributing to selection of a correct sampling strategy are: the aim of the study, the variability of the studied area, costs and parameters to be determined. Fig.1 is a schematic diagram defining the steps necessary to design a correct sampling strategy and variables that must be defined to reach the aim.

The purpose of sampling sea components to describe their composition and characteristics as functions of space and/or time. The principal question in this case regards the number of samples that are necessary to describe the system in sufficient detail.

To define the sampling frequency for internally correlated non-homogeneous samples, as is normally observed in marine systems, the theorem of Nyquist could be used [1]. This tells us that the sampling frequency must be at least twice as high as the highest frequency of variability that occurs in the studied system. Although this theorem works well in physics, in chemistry the approach is more difficult because to evaluate system variability it is necessary to know the exact composition of the object with a very high resolution. Pressure from economic and time factors may restrict the execution of an extensive sampling campaign, but knowledge of general information about the studied area can help to reduce sampling frequency. For example, measurements of parameters such as salinity, temperature and density can give information about the homogeneity of the seawater area to be studied, measurements of fluorescence *in situ* provide information about the depth and zone where the phytoplankton biomass reaches maximum value.

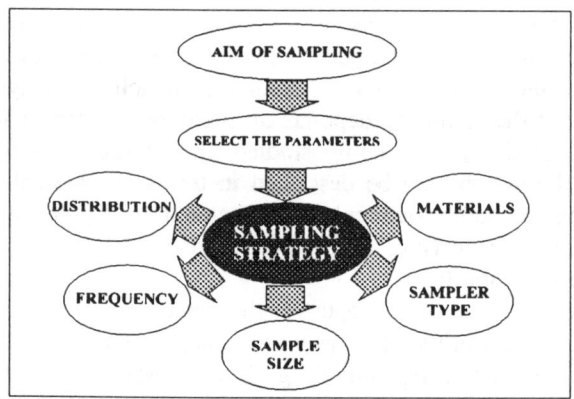

Figure 1. Schematic diagram defining the steps necessary in designing a sampling strategy.

As a general rule variability in water composition as a function of time is much higher than in sediments but normally presents a higher spatial homogeneity. The sediment composition integrates the change in water composition; while it is scarcely affected by sudden changes in water characteristics; its spatial variability is strongly affected by the principal hydrological characteristics of the studied area (mean seasonal currents, mean composition of effluents).

The sampling strategy will be strongly dependent on the area studied. Considering that in the marine environment three zones can be identified (bays, harbours and inlets; coastal areas; open sea) some general rules about frequency and density of sampling can be suggested:

1) Bays, inlets and harbours are strongly affected by local inputs. They are characterised by low homogeneity, especially in the vicinity of effluents, both for water and sediments: the sampling strategy therefore tends to increase the sampling density. Sudden changes in water characteristics are determined by the tidal regime during

the day, so frequent sampling is necessary to describe the water composition and its variability.
2) Coastal waters are substantially affected by human activities and natural effluents. Generally these zones present higher nutrient content compared with open sea areas. The characteristics of coastal water may be subject to seasonal variability due to the input of freshwater deriving from effluents. Sampling sites are normally located at the vertex of a grid whose mesh dimension is defined by knowledge of the hydrological regime and related to water quality and variability.
3) Open sea and particularly oceanic areas present high water stability and horizontal homogeneity, the vertical structure of water is defined by the hydrodynamic regime and only seasonal variations can be observed. Weather affects only the surface and subsurface layers changing the characteristics and the thickness of the mixed layer. The sampling density can be limited to a few significant sites, while the vertical dishomogeneity of oceanic areas means that the sampling must be carried out at many depths.

For the collection of sea components a very large range of samplers is available. Choice depends on the component to be sampled, the purpose of sampling, the parameters to be determined and, in some cases, on the analytical procedure to be applied. It should be borne in mind that is essential to collect an adequate amount of sample mantaining the characteristics and information necessary to achieve the aim. For example, if we wish to study the process of exchange of components between water and sediments or the processes of transformation taking place in the sediment, it is essential to collect a sufficient amount of sediment to determine the components of interest and the samples collected must maintain the vertical structure to emphasize the vertical distribution of components; moreover the sample composition should not be changed during the collection. This means that devices used to collect samples must be of correct volume, must be made of a suitable material and must ensure the avoidance of any alteration or contamination of samples.

2. Components of the sea

The sea is a large mass of water containing large amount of salts in a fairly uniform mixture. Frequently it would be a considerable challenge just to establish accurately the composition of the chemical properties of an isolated sample of sea-water, but the sea is not isolated so knowledge of the constituents of the system and the interfaces of contact with the sea-water is necessary to an understanding of the properties of the system.

The sea can be schematically represented as illustrated in fig.2, where we can distinguish 4 components (atmosphere, bulk seawater, biota and sediments) separated by 3 interfaces (atmosphere-seawater, seawater-sediment and seawater-biosphere) which are involved in active transport of matter and energy. Each component and interface has its special properties and affects the properties of others. Here we will examine the sediment, the water and the water-atmosphere and water-sediment interfaces.

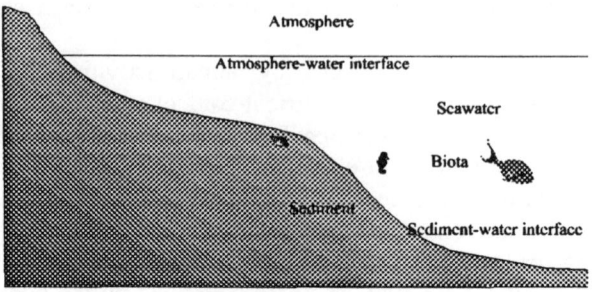

Figure 2. Sea components

Water-atmosphere interface. This part is responsible for the active transport of gases in both directions, the dispersion of aerosols originated by the sea and the collection of atmospheric particulate.

Processes determining the composition of surface water have already been described [2]. Material arrives at the surface from the bulk seawater or by dry or wet atmospheric depositions. Processes contributing to transport from bulk to surface are diffusion, convection, upwelling and rising gas bubbles, the last of which is probably the most important process. Material transported by the air can contribute substantially to the surface water composition, reaching the air-water interface through rainfall or dry deposition (gas, aerosol, particulate).

The material leaves the water surface toward the main water body or the atmosphere. Surface material can be dispersed by sinking particles, which may transport adsorbed matter on their surface, and by dissolution of water-soluble molecules. Dispersion of sea surface microlayer toward the atmosphere can be due to wind-generated aerosols or via the bursting of bubbles or by evaporation of volatile compounds. Processes operating in the surface microlayer (biological decomposition or photochemical reaction) may generate products that can be dispersed to the atmosphere or bulk seawater.

Bulk seawater. This is the water mass included between the water-atmosphere interface and the water-sediment interface and represents the larger part of the water mass. Its composition is affected by natural processes of transport throughout the surface, the exchange between the sediment-water interface and input from river estuaries or other water effluents of natural or anthropic origin [3].

Sediment-water interface. The thickness of this layer is related to the characteristics of the sediment, the hydrology and hydrodynamics of the studied area. It is composed of micro-particles and organisms determining the compaction and mineralization of sediments. The principal processes affecting this layer are degradation of organic matter and adsorption of inorganic components as a function of red-ox processes [4].

Sediment. This represents the collector of settling particles from the complete water column, the site where the burial and metamorphosis of residual particulate matter takes place. It features a strong stratification and its characteristics change from surface to deeper layers as a consequence of physical and chemical processes collectively referred to as diagenesis, catogenesis and metagenesis [3].

3. Sampling sea components

3.1. Seawater

3.1.1. Bulk seawater

The ideal procedure for analysing seawater is to make measurements *in situ* by a probe which could be lowered from the surface to the bottom to give a continuous profile of the water column. The application of probes able to carry out *in situ* measurements reduces problems related to the handling of samples (problems deriving from sample transformation or contamination during collection or storage), it permits real time analysis and therefore detailed description of complete ecosystems and interfaces [5,6]. Devices capable of precise measurements are available for only a few parameters (salinity, temperature, pH, PO_2 etc.) and only recently have sensors to determine some trace elements been introduced; however the application of this latter procedure is limited because of the lack of devices with adequate sensitivity and systems that permit sufficiently accurate determinations. The more convenient sensors for *in situ* determinations are based on optical, potentiometric or voltammetric measurements. Optical sensors based on fluorescence would be appear to be the most sensitive [7,8]. Measurements carried out by ion-selective electrodes are selective to certain forms of elements; they are therefore useful for speciation studies but, with few exceptions, are not sensitive enough to trace metals at environmentally significant concentrations. In the last few years progresses hare been made on the *in situ* determination of trace metals by voltammetric measurements. A submersible probe for trace metal determination in the water column has been described by Tercier et al. [9]; the device described has been used to determine Cu, Pb, Cd and Zn concentration in sea water at natural level. A flow system using voltammetric determination has been developed for continous trace metal monitoring; it consists of a computer-controlled electrochemical instrument fitted with units for pretreatment of samples and introduction of internal [10]. Brainina [11] integrated a similar system into a flow-through immersion device.

Because of problems related to the non-availability of instruments to make *in situ* measurements for all parameters of interest, chemical oceanographers are still dependent on the collection of water samples at the sites and at the depths of interest.

Procedures of water sampling may be instantaneous or continuous, the choice of procedure being made on the basis of analytical requirements, the aim of the study and the characteristics of the system. In oceanographic studies the sampling is normally carried out by general purpose devices, but sometimes, especially when low concentration species must be determined, specially designed samplers responding to specific requirements are adopted [12,3].

Various procedures can be used to collect seawater samples (see fig.3) but they all fall within three general approaches: i)pumping water to the surface from the requested depth; 2)sampling by bottles lowered to an appropriate depth and closed by a signal from surface; 3)adsorbing the elements or compounds of interest on an appropriate material lowered to the desired depth.

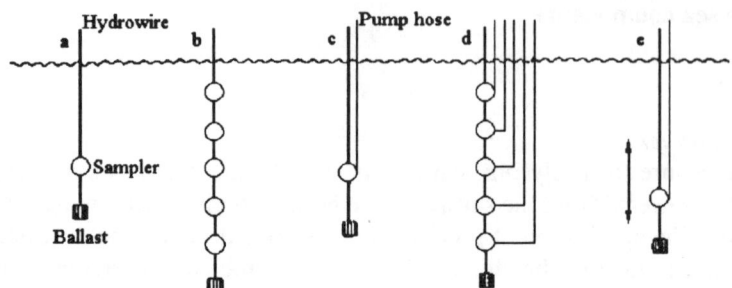

Figure 3.Different approaches to gathering water. a) Lowering one sampler at a fixed depth, b) multiple samples collected by samplers at different depths, c) lowering one pump at a fixed depth a collecting water through a hose, d) by a multiple pumping system at different depths, e) lowering one pump and changing the sampling depth.

Pumping systems have been used less frequently than the discontinuous sampling devices, usually only to sample shallow water, or when large volume samples or continuous profiles are required [14,15]. Collection of very large samples at different depths is necessary to detect natural or anthropogenic radionuclide elements and trace organic pollutants. Measurements by means of CTD and other probes have shown a general fine structure in the water column, missed if samples are collected by bottles at a limited number of depths. Sampling by pumping was used to collect large volumes of water (>20 tons) from depths of up to 3000 m [3] and in conjunction with continuous analyzers permits vertical and horizontal profiling [3]. Samples from depths of up to 75 m can be delivered at a rate of about 20 l min^{-1} by multistage submersible electric pumps.

Instantaneous samples are collected by bottles along the water column at different depths. Single or multiple instantaneous samples are collected by bottles bound to a cable, depths of sampling are obtained by measurement of the length of unrolled hydrowire, closure of samplers at desired depths is obtained in cascade by messengers running along the cable. An approximately simultaneous collection of multiple samples at different depths is obtained by bottles connected in a *rosette* sampler and closing the bottles at the desired depths by signals transmitted by a conducting cable. The latter procedure allows a general characterisation of the water column by physical and chemical measurements to be carried out at the same time as the sample collection by a multiparametric probe. Samples gathered by the previous systems are used to determine the majority of chemical and physico-chemical parameters (nutrients, salinity, particulate) and if the volume is sufficient to guarantee the representativeness of samples they can be used to determine some biological parameters such as phytoplankton.

For many general purposes the sampling is carried out at the standard depths reported by the "International Association on Physical Oceanography (see Table 1). However, deviations from such a general depth schedule can be applied if the sampling is carried out for special requirements (e.g. if the local hydrodynamic regime determines a particular vertical profile, or if only the photosynthetic zone is being studied).

Economic and time factors lead to the depth sampling scheme being simplified in many investigations.

Many devices are available for gathering water samples and the choice of the most appropriate is carried out on the basis of the nature and concentration of the components to be determined. However some general criteria must be followed when making the selection:

TABLE 1. Standard depths (m) reported by the International Association on Physical Oceanography.

0	50	200	500	1000	2500	6000
10	75	(250)	600	1200	3000	7000
20	100	300	(700)	1500	4000	
30	150	400	800	2000	5000	etc.

- Exchange between the surrounding water and that within the sampler should be rapid and complete. This can be obtained by devices with good flushing characteristics, comprising a tube connected to the hydrological wire parallel to its axis and sealed at the two ends by means of wide aperture valves.
- The closing mechanism should be reliable and after it has been activated the sample must be sealed completely to ensure that no exchange of water occurs as it is brought to the surface.
- The sampler should be constructed of materials resistant to corrosion and should not modify the sample composition. Many examples can be given of changes in sample characteristics due to inappropriate materials being used to construct samplers, especially when trace components are determined, but also parameters such as pH and conductivity can be considerably changed, see materials paragraph below.
- It should be light and easy to handle.

Integrated measurements of trace elements can be obtained with *in situ* sampling devices consisting of a pump drawing water through a preconcentrator column for a fixed time. Studies carried out using columns consisting of chelating or cation exchange resins showed that the recovery efficiency was higher than 90% if metals (Cd, Cu, Pb and Zn) were present in the ionic form [17-19]. All the procedures showed a considerable reduction of metal recovery if they were complexed by organic ligands; Zhang and Florence [19] used an aluminium hydroxide-coated cationic resin to evaluate the bioavailable metal species (labile metal forms).

Iwata et al. [20] used an adsorbent column of Amberlite resin to collect and preconcentrate persistent organochlorines. About 150-400 l of seawater were pumped from a seawater foucet through an XAD2 resin at a rate of less than 0.3 l/min, the chlorocompounds adsorbed were eluted by 300 ml of ethanol and concentrated until 5 ml. The extract was used to determine hexachlorocycloexane, DDT and its metabolites, chlorane compounds and PCBs in Pacific, Atlantic and Indian Oceanic waters.

3.1.2. Trace components

Metal distribution in marine areas has been studied for several decades, but early data were unreliable because of sample contamination during collection, storage, treatment and analysis. Advances in analytical techniques and improved laboratory practice have eliminated many sources of error that originate during analytical measurement, but they have also emphasized that sampling and conservation procedures frequently result in high sample contamination. In such cases the metal distribution patterns are biased and the interpretation of data becomes misleading and inconsistent with respect to known oceanographic parameters such as hydrodynamic flow, metal sources and biogeochemical cycles. The requirements for contamination control vary from one trace element to another as a function of their concentration in seawater and their ubiquitousness. Sources of contamination may derive from research vessels, which can be a significant source of contamination for some trace elements, and from unsuitable sampling and storage procedures. For some elements contamination control is relatively easy (Ba, As), while for ubiquitary trace elements (e.g. Zn, Fe, Pb) reliable results are available only if the entire procedure is subjected to meticulous control [21].

The present accepted lead concentration in seawater has decreased by about four orders of magnitude compared with the values found forty years ago as result of an artefact deriving from improvements in the reduction and control of contamination. Patterson and coworkers [22] were able to obtain the first reliable lead profiles by developing a new deep-water sampler by means of which they gathered samples free from contamination by the ship and hydrowire.

Although several deep-water samplers have been specifically developed to study areas presenting very low concentrations of lead [12,13], it has been shown that commercially available samplers can be used to collect uncontaminated samples for trace elements, and lead can be determined in these samples if its concentration is higher than that found in deep water in oceanic areas. Intercomparison exercises with sampling devices have provided assessments of different procedures in gathering seawater samples. Bruland and al. [23] reported a comparison between samples collected with the Patterson sampler and with a 30-litreTeflon-coated Go-Flo sampler (General Oceanics) mounted on a non-metallic hydroline. They found no significant differences for Zn, Cu, Ni, and Cd, while Spencer et al. [24] show significant differences in zinc and lead concentration in surface seawater samples collected by hand in 1 l FEP bottles and 30 l Teflon coated Go/Flo samplers. The results showed Zn concentration as 10 times higher in the samples collected with the commercial device. Capodaglio et al. [25] compared four different samplers (Go-Flo, Niskin, van Dorn and Rutner) with different capacities. The results showed that the volume/surface contact ratio can have a considerable effect on sample contamination also when Teflon-coated Go-Flo bottles and volume higher than 10 l were used. The problems of contamination of samples can be kept under control if sufficient attention is devoted to the cleaning and conditioning procedures.

3.1.3. Surface microlayers
The composition of the micro-layer in the surface of the sea is enriched by natural and anthropogenic surface-active substances which form films and affect the exchange of matter (water, gas and dissolved components) between water and atmosphere. Pollutants such as oil or other lipophilic substances affect not only the transport processes but also biological activity, which is concentrated in the sea surface layers.

Estimates of the surface layer thickness may be inferred from the results of experimental studies at air-water interfaces. Measurements of elliptical polarization of light, as a consequence of the transition layer at the water surface, showed layer thickness to be in the range 0.5-11 nm [26]. Measurements at temperature gradients under calm conditions appeared to show a thickness of about 100 μm and the thickness increases with the turbolence. Measurements of gas exchange at the seawater surface give a value of approximately 30 μm for the microlayer. An alternative approach can derive from considerations of the depth at which the water molecules are perturbed from their normal distribution in the bulk of the liquid. Horne [27] reviewed theories in this field and it was concluded that the perturbation extends several hundred nanometers from the water surface.

It is evident that different approaches produce a surface microlayer thickness which varies by many orders of magnitude as function of the parameters used to describe it. However the majority of results have been obtained with samples collected from the sea surface by devices able to collect approximately the top 100 μm, this layer appears to be a thick slice compared with the dimension of one surface active molecule presenting dimension of about 3×10^{-3} μm. An important objective for future research may therefore be the development of techniques to sample thinner slices of the surface water in order to determine the extent to which they differ in chemical, physical and biological properties.

Sampling of the surface material can be carried out with nets and adsorbing matter. Garret [28] sampled the top 150 μm of the water with a 16-mesh screen. He immersed and withdrew a screen horizontally and drained the water adhering to it into a sample bottle. The irreversible adsorption of some material on the mesh reduced the efficiency of sampled surface water recovery to about 75%. By repeating the procedure 200/250 times they were able to collect approximatelly 100 m^2 of surface water (~20 l). Piotrowicz et al. [29] used a similar system, consisting of a 20-mesh polythene screen, to collect microlayer samples for trace metal analysis. Samplers of this type have been used in most of the studies of sea surface material so far conducted. The principal deficiencies of this screen sampler are the long time it takes to get a reasonable volume of water and doubt that a sufficiently thin and unmixed layer is sampled.

Harvey [30] tried to overcome the above problems by adsorbing the microlayer (60-100 μm) on a rotating stainless steel drum covered by ceramic and continually removed by a wiper. The thickness of the sampled layer depends on the rotation speed and the water temperature. Under average conditions they were able to collect a considerable volume at a sampling rate of about 300 ml/min. Applicazione[31]

Harvey and Burzell [32] used a glass plate immersed and withdrawn vertically through the surface; a surface layer of about 60-100 μm adhered to the plate and was removed using a neoprene wiper blade.

Hamilton and Clifton [33] described a device in which a thin film of water is harvested by freezing it onto a solid collector with liquid nitrogen. They deployed the apparatus to sample dissolved and particulate material.

More recently, Turner and Liss [34] described an adaptation of the cryo-sampler previously described by Hamilton and Clifton to gather a microlayer for the measurement of gases present in trace amounts. They applied the sampler to determine the difference of contration between the surface microlayer and the subsurface layer of dimethylsulfide and low molecular weight halocarbons both natural (e.g. methyl iodide) and man-made (e.g. Freon-11). They calculated the enrichment and depletion factor for the trace gases and compared the results to those obtained by the Garrett screen [28]. They concluded that there are small differences between the two devices (evident differences are detected only for dimethylsulfide during phytoplankton blooms) and the cryo-sampler is not very satisfactory for gases with high concentration in the air phase. The authors hypothesized that cryo-sampling of the microlayer should prove preferable to most other methods for many reactive/short-lived substances of current interest.

3.2. Sediment-water interface

The zone affected by flow of mass and energy between sediment and bulk water can be identified in the water layer immediately in contact with surface sediment and the inerstitial water; the latter represents the mediator fluid in the exchange of components between sediment and water.

3.2.1. Water layer close to the sea floor

The water composition in the vicinity of the sea floor is modified by processes occurring with sediments. Two important criteria must be fulfilled in sampling this layer [35]: (i) the device must collect samples at a known and small distance from the seafloor, (ii) the procedure must ensure that the sample collected is representative of the water at the particular depth sampled. The samplers must be lowered very close to the bottom causing minimal disturbance of the water or it must be possible to leave it in position for sufficient time for the water composition to normalize.

Various sampling procedures have been used to gather water close to the bottom, some operating by horizontal devices using a variety of closure mechanisms [36,37], others pumping water to the surface through intakes placed on the bottom [38]. However, recent measurements of dissolved O_2, pH, nitrate and conductivity carried out by sensors have shown that the diffusion layer on the sea floor is frequently only a few millimeters thick [39-42]; therefore, collecting samples within a layer of 1-2 m by sampling bottles does not ensure that the samples gathered are representative of the sediment-water interface. These problems can be overcome by taking measurements of parameters of interest by *in situ* probes [6].

3.2.2. Interstitial water

Interstitial waters are acqueous solutions that occupy the free spaces between sediment particles. Their composition reflects the nature of the original fluids buried with the sediments, particle-fluid reactions and the migration of fluids by convection and diffusion [4].

Investigation of pore water chemistry is of great interest with regard to post-deposition reactions and diagenesis. It represents an important phase in the transport of trace elements from sediments to the overlying water and *vice versa*. The principal problem in studying the distribution of dissolved species in this medium lies in collecting samples without disturbing the physico-chemical conditions in the medium to be investigated; in particular, it has been shown that temperature, pressure and redox variations exert a strong influence on the composition of interstitial waters.

Two basic approaches are possible: squeezing of sediment cores or *in-situ* extraction. The first procedure may induce some changes of composition in the sample due to the contact of sediment with air during the transfer and handling of core, even when inert gases are utilized to prevent oxidation of water during extraction, the chemical and physico-chemical conditions of samples (CO_2 and H_2S content, pH, temperature and elements speciation) are often modified and hence the results obtained are frequently not representative of the studied media [4].

If analysis is carried out immediately after *in-situ* extraction it minimizes many of the previous sample alterations [43], but the method presents the limitation that the devices are suitable only for use in shallow water and it is difficult to take pore-water samples at depths greater than a few metres.

Application of *in situ* measurements by suitable probes seem the most convenient solution to problems of sample modification due to oxidation and alteration of content of gaseous components [6].

3.3. Sediment

3.3.1. Suspended matter

Suspended matter plays a fundamental role in the definition of the chemical composition of water because the particles are the carriers by which many chemical species are trasported from the surface to the deeps and to the sediments. Along the water column, the particulate undergoes a series of processes of transformation and compaction that determines their wide variability of density, size and chemical composition in the water [3,44], so, both the origin and the transformation undergone determine a very large dimensional distribution of particles ranging between a few millimeters down to 0.1 μm. The normal classification of water components tends to identify the border between colloidal matter and suspended matter around 0.1 μm [45], but if we consider the characteristics of the colloidal matter and that processes, controlled by dynamic equilibria connect all the water components, it is evident that this boundary between dissolved components and particulate matter cannot be clearly marked out; in general the differentiation between the two phases is dependent on the procedure used to determine the suspended matter.

Two different approaches can be used to quantify the suspended matter in the sea: i) gravimetric methods that involve the separation of the material by physical procedures (filtration, centrifugation), ii) optical methods without any need to separate the material from the dissolved phase. However, information about the chemical characteristics of suspended matter can be obtained only by analysis carried out on particles separated by physical procedure.

The optical methods, especially if applied *in situ*, provide the best information on the real distribution and nature of suspensoids because they avoid any alteration of particulate characteristics determined by the sampling techniques (agglomeration and disaggregation). In addition, optical methods *in situ* may provide a more detailed spatial description of particle distribution. The interpretation of results obtained by optical measurements is not always easy because optical phenomena depend on the number, size and shape of particles. Two different optical phenomena can be used to evaluate the suspended matter in seawater: light adsorption and light scattering. Instrumentation based on measurement of light adsorption (transmissiometry) presents a sensitivity (the ideal operating range is between 0.5-5 mg TSM l^{-1}) wihch is normally not sufficient to determine open-sea levels of total suspended matter (TSM), so the usefulness of this type of instrument is limited to measurements in coastal or other continental shelf and slope environments. Instruments using light scattering measurements (nefelometry) is normally preferred for *in situ* determination of TSM in open-sea water because of their higher sensitivity wich respect to the transmissiometry measurements [44].

In the last fifteen years studies have been carried out to quantify the suspended matter in seawater by Remote Sensing techniques; results have shown that the procedure can be applied only once a suitable algoritym is available to transform spectral data in suspended matter concentration [46].

Separation and subsequent weighing of suspended matter from seawater permits the determination of absolute concentration of TSM; however, a physical classification based on particle size does not define clear boundaries between different groups (particulate and dissolved), so classifications in the two groups are defined in an essentially operational manner. The separation can be accomplished either by centrifugation or filtration, but it must be remembered that particle density has a considerable effect on the efficiency of particle removement when centrifugation is used.

The most commonly used technique for the recovery of particulate which must be analysed is filtration through pre-weighed membrane filters for determination of both the inorganic and the organic component [47,48]. The filter material is normally mixed esters of cellulose or polycarbonate, especially if the dissolved phase is to be analysed for determination of trace elements; however, glass fibre filters are also used when organic material must be determined. The pore size ranges between 0.1 µm and a few µm but the pore size most used to separate dissolved and particulate components is 0.45 µm. The filtration can be carried out by exerting a pressure or by exerting a depressure, in both the cases the difference in pressure must not be too high (normally lower than 0.5 bar) to avoid crushing particles and cells.

When a large amount of particulate is necessary or the collection of large-volume samples is required to obtain a sufficient amount of particulate, continuous-flow centrifugation systems can be used [49]. When processes of sedimentation or processes of particulate material degradation are being studied, the sedimenting matter is collected by sediment traps placed along the water column [50,51].

3.3.2. Sea-floor
Because of the heterogeneity and complexity of solid matter in aqueous systems, care is necessary during sampling. The procedure has to consider the heterogeneity of the soil matrix both vertically and horizontal. Several devices have been developed for the extraction of sediments, interstitial water and suspended material. The type of sampler selected is related to the aim of the study [52]. Among the scientific problems that can be faced when studying sediment composition and distribution are: i) the influence on the sediment composition of physical, chemical and biological variables of surface water and

along the water column on the sediment composition; ii) the effects of the flow regime and the physico-chemical characteristics of bottom waters on deep-sea sediments; iii) the influence of benthic fauna on the sediment; iv) the relationship of sediment composition, and surrounding physical and chemical conditions, to the geotechnical properties of deep-sea sediemnts.

The simplest devices are grabs; application of the system causes the mixing of material, so it can be used only when knowledge of the vertical structure is not necessary; the ease of use, the effectiveness in all sediment types and the large sample volumes collected are the main advantages of grab samplers. When the principal requirement of the study is to know the vertical stucture of sediment the more useful device is the corer. There are different types of corer suited to different studies and a detailed description is available [52]. Gravity corers are used in the study of sediment textures and evaluation of paleoclimatic changes, for which cores of 2-3 m of sediment material are required; the major disadvantage arises from limited penetration and the disturbance deriving from the rolling of the ship, which can cause the corer to bounce one or more times along the sea floor determining a loss of the vertical structure. Cores of up 30 m in length have been taken by piston corers. Some corers have been introduced to collect the first metre of sediment, maintaining the structure and the stratification of the surface layer unaltered [53]. In this case cores longer than 2-3 metres have to be hoisted on board ship in a horizontal position, leading to disturbance of the core top. These devices collect sediment together with the overlying water layer; this permits study of the processes of water-sediment exchange. Using box corers it is possible to collect large sections of suface sediment to study the sedimentary structure and pore-water chemistry.

As reported above, sediment is caracterized by low homogeneity. Although vertical stratification can be evaluated by corer samplers, horizontal variability, that may be present at small-scale level (of the order of a few centimetres), cannot be estimated using samplers collecting only a few tens of square centimetres of sediment surface. Fowler and Kulm [54] introduced the use of multiple-barrelled corers that not only collect more surface sediment, but also allow the small-scale variability to be evaluated by comparing the composition of samples collected by 5-6 individual barrels simultaneously deployed.

Study to determine sedimentation rates or to define the characteristics and composition of settling particles is carried out by collecting samples using bottom sediment traps. Devices are normally made of plastic material and ballasted on the bottom by arms and weights that must limit disturbances [55].

4. Materials

The selection of the correct materials to be in contact with the samples is an essential part of the sampling strategy; use of the wrong materials to construct the samplers or bottles used to collect and store samples respectively will lead to unrepresentative samples being analysed; Park [56] showed that the use of samplers made of brass or other copper alloys whch are uickly corroded by sea water can determine a decrease in dissolved oxygen and the transformation of bicarbonate into carbonate, will lead to an increase in pH and alkalinity and changes of conductivity (this will give rise to a mistake in the value of salinity if measured by a salinometer). Consideration in the selection of the materials should include chemical composition, chemical resistence, thermal stability, permeability, adsorption and desorption behaviour and the treatments and conditioning procedures necessary before use [57].

General purpose samplers for gathering seawater are normally made of plastic materials (polycarbonate, polyvinyl chloride or in same cases polymethyl methacrylate) though metallic samplers are suitable for the collection of samples to determine dissolved organic components. The determination of some chemical parameters, for example trace species (organic compounds and trace metals), requires the use of a dedicated sampler made of controlled material to avoid modification of the sample (see Tab. 2). The gathering of samples to determine dissolved gases requires systems excluding any contact with the atmosphere and materials which are permeable. The collection of water to determine bacteria composition of course requires the use of sterile samplers.

TABLE 2. Selection of the materials to be in contact with the samples.

Parameters	Avoided materials	Prefered materials
DO, pH, Alkal., conductiv.	Corroded metals	Plastics materials
Organic components, DOC	Polyethylene, polypropilene	Teflon, Stainless steel, glass
Trace elements	Glass[a], metals	Teflon, polyethylene[b], fused silica[a]
Gas analysis	Plastics, permeable materials	Glass, stainless steel
Microbiological analysis	Any not sterile material	Sterile polyethylene bags

[a]Prefered materials when mercury is determined; [b]low density polyethylene and low fillers plastic materials.

For trace components, inorganic (glass, stainless steel, etc.) or plastic materials (polyethylene, polyfluorocarbon, etc.) are preferred as a function of components to be determined to avoid sample contamination. However it must be kept in mind that each material contains undesiderable impurity, and their concentration varies greatly with manufacters, grade and even lot numbers. All the materials should be carefully cleaned before use. Procedures for cleaning plastic materials used to collect and store seawater samples to be analysed for determination of trace elements are described [58,59]. Materials used in the collection treatment and storage of samples that will be used to determine trace organic components are normally subjected to several steps of cleaning by ultrapure organic solvents (hexane, acetone, chloroform, etc.) and rinsed by distilled water [47].

5. References

1. Kateman G. and Pijpers F.W. (1981) *Quality Control in Analytical Chemistry*, J. Wiley and Sons, New York.

2. Liss P.S. (1975) in J.P. Riley and G. Skirrow(eds.), *Chemical Oceanography*, vol. 2, chapt. 10, Academic Press, London, pp. 193.
3. Chester R. (1990) *Marine Geochemistry*, Unwin Hyman, London.
4. Manheim F.T. (1976) in J.P. Riley and R. Chester (Eds.) *Chemical Oceanography*, vol. 6, chapt. 32, Academic Press, London, pp. 115.
5. Buffle J., Wilkinson K.J.,. Tercier M.L and Parthasarathy N. (1997), *Ann. Chim.*, **87**, in press.
6. Tercier M.-L- and Buffle J.(1993) *Electroanalysis,* **5**, 187.
7. Taglacier A., Muller-Ackermann E., Kammerlover H. and Niesner R., (1996) Euroanalysis IX, Bologna, Sept. 1-7, Abst. TuL 26.
8. Spichicher U., Simon W., Bakker E., Lerchi M., Bühlmann P., Hang J.P., Kuratti M., Ozawa S. and West S. (1993) *Sensors and Activators*, **B11**, 1-8.
9. Tercier M.-L., Buffle J., Zirino A., De Vitre R.R. (1990) *Anal. Chim. Acta*, **237**, 429.
10. Achterberg E.P.and van den Berg C.M.G. (1994) *Anal. Chim. Acta*, **284**, 463.
11. Kh.Z. Brainina, Khanina R.M., Forshtadt V.M., Vilchinskaya E.A. and Gaponenko G.L. (1990) Proc. of J. Heyrovsky Centennial Congress on Polarography 41st Meeting of International Society of Electrochemistry, Prague, pp. 175.
12. Boyle E.A., Chapnick S.D., Shen G.T. and Bacon M.P. (1986) *J. Geophys. Res.*, **91**, 8573.
13. Patterson C.C. and Settle D. (1976) NBS Spec. Publ. (US), **422**, 321.
14. Bergen B.J., Nelson W.G. and Pruell R.J. (1993) *Environ. Sci. Technol.*, **27**, 938.
15. Gillain G. and Brihaye C. (1985) *Anal. Chim. Acta*, **167**, 387.
16. Jeffrey L.M., Fredericks A.D. and Hillier E. (1973) *Limnol. Oceanogr.*, **18**, 336.
17. Davey E.W. and Soper A.E. (1975)
18. Willie S.N., Sturgeon R.E. and Berman S.S. (1983) *Anal. Chim. Acta*, **149**, 59.
19. Zhang M.-P. and Florence T.M. (1987) *Anal. Chim. Acta*, **197**, 137.
20. Iwata H., Tanaka S., Sakai N. and Tatsukawa R. (1993) *Environ. Sci. Technol.*, **27**, 1080.
21. Bruland K.W. (1983) in J.P. Riley and R. Chester (Eds.) *Chemical Oceanography*, vol. 8, chapt. 43, Academic Press, London, pp. 157.
22. Schoule B.K. and Patterson C.C. (1981) *Earth Planet Sci. Lett.*, **54**, 97.
23. Bruland K.W., Coale K.H. and Mart L. (1979) *Mar. Chem.*, **17**, 285.
24. Spencer M.J., Betzer P.R. and Piotrowicz S.R. (1982) *Mar. Chem.*, **11**, 403.
25. Capodaglio G., Toscano G., Cescon P., Scarponi G. and H. Muntau (1994) *Ann. Chim.*, **84**, 329.
26. Kinosita K. and Yokota H. (1965) *J. Phys. Soc. Japan*, **20**, 1086.
27. Horne R.A. (1969) *Marine Chemistry*, J. Wiley and Sons, New York.
28. Garret W.D. (1965) *Limnol. Oceanogr.*, **10,** 602.
29. Piotrowicz S.R., Ray B.J., Hoffman G.L. and Duce R.A. (1972) *J. Geophys. Res.*, **77**, 5243.
30. Harvey G.W. (1966) *Limnol. Ogeanogr.*, **11**, 608.
31. Garabetian F., Romano R.P. and Sigoillot J.C. (1993) *Mar. Environ. Res.*, **35**, 323.
32. Harvey G.W. and Burzell L.A. (1972) *Limnol. Ogeanogr.*, **17**, 156.
33. Hamilton E.J. and Clifton L.J. (1979) *Limnol. Oceanogr.*, **24**, 188.

34. S.M. Turner and Liss P.S. (1989) in J.P. Riley and R. Chester (Eds.) *Chemical Oceanography*, vol. 10, chapt 63, Academic Press, London, pp. 379.
35. Riley J.P. (1975) in J.P. Riley and G. Skirrow (Eds.) *Chemical Oceanography*, vol. 3, chapt. 19, Academic Press, London, pp. 193.
36. Joyce J.R. (1973) *J. Mar. Biol. Ass. UK*, **53**, 741.
37. Sholkovitz E.R. (1970) *Limnol. Oceanogr.*, **15**, 641.
38. Smith K.L. (1971) *Limnol. Oceanogr.*, **16**, 675.
39. Archer D., Emerson S. and Reimers C. (1989) *Geochim. Cosmochim. Acta*, **53**, 2831.
40. Gundersen J.K. and Joergensen B.B. (1990) *Nature*, **345**, 604.
41. Wilson T.R.S., McPhail S.D., Braithwaite A.C., Koch B., Dogan A. and Disteche A. (1989) *Deep-Sea Res.*, **36**, 315.
42. De Beer D. and Sweerts J.P.R.A. (1989) *Anal. Chim. Acta*, **219**, 351.
43. Bertolin A., Rudello D. and Ugo P. (1995) *Marine Chem.*, **49**, 233.
44. Sackett W.M. (1978) in J.P. Riley and R. Chester (Eds.) *Chemical Oceanography*, vol. 7, chapt. 37, Academic Press, London, pp. 127.
45. Ferrari G.M. (1988) *Acqua-Aria*, 587.
46. Sturm B. and Tassan S. (1983) XXVIII General Assembly of Intern. Union of Geodesy and Geophysic (IVGG), Hamburg 15-17 August 1983.
47. Kelly A.G., Cruz I. and Wells D.E. (1993) *Anal. Chim. Acta*, **276**, 3.
48. Brzezinska-Paudyn A., Balicki M.R. and Van Loon J.C. (1985) *Water Air Soil Poll.*, **24**, 339.
49. Schuessler U. and Kremling K.(1993) *Deep-Sea Res.*, **40**, 257.
50. Khripounoff A. and Crassous P. (1994) *Deep-Sea Res.*, **41**, 821.
51. Anderson R.F., LeHuray A.P., Fleisher M.Q. and Murray J.W. (1989) *Geochim. Cosmochim. Acta*, **53**, 2205.
52. Moore T.C.Jr. and Heath G.R. (1978) in J.P. Riley and R. Chester (Eds.) *Chemical Oceanography*, vol. 7, chapt. 36, Academic Press, London, pp. 127.
53. Kogler F.C. (1973) *Meyniana*, **13**, 1.
54. Fowler G.A. and Kulm L.D. (1966) *Limnol. Oceanogr.*, **11**, 630.
55. Hargrave B.T. and Burns N.M. (1979) *Limnol. Oceanogr.*, **24**, 1124.
56. Park P.K. (1968) *Deep-Sea Res.*, **15**, 721.
57. Mizuike A. (1983) *Enrichment Techniques for Inorganic Trace Analysis*, Springer-Verlag, Berlin.
58. Mart L. (1979) *Fresenius Z. Anal. Chem.*, **296**, 350.
59. Scarponi G., Capodaglio G., Barbante C. and Cescon P., in S. Caroli (ed.) (1996) *Element Speciation in Bioinorganic Chemistry*, Chemical Analysis Series, vol. 135, chapt. 11, J. Wiley and Sons, New York, pp. 363.

TRACE AND ULTRATRACE METAL ANALYSIS: MATRIX REMOVAL AND SAMPLE PRECONCENTRATION

CORRADO SARZANINI
Department of Analytical Chemistry, University of Torino
Via P. Giuria, 5 - Torino, Italy

1. Introduction

Significant advances have recently been made in the detection power of analytical instrumentation possessing multi-element capability; preconcentration and matrix separation techniques however are still often needed for fresh water and sea water samples in order to achieve the necessary detection limits for the accurate determination of trace elements at ng/Kg (ppt) levels. The chemical techniques used in preconcentration must provide analyte isolation as well as enrichment through minimal sample manipulation in order to avoid contamination and to obtain low sample blanks. Metals present in sea water can be subdivided into a dissolved and an undissolved state (conventionally unretained and retained after filtration through 0.45 µm membrane filter) in an operational manner. But, this means that inorganic and organic colloids with a diameter below 0.45 µm in addition to labile and weak complexes originated by inorganic (Cl^-, OH^-, CO_3^{2-}, HCO_3^-, SO_4^{2-}) and organic (dissolved organic matter) ligands are counted as dissolved. This refers to speciation that must be mentioned with reference to preconcentration methods which allow the evaluation of total or of a fraction of metal.

Separation techniques are one of the three primary ways to achieve or enhance selectivity, the other two are chemical reactivity and signal discrimination. An analytical separation can be regarded as an operation involving the dividing of a sample into at least two parts of different composition in order to enrich one of the fraction with one component in relation to the rest.

2. Separation Techniques

Analytical separations can be roughly classified according to basic criteria:
a) the type of interface across which mass transfer takes place
   solid-liquid, solid-gas, liquid-solid, liquid-liquid, gas-liquid
b) the forces involved in the process
   purely mechanical (filtration), physical (distillation, zone fusion), solubility-based partitioning processes (liquid-liquid, gas-liquid chromatography), based on the behaviour in an electric field (electrophoresis) or chemical (precipitation, ion-exchange).

In addition analytical separation can be thermodynamically or kinetically controlled, or both.

Other classifications of analytical separation techniques are based on dynamic or operational aspects: mass transfer takes place across a well defined interface or in the bulk phase.

2.1 STATIC AND DINAMIC SEPARATIONS

Different approaches can be followed in order to remove the matrix and preconcentrate the analytes: static and dynamic separations can be performed. Such procedures are usually defined off-line or on-line, involving sample treatments which are performed indipendently-of or in direct connection with the analytical instrumentation.

The main preconcentration methods, from a hystorical point of view, are : Evaporation, Precipitation, Coprecipitation, Flotation, Extraction (homogeneous or heterogeneous), Sorption (adsorption, chelation on functionalized supports, ion-exchange), Electrochemical methods and Special Techniques. An overall classification proves very difficult because more than one physical and/or chemical principle may be involved in the techniques considered and may be taken as a basis for classification. Preconcentration methods based on:
- Precipitation and related techniques
- Flotation
- Extraction
- Special Techniques

will be hereafter considered in order to give an overview of different approaches and tools now available with particular regard to seawater samples.

Particular attention will be devoted to the generically named *liquid-solid extractions*, probably the most effective of the techniques employed nowadays.

2.2. PRECIPITATION

The solubility of many compounds is sufficient to prevent quantitative precipitation at trace level concentrations. In these cases the use of high purity collectors (carriers), in the form of precipitates, such as hydroxides is most frequently used, in order to obtain a quantitative coprecipitation.

Preconcentration procedures utilizing the collection of trace elements via coprecipitation have often been used to enhance the sensitivity and/or selectivity of a given analytical method. Their major advantage over preconcentration techniques lies in simplicity.

In order to fully utilize the above advantages, the carrier precipitate used should not have any adverse effect on the analytical signal (no secondary matrix effect) and the coprecipitation of the elements of interest should be quantitative. From a practical point of view an easy filtration of the precipitate is also required.

Group precipitants are reagents which react with collectors to give hydroxides or sulphites. Common hydroxide-type collectors include $Fe(OH)_3$, $La(OH)_3$, $Cd(OH)_2$, $Zn(OH)_2$, $Bi(OH)_3$ and $Al(OH)_3$ or mixed collectors such as equimolar amounts of Fe(II) and Fe(III) hydroxides which completely precipitate traces of aluminium, titanium, chromium, vanadium, zirconium and antimony; manganese dioxide has also been widely used. Gallium hydroxide has been employed as coprecipitant for the enrichment (200x) of Al, Co, Cr, Cu, Fe, Mn, Ni, Pb, Ti and Zn traces (10-500 ng/l) from seawater [1-3]. Coprecipitation with Zr was used to obtain 20-fold preconcentration for the determination of 17 trace metals from seawater by ICP AES [4] and more recently for the determination of trace amounts of Be, Cr, Fe, Co, Ni, Cu, Cd and Pb in natural water samples [5].

Although inorganic reagents have been used more frequently than their organic counterparts, the precipitation of trace elements with organic reagents has found extensive use in analytical chemistry.

Typical organic ligands that have been used as precipitants are: cupferron (Fe carrier), 5,7-dibromo-8-hydroxyquinoline, 8-hydroxyquinoline (Fe carrier), ammonium pyrrolidine dithiocarbamate (APDC, Co, Pb or Zn carriers) [6,7], and 1-(2-pyridylazo)-2-naphthol (PAN) [8]. If samples with high acid contents are to be analyzed, it is advatageous to carry out the precipitation in the original acidic solution in order to avoid contamination and losses during the neutralization or acid evaporation steps. One of the reagents enabling preconcentration from acidic media is dithizone (2-phenylhydrazide-phenylazo-thioformic acid, DTZ) which forms stable complexes with some metal ions.

The choice of a collector is very important and, as mentioned above, valid alternative to inorganic ones are the organic coprecipitants which are often combined with inert coprecipitants of the naphthalene type.

Particular approaches of the precipitation technique are:
- Continuous precipitation, involving a precipitation step with sample collection and matrix removal followed by on-line sample dissolution and driving to the detection system.
- Reductive precipitation, where analytes are enriched from large sample volumes by precipitation through a reduction to a zero oxidation state.

2.2.1 *Continuous Precipitation*

Recently automation of precipitation and coprecipitation procedures has been achieved by using flow injection (FI) techniques [9]. On-line preconcentration by precipitation or by coprecipitation are equally feasible, but coprecipitation methods are preferred in consideration of the greater availability of appropriate carriers enabling trace analyte precipitation through adsorption, occlusion and formation of isomorphic crystals. Difficulties arising from the manipulation of large amounts of precipitate in a continuous flow system have been overcome using knotted reactors as filterless precipitate collectors [9]. The basic configuration of the FI on-line coprecipitation-dissolution flame atomic absorption spectroscopy (FAAS) system and the steps involved are illustrated in Figure1. The sample, at a proper pH, and the precipitating reagent solution are continuosly pumped and mixed through a reaction coil. After the precipitate has been filtered out and the matrix driven to waste, by simultaneously switching the two selecting valves the eluent stream flows through the precipitate to dissolve it and to sweep the analyte to the detector. To give an example, a Se(IV) on-line preconcentration procedure can be mentioned which involved precipitation with lanthanum hydroxide from water samples, and whose detection limit was 1ng/l with a precision of 0.7% [10].

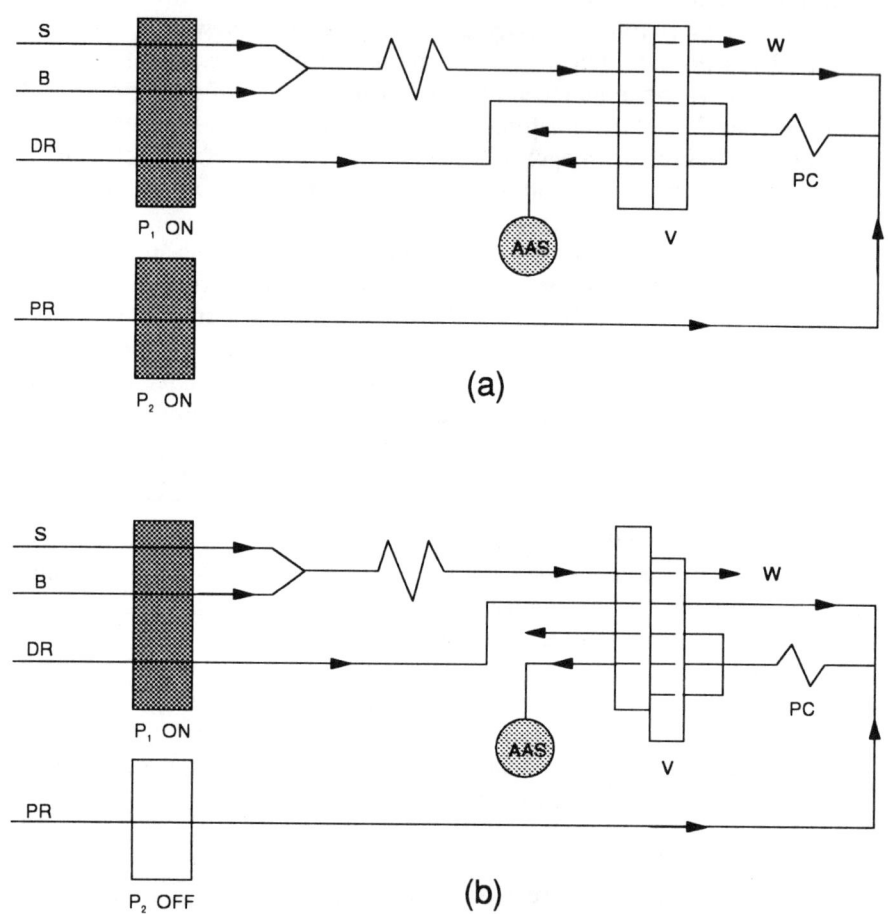

Figure 1. FI on-line coprecipitation preconcentration FAAS system using a knotted reactor precipitate collector: (a) precipitation (loading) sequence; (b) dissolution sequence. P_1, P_2, pumps; V, multifunctional valve; S, sample; B, Buffer solution; DR, dissoution reagent; PR, precipitation reagent; PC, precipitate collector; W, waste.

2.2.2 Reductive precipitation

A method enabling the simultaneous collection and preconcentration of a wide variery of elements, including hydride-forming elements, is reductive precipitation. Sea-water or liquid samples are adjusted to the proper pH (8-9, usually a high purity ammonia solution), to which iron and palladium (catalyst and coprecipitating agents) are added. A portion of sodium terahydroborate ($NaBH_4$) is added to the sample and the bottle swirled (1-2 min); the mixture is usually left to stand about 15 h, and the sample is filtered through a 25-mm diameter cellulose nitrate 0.45 μm membrane filter. A reduced volume of a mixture of high purity concentrated nitric and hydrochloric acids is used to wash the empty bottle and to dissolve the precipitate from the walls, collecting the liquid obtained in a volumetric flask for the next determination. Borohydride reduction, according to the following reactions:

$$BH_4^- + 3H_2O \rightarrow H_3BO_3 + 7H^+ + 8e^-$$

$$M^{n+} + ne^- \rightarrow M$$

$$M^{n+} + nB^- \rightarrow MB_n$$

gave recoveries, in form of element (M) or boride (MB_n), greater than 90 % for the 14 species considered [11]. The advantages of the method include its applicability to 23 elements and the exclusion of the major and minor elements (*e.g.* Ca, Mg, Na and K) with the analytical interferences associated with them. The whole procedure, applied to near-shore and open ocean sea water reference materials (CASS-1 and NASS-2, salinity 31.8 and 35.07 $^o/_{oo}$ respectively), gave results in good agreement with accepted values at ng/ml concentration levels; reproducibility was good and detection limits (3 times r.s.d.) varied from 0.3 to 19 ng/l for As, Cd, Co, Cr, Mn, Ni, Pb, Sb, Se, and Zn [12]. The method has also been successfully applied to Antarctic seawater samples [13].

2.3. FLOTATION

Flotation methods essentially consist in collecting single particles on the surface of gas bubbles with the aid of collectors, surfactants or organic solvent droplets. Arsenic, molybdenum, uranium and vanadium were preconcentrated from sea water by colloid flotation based on hydrous iron(III) oxide and sodium dodecyl sulphate [14]. A typical application is the flotation of metal quinolin-8-olates with phenolphthalein or 2-naphthol as collector and octadecylamine as surfactant (Ni, Co, Cd recovery yields > 90% at ng ml^{-1} concentration) [15]. A new type of flotation without surfactant and gas bubbling has been proposed by Marczenko [16]. Ion-associations of multicharged anionic complexes, formed by the trace elements to be determined with hydrophobic dyes, are accumulated, under shaking, at the phase boundary of a low polarity solvent added to the solution and collected.

2.4. CLOUD POINT

A preconcentration technique based on phase separation which could be considered as a mixture of liquid-liquid and liquid-solid extraction procedures is the cloud point methodology. Aqueous solutions of non-ionic surfactants heated to a temperature known as the *cloud point* become turbid. Above this temperature a phase separation occurs and two phases in equilibrium are originated: a surfactant-rich phase (with a small volume of water) and an aqueous phase (with a surfactant concentration close to *cmc*). The separation mechanism has still to be fully elucidated [17-19]; this process has however been successfully used for the extraction and preconcentration of metal cations after formation of complexes with hydrophobic ligands [20, 21]. As an example, a cloud point preconcentration procedure recently developed for Cd in sea water can be mentioned [22]. The analyte is complexed with 1-(2-pyridylazo)-2-naphthol (PAN) in the presence of Triton X-114 surfactant and the sample heated to 40°C for 5 min. On cooling, the surfactant-rich phase becomes viscous and the aqueous phase separates. The surfactant-rich phase is dissolved (with a mixture of methanol and HNO_3) and the analyte determined at µg l^{-1} concentration level.

2.5. EXTRACTION

Extraction procedures can be mainly divided into liquid-liquid or liquid-solid (solid-phase extraction, SPE).

2.5.1. Liquid-Liquid-Extraction

The most general procedure, for liquid-liquid extraction, can be summarised as follows: if a water-immiscible solvent and an aqueous sample containing a hydrophobic species are brought into contact, the chelate is transferred into the organic phase.

From a practical point of view, the liquid-liquid extraction can be performed following two procedures:
- ionizable compounds can be extracted into organic solvents as "neutral" ion-pairs in the presence of an appropriate ion-pairing agent.
- hydrophylic compounds, which are difficult to remove from the aqueous phase, can be extracted by forming a hydrophobic complex with an appropriate complexing agent and the reaction may be used for a selective isolation.

The most widely employed extractants form neutral chelates which have greater affinity for organic solvents than for the aqueous phase. The metal determination following this approach can be performed by direct analysis of the organic phase or after an acid back-extraction. For simultaneous trace metal determinations in seawater, extraction with carbon tetrachloride, methyl isobutyl ketone (MIBK), chloroform, with or without acid back extraction, has been used after chelation with diethyldithiocarbamate (DDC), ammonium pyrrolidinedithiocarbamate (APDC) or a mixture of the two reagents [23,24,25]. An alternative double chelate extraction (APDC-8-quinolinol-MIBK) was performed for the determination of Cd, Zn, Pb, Cu, Fe, Mn, Co, Cr, and Ni in coastal seawater, allowing recovery of the Mn-oxinate [26].

The extraction methods may be used for selective or group separation of trace elements,collected in the extract or for matrix removal. These methods have the advantage of simplicity and rapidity, but the low concentration factors achieved is one of their major drawbacks. In some cases higher preconcentration factors are obtained by solvent volatilization.

The solvent, in addition to being non toxic, not highly flammable, easily recoverable, must be: water immiscible, have convenient specific gravity, be volatile, stable, transparent to UV, not emulsifying during extraction and as selective as possible.

The other more promising solution is the solid phase extraction.

2.5.2. Liquid-Solid Extraction

The extraction of compounds from a solution onto a solid phase (SPE, also namely liquid-solid or sorbent- extraction) using silica, alumina, celite, charcoal, polymeric or ion-exchange resins (see below) has long been practised. Silica gels have been bound with a wide variety of functional groups (*e.g.* alkyl, phenyl, amino, ciano, diol, alkylsulfonate and quaternary ammonium gropus) to provide a specific interaction with analytes. More recently chelating functional groups (*e.g.* quinolines, iminodiacetic) have been bound in order to obtain specific metal ion retention.

Although ion-collecting filters (*e.g.* cellulose filters modified by the insertion of chelating groups, see below) have been proposed and used, from a practical point of view liquid-solid extraction is commonly performed by driving the sample through a column (microcolumn, cartridge, microtube or other kind of container) packed with the proper stationary phase. A typical, off- or on-line, procedure involves the following steps:

- the column is packed with the selected stationary phase,
- the column is equilibrated with water or a buffer of controlled pH, ionic strength and with or without a chelating or ion-pairing agent at the proper molarity,
- the sample is loaded into the column directly or after pH adjustment and, if required, after complexation or ion-pair formation of analytes,
- the column is washed with an appropriate eluent to remove impurities,
- the analyte is selectively eluted with a small volume of a solvent mixture optimized for its recovery.

One of the most relevant advantages of connecting the preconcentration manifold directly with the separation and/or detection instrument, is the reduction of sample contamination particularly when trace levels must be detected.

Extraction Mechanisms. The acting mechanisms, for the retention of metal ions, coupled with the different kinds of materials are:

- *adsorbtion of neutral species* obtained *via* interaction with hydrophobic groups formed by:
  - adding a proper ligand able to originate neutral complexes with the analytes
  - adding an ion interaction reagent able to originate ion-pairs, with hydrophobic properties, with metal ions.
- *chelation* obtained by
  - reaction between metal ions and coordinating groups grafted on the solid
- *ion-exchange* which can be performed following two ways:
  - cation exchange, which involves a direct exchange of the metal ion with the counter ion of the resin and its retention
  - anion exchange, which requires the preliminary addition of a ligand (usually sulphonated or phosphonated) able to form a negatively charged complex with the analyte.

The study and development of a SPE procedure must take into account the nature of the analytes (ionic or not) and of the matrix to be removed: in fact the solid phase and the liquid-solid interaction mechanism must be properly selected on this base.

The main approaches for metal ions consist either in the retention of charged species or on modifying their nature before or during the retention step. A few words must be spent to describe the manifold structure and the materials used.

Materials. The commonly used materials are:
- sorbents: foams or synthetic polymers such as Amberlite XAD family supports (polistyrene-divinylbenzene, polyacrylic, etc.), silica based (C_8, C_{18}, CN, etc.) supports,
- chelating materials: polymers (Chelex 100) or modified silica (controlled pore glass, CPG) with ligands adsorbed or chemically bound onto the surface,
- ion-exchangers as such or properly modified by inserting ligands.

Sorbents can be of inorganic or organic nature. No attempts have been made hereafter to define foams or resins and their typology (macro-, micro-, porous, poly-esthers or ether, mixed or not, styrene-divinylbenzene copolymer, polyuretane foams, acrylic, and so on) in detail. As above mentioned, the Amberlite XAD family supports (polystyrene-divinylbenzene, polyacrylic, etc.), metacrylic resins and silica based (C_8, C_{18}, CN, etc.) supports are the most commonly used adsorbents. If silica-based supports are used, the residual silanol sites provide, for the analyte, a second type of binding sites of a different nature to that of, *e.g.* C_{18}, binding sites, so it is advisablee to use as supporting material in the column either a silica C18 end-capped type or a polymeric-based C18 derivatized support.

As mentioned above the approaches for performing a preconcentration with this kind of materials consist in loading the solid phase with a suitable ligand or adding it to the sample before elution.

Chelating materials have been obtained: from different kinds of cellulose (microcrystalline, amorphous) and cotton wool by direct bonding or by diazo-coupling *e.g.* iminodiacetic acid-ethyl-cellulose (IDA-cellulose)[27], 1-(2-pyridylazo)-2-naphthol (PAN) [28]; by synthetizing resins *e.g.* hydroxamic [29, 30], poly(vinylpyrrolidoxime) [31], poly(acrylamidoxime) [32, 33], dithiocarbamate [34] and loading foams and resins *e.g.* XAD-2 *plus* diphenylcarbazone [35], XAD-4 *plus* propylenediaminotetraacetic acid [36], 7-dodecenyl-8-oxine [37], 8-oxine [38], several ligands [39], bis(carboxymethyl)dithiocarbamate [40]. However polymers or modified silica coupled with ligands are the most commonly used stationary phases. Among synthetized resins, Chelex 100, which is the purified form of the ion-exchanger Dowex A1 (and usually included in ion-exchanger products, see below) has received much attention as well as 8-hydroxyquinoline (oxine) derivatives and silica based supports. The latter technique, originally based on the complexation of metal ions with 8-hydroxyquinoline followed by adsorption on C18-bonded silica [41] resulted in an enrichment factor of 50-100 and total recovery for Cd, Cu, Pb, Mn, Fe, Ni, Cr and Co from sea water [42]. A more efficient preconcentration was achieved with silica immobilized 8-hydroxyquinoline (I-8-HOQ) [43, 44] followed by acid elution which made it possible to determine metal ions in sea water and certified samples [45].

Ion exchangers are very widely used and provide the most universal preconcentration technique. Their application is obviously limited by the presence of the element of interest, in the system considered, in the form of ions or well-defined states suitable for being retained by the exchangers. Usually with acidification of the samples or addition of complex forming agents capable of giving origin to charged compounds ion exchange techniques can be applied to trace element preconcentration. An attractive alternative technique to classic ion-exchange consists in the use of "reverse ion-exchange" for cation preconcentration. The method is based on the use of ligands which enable both metal complexation and anion exchange. The procedure can be realised in two ways: a) the resin is loaded with the proper ligand; b) the metal ions considered are converted to anionic complexes by means of a specific chelating agent provided with anionic groups not involved in the metal coordination and able to perform the exchange on the resin. A good separation can be achieved in this case from a highly saline matrix because alkaline-earth ions have low affinity for chelating agents and in addition they are rejected by the charge of the ion-exchanger surface.

Selective ion-exchangers are produced for special purposes and, as mentioned above (Chelex 100), an incorrect definition is sometimes used to identify specific products. Chelex 100 is characterized by presence on its surface of iminodiacetic functional groups, $-N(CH_2COOH)_2$, able to complex and retain the metal ions. Recently prepacked microcolumns, namely MetPac CC1 (Dionex, Sunnyvale), containing a resin similar to Chelex 100 and enabling metal ion preconcentration have been produced. The first investigation of the uptake of trace elements from sea water showed good retention efficiency for Chelex 100 [46]. In recent years a number of papers have been published on the performance of this resin as a function of different counter-ions [47, 48, 49], buffering systems [50, 51], flow rates and pretreatments [52], and its suitability for sea-water was demonstrated [53, 54].

Cation exchange resins are not usually employed for trace cation concentration and the above mentioned "reverse ion-exchange" is preferred and more suitable in this field. Some of the studies, for developing the preconcentration procedures for trace metal ions determination in complex matrix and sea-water samples, concern the behaviour of the sulphonated form of 2-benzoylpyridine-2-pyridylhydrazone BPPH-S [55, 56], 8-hydroxyquinoline-5-sulphonic acid and derivatives [57, 58], sulphonated hydroxyazo-dyes and similar ligands [59-64].

Classic ion-exchangers (PS-DVB based) and new polymeric materials have mainly been considered for matrix removal and metal ion preconcentration. The former kind of material is particularly suitable for this application since preconcentration procedures, even when performed on-line (see below), imply, for sample loading, very reduced flow rates (*i.e.* backpressure) so as not to disorganize the resin skeleton.

Practical Aspects. Obviously, in consideration of the mechanism and materials selected the parameters to be kept under control will be:
- proper pH for complex formation (precipitation must be avoided)
- ligand or ion interaction reagent concentration and solubility (sufficient to originate complex or ion-pair but not so high as to compete for resin sites)
- ionic strength (mainly for ion exchange competition)

and, from a practical point of view, elution velocity of the sample must be the maximum allowed by adsorbtion, chelation or ion-exchange kinetics utilized for metal ion retention.

Column design (tubular, conical) strongly also influences the performance of preconcentration systems [9]; the optimum column design for achieving high efficiency depends on several factors, *e.g.* sample loading rate and volume, sorbent capacity for the analytes, particle size of the packing material and its specific properties.

After metal ion retention and matrix removal, the optimization of preconcentration involves the selection of an eluent suitable for the recovery of metal ion(s) or complex(es) from the solid phase.

Developing this step, attention must be paid to obtain the total recovery of the analytes by using the lowest possible volume and the most rapid elution rate.

2.5.3. Continuous Configurations

To reduce sample manipulations, loss of analytes and contamination, on-line preconcentrations based on *solid phase extraction* procedures have also been developed. The term "on-line" refers to the fact that the preconcentration step, usually based on an enrichment cartridge, is performed in series with the system of detection. The process can be continuous or not, depending on the kind of detection system coupled with the enrichment manifold (*e.g.* flame AAS, ICP-AES or graphite furnace AAS). These procedures, regarded as an evolution of Flow-injection Analysis (FIA) [65] are often termed *flow injection* (FI) methods.

The basic and simplest manifold includes two valves (*a,b*) and a microculmn filled with the proper sorbent. Valve *a*) is used for connecting the driving system(s) with the preconcentration column, for washing it, loading the sample, and finally for recovering the preconcentrated, matrix-free, analyte. Valve *b*) alternatively connects the eluates to waste system or to the detector.

When complexation is required in on-line systems, the sample of interest is injected into the manifold and merged with the chelating agent at a confluence point downstream (see below). After passing through a mixing coil, which allows enough time for the chelate to form, the metal chelate is preconcentrated on the column. This procedure is superior to the direct addition of the chelating agent to the sample since in this case a purification cartridge can be added before mixing sample and ligand.

For analyte recoveries in some cases a satisfactory solution can be to release the preconcentrated species by applying a stripping agent in the opposite direction to that used for loading. In addition contamination and/or modification of the solid phase must be avoided in order to be able to perform a series of measurements without changing the preconcentration column.

The following are some examples concerning the above mentioned techniques. The first on-line FI analysis trace enrichment element, using a microcolumn of Chelex-100 and flame AAS was developed by Olsen *et al.* [66]

A closed-loop on-line enrichment procedure, based on *reversed ion-exchange*, in combination with an ICP-AES spectrometer was developed for the analysis of trace metal ions in seawater samples. The procedure utilised 8-hydroxy-7-

iodoquinoline-5-sulphonic acid coupled with an anion-exchange resin and enabled the determination of Cd, Cu, Fe, Mn, Ni, Pb and Zn [67].

Quinolin-8-ol was immobilised on Sepabeads (polyvinyl polymer with alkyl amino groups) or Capcell-NH2 (silicone-coated silica gel with alkyl amino groups) to obtain clean adsorbents for the concentration of metals from seawater [68] as well as silica immobilised quinolin-8-ol was used for their on-line separation and GF AAS, ICP-AES or ICP MS determination [69-71].
On-line chelating procedures were also developed e.g. by loading a microcolumn (XAD-2 resin) with 1-(2-thiazoliylazo)-2-naphtol before seawater sample elution; by retention on C18 microcolumn of Na diethyldithiocarbamate metal complexes; or by retention of metal traces on Prosep IDA (iminodiacetate chelating reagent immobilized on CPG support) and coupling ICP-AES, GF AAS or ICP MS determination [72-74]. A comprehensive review on flow-injection methods for the determination of cationic species and trace elements in sea water has also been published [75].

Finally, in some cases the on-line preconcentration manifold coupled with GF AAS was developed by inserting a miniature XAD-2, XAD-7 or silica C18 column, at the tip of the autosampler arm, enabling the retention of pyrrolidinedithiocarbamate metal complexes [76, 77].
The schematic diagram of the system [76] and the enrichment column built with concentric tygon and PTFE tube are shown in Figure 2.

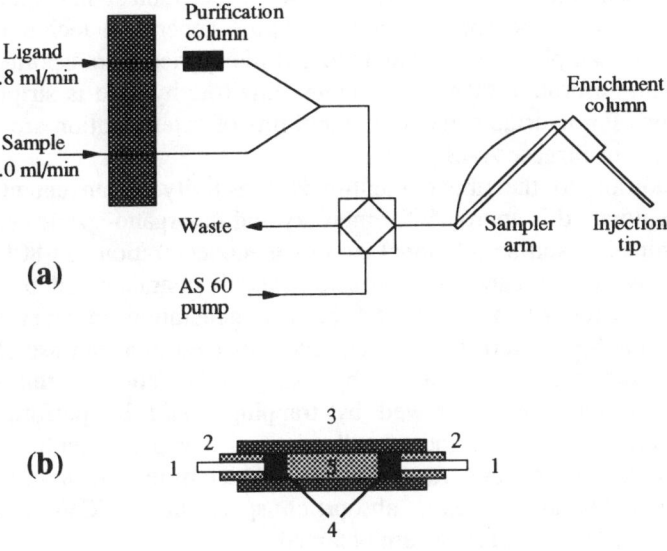

Figure 2. (a) Schematic diagram of the preconcentration manifold. (b) Section of the enrichment column: 1, PTFE tube (0.15 mm i.d.); 2, Tygon tube (1.3 mm i.d., 2.0 mm o.d.); 3, Tygon tube (2.1 mm i.d.); 4, PTFE; 5, resin

The direct injection of the eluate into the graphite furnace gave high preconcentration factors (20-225) which enabled the determination of Cd, Pb, Cu, Ni, Co and Fe in Antarctic seawater. In addition through an on-line purification of the ligand (see scheme) blank levels were very low and the detection limits ranged from 0.4 ng l^{-1} (Cd) to 25 ng l^{-1} (Fe).

2.6. SPECIAL TECHNIQUES

Some analytical techniques of detection involve chemical reactions or instrumental procedures which are themselves an approach to matrix removal or analyte preconcentration. Two basic examples concern hydride generation and electrochemical techniques.

2.6.1. *Hydride Techniques*

Elements such as As, Se, Sb, Bi, Ge, Sn, Te and Pb form hydrides in acidic solutions with NaBH$_4$, for example the reaction of As(III) can be represented as follows:

$$3BH_4^- + 3H^+ + 4H_3AsO_3 \rightarrow 3H_3BO_3 + 4AsH_3 + 3H_2O$$

and when a basic solution of borohydride is added to an acidic solution, excess hydrogen is also produced according to:

$$BH_4^- + 3H_2O + H^+ \rightarrow H_3BO_3 + 4H_2$$

The hydride technique proceeds through several steps: the hydride is generated by a chemical reaction; and is swept out of the solution into the atomizer by a carrier gas; it is then decomposed in the atomizer and the atomic absorption signal is measured. Two modes of operation can be applied for the hydride generation technique: the batch system (the whole sample is reduced and the hydride formed is transported by a carrier gas stream to an absorption tube), continuous flow (the hydride is stripped in a gas-liquid separator). For hydride generation, the limits of determination are 10-100 times lower than those for furnace AAS.

In addition to the above mentioned sensitivity enhancement and matrix interference removal, this approach for mercury and its organo-species enables a cold vapour determination and an additonal step of preconcentration [78-80]. In this case metallic mercury is originated after reduction and is characterized by a significant vapour pressure at room temperature. So, after its generation, mercury can be swept out of the solution by an inert gas (*e.g.* N$_2$) and collected in a trap usually filled with metallic Au, where mercury is retained, by *amalgam* formation. In this way mercury reduction and volatilization followed by trapping could be performed on large volumes of sample and, at the end of collection, all the preconcentrated mercury is removed from the trap by electrical heating and determined in a small volume of carrier gas by cold-vapour atomic absoprtion spectrometry (CV-AAS). Detection limits in the sub ng/ml to ng/l range are obtained.

A reaction of ethylation with subsequent trapping of Pb(Et)$_4$ in a graphite furnace at 400°C has also been developed and applied to marine reference materials [81]. The method, avoiding contamination of the sample, gave a concentration factor of 500 in 3-4 min (10 ml sample) and detection limits of 1 pg ml^{-1} were obtained.

2.6.2. Electrochemical Preconcentration

Electrolytic deposition represents an efficient way for the enrichment and isolation of trace components. It has the advantage that little or no reagent addition is required, thus minimizing contamination risks, in addition seawater matrix does not interfere with the determination. The electrochemical preconcentration scheme can be coupled with a range of instrumentation techniques, a brief overview is given hereafter.

Stripping Analysis. Stripping analysis is an analytical method that incorporates an electrolytic preconcentration step. The technique includes extremely low detection limits, 10^{-10}-10^{-11} M, multielement and speciation capabilities, and stability for on-line measurements.
Two main steps are involved: a preconcentration step and a stripping step. In the first step the analyte is accumulated onto or into the working electrode and in the second step the accumulated species is oxidized or reduced back into the solution. The response is proportional to the concentration of the analyte that is in or on the electrode as a function of the nature of preconcentration and stripping steps. There are different versions of stripping analysis, namely:

Anodic Stripping Voltammetry (ASV). The metal ion is reduced to the metal which dissolves the in mercury, electrode; then a positive potential scan oxidizes the species which are stripped from the electrode as a function of their standard potentials and anodic peak currents are measured.

Potentiometric Stripping Analysis (PSA). The preconcentration step in PSA is the same as for ASV: the metal is elctrolytically deposited (reduction) onto the electrode (usually a mercury film). The stripping step is obtained by chemical oxidation in the presence *e.g.* of mercuric ions or oxygen.

$$M(Hg) + oxidant \rightarrow M^{n+} + Hg$$

or by applying a constant anodic current. The experimental curve monitored, reporting electrode potential as a function of time, contains qualitative and quantitative information.

Cathodic Stripping Voltammetry (CSV). Cathodic stripping voltammetry, best suited for inorganic anions and metal complexes or organic sulfur compounds able to give insoluble salts with the electrode, involves formation of an insoluble species at the electrode followed by a stripping step obtained with a potential scan in the negative direction.

Adsorptive Cathodic Stripping Voltammetry (AdSV). This is a very sensitive method for trace metals, it is based on the formation of an appropriate metal chelate that is accumulated onto the working electrode. Successively, by the application of a negative-going potential scan the reduction of the metal or ligand in the complex is monitored.

For specific accumulation onto electrodes before voltammetric quantitation, chemically modified electrodes are also used. The proper choice of a modifier results in an enhancement of selectivity and sensitivity; usually the bound agent is incorporated into the surface of the electrode by mixing it with carbon paste or by forming derivatives of polymeric films. In these cases selective accumulation must be obtained, surface saturation prevented and the surface must be regenerable (analyte-free).

The most commonly used electrodes are the hanging mercury drop electrode (HMDE) and the mercury film electrode (MFE), but gold and carbon electrodes have also been used for the determination of trace metals with an oxidation potential more positive than mercury (*e.g.* Ag, As, Au, Se and Te).

3. References

1. Akagi,T., Fuwa, K. and Haraguchi, H. (1985) Simultaneous multi-element determination of trace metals in sea-water by inductively coupled plasma atomic-emission spectrometry after co-precipitation with gallium, *Anal. Chim. Acta* **177**, 139-151.
2. Akagi, T. and Haraguchi, H. (1990) Simultaneous multi-element determination of trace metals using 10 ml of sea-water by inductively coupled plasma atomic-emission spectrometry with gallium co-precipitation and micro-sampling technique *Anal.Chem.* **62**, 81-85.
3. Sawatari, H., Fujimori, E. and Haraguchi, H. (1995) Multi-element determination of trace elements in sea water by gallium coprecipitation and inductively coupled plasma mass spectrometry, *Anal. Sci.* **11**, 369-374.
4. Akagi, T., Nojiri, Y., Matsui, M. and Haraguchi, H. (1985) Zirconium co-precipitation for simultaneous multi-element determination of trace metals in sea-water by inductively coupled plasma atomic-emission spectrometry, *Appl. Spectrosc.* **39**, 662-667.
5. Nakamura, T., Oka, H., Ishii, M. and Sato, J. (1994) Direct atomization atomic-absorption-spectrometric determination of beryllium, chromium, iron, cobalt, nickel, copper, cadmium and lead in water with zirconium hydroxide co-precipitation, *Analyst* **119**, 1397-1401.
6. Bloom, N.S. and Crecelius, E.A. (1984) Determination of silver in sea-water by co-precipitation with cobalt pyrrolidinedithiocarbamate [pyrrolidine-1-carbodithioate] and Zeeman graphite-furnace atomic-absorption spectrometry, *Anal. Chim. Acta* **156**, 139-145.
7. Lan, C.R., Tseng, C.L., Yang, M.H. and Alfassi, Z.B. (1991) Two-step co-precipitation method for differentiating chromium species in water followed by determination of chromium by neutron-activation analysis, *Analyst* **116**, 35-38.
8. Bem, H. and Ryan, D.E. (1984) Determination of seven trace elements in natural waters by neutron-activation analysis after pre-concentration with 1-(2-pyridylazo)-2-naphthol, *Anal. Chim. Acta* **166**, 189-197.
9. Fang, Z. (1995) *Flow Injection Atomic Absorption Spectrometry*, John Wiley & Sons, N.Y.
10. Tao, G.H. and Hansen, E.H. (1994) Determination of ultra-trace amounts of selenium(IV) by flow-injection hydride-generation atomic-absorption spectrometry with online pre-concentration by co-precipitation with lanthanum hydroxide. *Analyst* **119**, 333-337.
11. Skogerboe, R.K., Hanagan, W.A. and Taylor,H.E. (1985) Concentration of trace elements in water samples by reductive precipitation, *Anal. Chem.* **57**, 2815-2818.
12. Nakashima, S., Sturgeon, R.E., Willie, S.N. and Berman, S.S. (1988) Determination of trace elements in sea-water by graphite-furnace atomic-absorption spectrometry after pre-concentration by tetrahydroborate reductive precipitation, *Anal. Chim. Acta* **207**, 291-299.
13. Mentasti, E., Porta, V., Abollino, O. and Sarzanini, C. (1989) Trace metal determination in Antarctic seawater, *Ann. Chim.* (Rome) **79**, 629-637.
14. Murthy, R.S.S. and Ryan, D.E. (1983) Determination of arsenic, molybdenum, uranium and vanadium in sea-water by neutron-activation analysis after preconcentration by colloid flotation, *Anal. Chem.* **55**, 682-684.

15. Caballero, M., Lopez, R., Cela, R. and Perez-Bustamante, J.A. (1987) Pre-concentration and determination of trace metals in synthetic sea-water by flotation with inert organic collectors, *Anal. Chim. Acta* **196**, 287-292.
16. Marczenko, Z. (1985) New type of flotation of ion-association compounds of complexes of multicharged anions with basic dyes, *Pure and Appl. Chem.* **57**, 849-854.
17. Degiorgio, V., Piazza, R., Corti, M. and Minero, C. (1985) Critical properties of nonionic micellar solutions, *J. Chem. Phys.* **82**, 1025-1031.
18. Blankschtein, D., Thurston, G.M. and Bebedek, G.B. (1986) Phenomenological theory of equilibrium thermodynamic properties and phase separation of micellar solutions, *J. Chem. Phys.* **85**, 7268-7288.
19. Lindman, B. and Wennerstrom, H. (1991) Nonionic micelles grow with increasing temperature, *J. Chem. Phys.* **95**, 6053-6054.
20. Watanabe, H. and Tanaka, H. (1978) A non-ionic surfactant as a new solvent for liquid-liquid extraction of zinc (II) with 1-(2-pyridylazo)-2-naphthol, *Talanta* **25**, 585-589.
21. Buhai, L. and Rigan, M. (1990) Liquid-solid extraction system based on tween 40-salt-H_2O without organic solvents, *Talanta* **37**, 885-888.
22. Garcia Pinto, C., Pérez Pavon, J.L., Moreno Cordero, B., Romero Beato, E. and Garcia Sanchez, S. (1996) Cloud point preconcentration and flame atomic-absorption spectrometry: application to the determination of cadmium, *J. Atom. Abs. Spectroscopy* **11**, 37-41.
23. Chakraborti, D., Adams, F., Van Mol, W. and Irgolic, K.J. (1987) Determination of trace metals in natural waters at nanogram per litre levels by electrothermal atomic-absorption spectrometry after extraction with sodium diethyldithiocarbamate, *Anal. Chim. Acta* **196**, 23-31.
24. Kremling, K. and Petersen, H. (1974) APDC-MIBK extraction system for the determination of copper and iron in 1 cm^3 of sea water by flameless atomic-absorption spectrometry, *Anal. Chim. Acta* **70**, 35-39.
25. McLeod, C.W., Otsuki, A., Okamoto, K., Haraguchi, H. and Fuwa, K. (1981) Simultaneous determination of trace metals in sea-water using dithiocarbamate pre-concentration and inductively coupled plasma emission spectrometry, *Analyst* **106**, 419-428.
26. Sturgeon, R.E., Berman, S.S., Desaulniers, A. and Russell, D.S. (1980) Preconcentration of trace metals from sea-water for determination by graphite furnace atomic-absorption spectrometry, *Talanta* **27**, 85-94.
27. Horvath, Zs., Lasztity, A., Szakacs, O. and Bozsai, G. (1985) Iminodiacetic acid - ethylcellulose as a chelating ion exchanger. I. Determination of trace metals by atomic-absorption spectrometry and collection of uranium, *Anal. Chim. Acta* **173**, 273-280.
28. Taguchi, S., Yamazaki, S., Yamamoto, A., Urayama, Y., Hata, N., Kasahara, I. and Goto, K. (1988) Acid-soluble membrane filter for the pre-concentration and electrothermal-atomization atomic-absorption spectrometric determination of trace levels of cadmium in water, *Analyst* **113**, 1695-1698.
29. Vernon, P. and Eccles, H. (1973) Chelating ion-exchangers containing 8-hydroxyquinoline as the functional group, *Anal. Chim. Acta* **63**, 403-414.
30. Vernon, P. and Zin, W. (1981) Chelating ion exchangers containing N-substituted hydroxylamine functional groups. VI. Sorption and separation of gold and silver by a polyhydroxamic acid, *Anal. Chim. Acta* **123**, 309-313.
31. Willie, S.N., Sturgeon, R.E. and Berman, S.S. (1983) Comparison of quinolin-8-ol-bonded polymer supports for pre-concentration of trace metals from sea-water, *Anal. Chim. Acta* **149**, 59-66.
32. Colella, M.B., Siggia, S. and Barnes, R.M. (1980) Synthesis and characterization of a poly(acrylamidoxime) metal-chelating resin, *Anal. Chem.* **52**, 967-972.
33. Colella, M.B., Siggia, S. and Barnes, R.M. (1980) Poly(acrylamidoxime) resin for determination of trace metals in natural waters, *Anal. Chem.* **52**, 2347-2350.
34. Yamagami, E., Tateishi, S. and Hashimoto, A. (1980) Application of chelating resin to determination of trace amounts of mercury in natural waters, *Analyst* **105**, 491-496.
35. Osaki, S., Osaki, T. and Takashima, Y. (1983) Determination of chromium(VI) in natural waters by sorption of chromium - diphenylcarbazone with XAD-2 resin, *Talanta* **30**, 683-686.
36. Moyers, E.M. and Fritz, J.S. (1977) Preparation and analytical applications of a propylenediaminetetraacetic acid resin, *Anal. Chem.* **49**, 418-423.
37. Isshiki, K., Tsuji, F., Kuwamoto, T. and Nakayama, E. (1987) Pre-concentration of trace metals from sea-water with 7-dodecenylquinolin-8-ol-impregnated macroporous resin, *Anal. Chem.* **59**, 2491-2495.
38. Isshiki, K., Sohrin, Y., Karatani, H. and Nakayama, E. (1989) Pre-concentration of chromium(III) and chromium(VI) in sea-water by complexation with quinolin-8-ol and adsorption on macroporous resin, *Anal. Chim. Acta* **224**, 55-64.

39. Isshiki, K. and Nakayama, E. (1987) Selective concentration of cobalt in sea-water by complexation with various ligands and sorption on macroporous resins, *Anal. Chem.* **59**, 291-295.
40. Plantz, M.R., Fritz, J.S., Smith, F.G. and Houck, R.S. (1989) Separation of trace metal complexes for analysis of samples of high salt content by inductively coupled plasma mass spectrometry, *Anal. Chem.* **61**, 149-153.
41. Watanabe, H., Goto, S., Taguchi, S., McLaren, J.W. and Berman, S.S., Russell, D.S. (1981) Pre-concentration of trace elements in sea-water by complexation with quinolin-8-ol and adsorption on C18-bonded silica gel, *Anal. Chem.* **53**, 738-739.
42. Sturgeon, R.E., Berman, S.S. and Willie, S.N. (1982) Concentration of trace metals from sea-water by complexation with 8-hydroxyquinoline [quinolin-8-ol] and adsorption on C18-bonded silica gel, *Talanta* **29**, 167-171.
43. Sturgeon, R.E., Berman, S.S., Willie, S.N. and Desaulniers, J.A.H. (1981) Pre-concentration of trace elements from sea-water with silica-immobilized quinolin-8-ol, *Anal. Chem.* **53**, 2337-2340.
44. Sturgeon, R.E., Berman, S.S., Willie, S.N., Desaulniers, J.A.H. and Russell, D.L. (1983) Preconcentration of trace metals from seawater using silica immobilized 8-hydroxyquinoline, *Marine Chemistry* **12**, 219-225.
45. Nakashima, S., Sturgeon, R.E., Willie, S.N. and Berman, S.S. (1988) Determination of trace metals in sea-water by graphite-furnace atomic-absorption spectrometry with pre-concentration on silica-immobilized quinolin-8-ol in a flow system, *Fresenius' Z. Anal. Chem.* **330**, 592-595.
46. Riley, J.P. and Taylor, D. (1968) Chelating resins for the concentration of trace elements from sea water and their analytical use in conjunction with atomic absorption spectrophotometry, *Anal. Chim. Acta* **40**, 479-485.
47. Abdullah, M.I., El-Rayis, O.A., and Riley, J.P (1976) Re-assessment of chelating ion-exchange resins for trace metal analysis of sea water, *Anal. Chim. Acta* **84**, 363-368.
48. Figura, P. and McDuffie, B. (1977) Characterization of the calcium form of Chelex-100 for trace metal studies, *Anal. Chem.* **49**, 1950-1953.
49. Lamathe, J. (1979) Methode d'elution selective pour l'extraction des metaux lourds de l'eau de mer sur resine chelatante, *Anal. Chim. Acta* **104**, 307-317.
50. Kingston, H.M., Barnes, I.L., Brady, T.J., Rains, T.C. and Champ, M.A. (1978) Separation of eight transition elements in estuarine and seawater with chelating resin and their determination by graphite furnace atomic absorption spectrometry, *Anal. Chem.* **50**, 2064-2070.
51. Buckley, J.A. (1985) Preparation of Chelex-100 resin for batch treatment of sewage and river water at ambient pH and alkalinity, *Anal. Chem.* **57**, 1488-1490.
52. Paulson, A.J. (1986) Effects of flow rate and pre-treatment on the extraction of trace metals from estuarine and coastal sea-water by Chelex-100, *Anal. Chem.* **58**, 183-187.
53. Sturgeon, R.E., Berman, S.S., Desaulniers, J.A.H. and Russell, D.L. (1980) Preconcentration of trace metals from sea-water for determination by graphite furnace atomic-absorption spectrometry, *Talanta* **27**, 85-94.
54. Sturgeon, R.E., Berman, S.S., Desaulniers, J.A.H., Mykytluk, A.P., McLaren J.W. and Russell, D.L. (1980) Comparison of methods for determination of trace elements in sea-water, *Anal. Chem.* **52**, 1585-1588.
55. Going, J.E. and Sykora, C. (1974) Photometric and potentiometric study of metal complexes of 2-(3'-sulfobenzoyl)pyridine-2-pyridylhydrazone, *Anal. Chim. Acta* **70**, 127-132.
56. Going, J.E., Wesemberg, G. and Andreiat, G. (1976) Preconcentration of trace metal ions by combined complexation-anion exchange. Part I. Cobalt, zinc and cadmium with 2-(3'sulfobenzoyl)-pyridine-2-pyridylhydrazone, *Anal. Chim. Acta* **81**, 349-360.
57. Berge, D.G. and Going, J.E. (1981) Pre-concentration of trace-metal ions by combined complexation - anion exchange. II. Cobalt, zinc, cadmium and lead with 8-hydroxy-quinoline-5-sulphonic acid, *Anal. Chim. Acta* **123**, 19-24.
58. Abollino, O., Mentasti, E., Porta, V. and Sarzanini, C. (1990) Immobilized 8-oxine units on different solid sorbents for the uptake of metal traces, *Anal. Chem.* **62**, 21-26.
59. Sarzanini, C., Mentasti, E., Gennaro, M.C. and Marengo, E. (1985) Enrichment of aluminium traces in liquid samples, *Anal. Chem.* **57**, 1960-1963.
60. Sarzanini, C., Gennaro, M.C., Porta, V. and Mentasti, E. (1987) Comparison of ligands for the recovery of trace aluminium(III) by anion exchange, *Anal. Chim. Acta* **198**, 191-196.

61. Sarzanini, C., Abollino, O., De Luca, M. and Mentasti, E. (1992) Ion exchange for the determination of stability constants of metals-Plasmocorinth B complexes and preconcentration procedure, *Anal. Sciences* **8**, 201-206.
62. Sarzanini, C., Abollino, O. and Mentasti, E. (1992) Hydroxyazo-dyes in metal ions preconcentration by ion exchange, in M.J. Slater (ed.), *Ion Exchange Advances*, Elsevier Appl. Science, London, pp. 279-286.
63. Abollino, O., Sarzanini, C., Mentasti, E. and Liberatori, A. (1993) Trace metal ions preconcentration with sulphonated azo-dyes and ICP/AES determination, *Spectrochim. Acta* **49 A**, 1411-1421.
64. Abollino, O., Sarzanini, C., Mentasti, E. and Liberatori, A. (1994) Evaluation of stability constants of metal complexes with sulphonated azo-ligands, *Talanta* **41**, 1107-1112.
65. Ruzicka, J. and Hansen, E.H. (1981) *Flow Injection Analysis*, J. Wiley & Sons, New York.
66. Olsen, S., Pessenda, L.C.R., Ruzicka, J. and Hansen, E.H. (1983) Combination of flow-injection analysis with flame atomic absorption spectrophotometry: determination of trace amounts of heavy metals in polluted sea-water, *Analyst* **108**, 905-917.
67. Porta, V., Sarzanini, C. and Mentasti, E. (1989) Online pre-concentration and ICP determination for trace metal analysis, *Mikrochim. Acta* **III**, 247-255.
68. Kasahara, I., Willie, S.N., Sturgeon, R.E., Berman, S.S., Taguchi, S. and Goto, K. (1993) Preparation of quinolin-8-ol immobilized adsorbents with minimum contamination for the pre-concentration of trace metals in water, *Bunseki-Kagaku* **42**, 107-110.
69. Azeredo, L.C., Sturgeon, R.E. and Curtius, A.J. (1993) Determination of trace metals in sea-water by graphite-furnace atomic-absorption following on-line separation and pre-concentration, *Spectrochim Acta* **48B**, 91-98.
70. Lan, C.R. and Yan, M.H. (1994) Synthesis, properties and applications of silica-immobilized 8-quinolinol. II. On-line column pre-concentration of copper, nickel and cadmium from sea-water and determination by inductively-coupled plasma atomic-emission spectrometry, *Anal. Chim. Acta* **287**, 111-117.
71. Nelms, S.M., Greenway, G.M. and Hutton, R.C. (1995) Application of multi-element time-resolved analysis to a rapid online matrix separation system for inductively coupled plasma mass spectrometry, *J. Anal. At. Spectrom.* **10**, 929-933.
72. Porta, V., Sarzanini, C., Abollino, O., Mentasti, E. and Carlini, E. (1992) Preconcentration and ICP-AES determination of metal ions with on-line chelating ion-exchange *J. Anal. At. Spectrom.* **7**, 19-22.
73. Sperling, M., Yin, X. and Welz, B. (1992) Determination of ultra-trace concentrations of elements by means of on-line solid sorbent extraction graphite-furnace atomic-absorption spectrometry, *Fresenius' J. Anal. Chem.* **343**, 754-755.
74. Greenway, G.M., Nelms, S.M. and Koller, D. (1996) Application of a novel iminodiacetate chelating material to automated matrix separation for inductively coupled plasma mass spectrometry, *Anal. Commun.* **33**, 57-59.
75. Atienza, J., Herrero, M.A. and Puchades R. (1992) Flow-injection analysis of sea-water. II. Cationic species, *C.R. Anal. Chem.* **23**, 1-14.
76. Porta, V., Abollino, O., Mentasti, E. and Sarzanini, C. (1991) Determination of ultra-trace levels of metal ions in seawater with on-line pre-concentration and electrothermal atomic absorption spectrometry, *J. Anal. At. Spectrom.* **6**, 119-122.
77. Liu, Z.S. and Huang, S.D. (1993) Automatic on-line pre-concentration system for graphite furnace atomic-absorption spectrometry for the determination of trace metals in sea-water, *Anal. Chim. Acta* **281**, 185-190.
78. Hatch, W.R. and Ott, W.L. (1968) Determination of sub-microgram quantities of mercury by atomic absorption spectrophotometry, *Anal. Chem.* **40**, 2085-2087.
79. Welz, B. and Schubert-Jacobs, M. (1988) Cold-vapour atomic-absorption spectrometric determination of mercury using sodium tetrahydroborate reduction and collection on gold, *Fresenius'-Z. Anal. Chem.* **331**, 324-329.
80. Zachariadis, G.A. and Stratis, J.A. (1991) Optimization of cold-vapour atomic-absorption spectrometric determination of mercury with and without amalgamation by subsequent use of complete and fractional factorial designs with univariate and modified simplex methods, *J. Anal. At. Spectrom.* **6** ,239-245.
81. Sturgeon, R.E., Willie, S.N. and Berman, S.S. (1989) Atomic absorption determination of lead at picogram per gram levels by ethylation with in situ concentration in a graphite furnace, *Anal. Chem.* **61**, 1867-1869.

HYPHENATED INSTRUMENTAL METHODS FOR THE DETECTION OF HEAVY METALS IN MARINE ENVIRONMENT

R.Frache
University of Genoa - Department of Chemistry and Industrial Chemistry
Via Dodecaneso 31 - 16146 Genoa - Italy

1. Introduction

In the past, analytical methods for the quantitative determination of metals and non metals have mainly been developed with the aim of determining the total concentration of the individual elements. However it is increasingly recognized that information about the physicochemical forms of the elements is required for undestanding their chemical behaviour. This information is particularly crucial in studies related to environmental or toxicological investigations.

In marine chemistry the oceanographer needs to know the species in which the element is present in order to gain insight into the role of these compounds in sedimentary cycles, into the physical chemistry of the sea and into the nature of pollutants interactions as well as in the complexity of the biogeochemical cycle.

In particular the aim of the **speciation analysis** is often to provide information about the **bioavailability** of the elements. In fact, for example, considerable evidence now exists for assuming that the free ionic forms of metals like Cu, Pb and Cd are usually the most toxic forms to aquatic biota and that the complexation by natural ligands reduces the toxicity of these metal ions. However the relationship between metal species and bioavailability can be rather complicated. For example, the toxic effect of Cd towards a green alga showed considerable seasonal variations in an eutrophic lake with a toxicity far exceeding what would be expected according to the estimated free ion activity during summer: it is hypothesized that qualitative changes in composition of the dissolved organic matter during the production period are responsable of this effect. For organometallic compounds, these are usually much more toxic than the corresponding inorganic elements. Hg, Pb and Sn obey this general rule whereas As represents an exception because most organoarsenic compounds are less toxic than inorganic arsenic species. The toxicity of organometallic species varies with the organism monitored also. For example, triethyltin is more toxic for mammals, trimethyltin for insects and tributhyltin for fish, fungi and bacteria [1].

Speciation analysis involves the use of analytical methods that can provide such information and this field has become one of the major research and it is currently one of the most challenging tasks for analytical chemists.

2. Hyphenated techniques

The most commonly used direct speciation hyphenated techniques today are achieved by an initial separation of the different analyte species in the sample matrix by **chromatography** followed by specific detection and determination of one or more chosen elements in the compound. During the last years, many analytical systems composed of any possible instrumental combinations have been developed but the use of the high performance liquid chromatography (HPLC) and gas chromatography (GC) in combination with element-specific detectors in probably the one of the most promising approaches in speciation analysis.

Few of the instrumental methods available for elemental analysis have sufficient selectivity and sensitivity for the direct quantitative determination of individual-containing species. The very sensitive techniques of **atomic absorption** and **atomic emission** usually determine only the total amount of metal sample and must therefore be combined with a chromatographic or another separation technique before individual species can be identified.

3. Chromatographic techniques

The chromatographic techniques are especially useful for determining chemical species. However, the choice of a suitable detection system needs some consideration since the standard detectors used in HPLC and GC are often inadeguate for the determination of metals and non metals.

The GC approach requires that the species are sufficiently volatile and thermally stable to elute from the gas chromatograph. So the speciation studies in several cases requires derivatization reactions in order to obtain volatile compounds. The derivatization and clean-up procedures becomes problematic and a longer times are needed for analysis. The HPLC is more generally applicable than GC to speciation problems. For example for many organometallic compounds with boiling-points that are not sufficiently low for GC, LC becomes the preferred choice.

There are a number of advantages in using LC: i) minimum sample preparation, ii) high-boiling compounds can be separated, iii) more operational parameters can be used. Methods have been developed for the separation of most type of inorganic species, such as cations, anions and metal complexes, in addition to organometallic compounds. Since most metal-containing species are charged, some form of ion chromatography should be particularly useful.

3.1. Detection techniques in chromatography

3.1.1. Requirements of a typical detector

The funcion of any detector employed in high performance liquid chromatography is to monitor accurately the amount or concentration of the analytes eluted from the column. Generally the following requirements are necessary [2]: a) capability to detect 1 part of solute or less in 10^6 part of eluent; b) no re-mixing of components as they pass through the detector; c) wide linear dynamic range to ensure that quantitative analysis can be accomplished in a strighforward manner; d) low drift and noise levels so that small amounts of solute can be observed; e) fast response time to record accurately and rapidly eluting peaks; f) insensitive to flow-changes, pulsation and temperature; g) insensitive to changes in eluent composition so that gradient elution can be performed; h) ease of operation and reliability.

3.1.2. Classification of detectors

The increasing demands for data concerning the molecular structure and/or the characterization of species which are present in a variety of complex matrices require two criteria to be met. First, separation techniques must be available that provide sufficient resolution to permit analytes of differing molecular structures to be resolved and advances in GC and HPLC meet this requirement. The second criterion involves the reliable and selective detection of the particular analyte(s) of interest often in minute quantities amid complex eluents.

Our intention in this paragraph is to present the subset of detectors generally labeled as element-specific or element-selectice detectors (ESD). Simply stated a detector capable of reproducibly detecting a unique property of an element or a particular molecular form of that element as neutral or ionic species will be considered to have element-specific or element-selective characteristics. Detectors that measure atomic transitions are element specific by the aforementioned definition. These include chiefly conventional flame and graphite furnace (electrothermal) atomic absorption (FAAS, GFAAS) and atomic emission (AES).

In Table 1 the detection limits for a number of elements by a variety of AAS and AES methods are given [2]. It can be seen that plasma AES offers excellent sensitivity.

Table 1. Detection limits for elements using AAS and AES

D.l. $\mu g\ ml^{-1}$	Atomic absorption		Atomic emission	
	flame	AAS	flame	plasma
10^{-7}		Be,Cd,Cr,Fe,Mg,Mn Zn		
10^{-6}		Al,Ca,Co,Cu,Na,Pb		Ca
10^{-5}		Ba, Ni, V		Sr
10^{-4}	Mg	Si, Ti	Ca, Na	B,Ba,Be,Cd,Co,Cu,Fe,Mg,Mn,Na
10^{-3}	Ag,Al,Be,Ca,Cd,Co,Cr,Cu,Fe,Mn,Na,Ni,Zn		Ag,Ba,Cr,Mg,Mn,Sr	Ag,Al,Cr,Hg,Ni,Pb,Sn,Th
10^{-2}	Ba,Pb,Sn,Sr	P	Al,Co,Cu,Fe,Ni	P,Si,Tl
10^{-1}	Si, Tl, V		Pb, Tl	V
1	Hg		B, Sn	Cd
> 1	B, Th	Hg, Si, Zn		

An illustration of the analytical advantages of using ESD is shown in Fig.1. The employment of nonspecific detectors provides detectable signal for many both important and nonimportant analytes regardless their chemical content but the analytes containing the element of interest may remain undetected. ESD system provide an excellent probe for detecting all eluting species that contain the selected element of interest. In this case the element X in contained both in molecules bearing and no bearing an active chromophore. Since capability for simultaneous detection of many elements exists, valuable data on the elemental composition of each resolved species may also be collected [3].

The use of ESD in HPLC permits others advantages: a) minimizing interferences ; b) the use of a detection method that responds only to a particular element or elements leads to an overall simplification of the necessary chromatography; c) derivatization reagents containing an organometallic or inorganic derivatizing species offer an extension on the direct HPLC/ESD detection system.

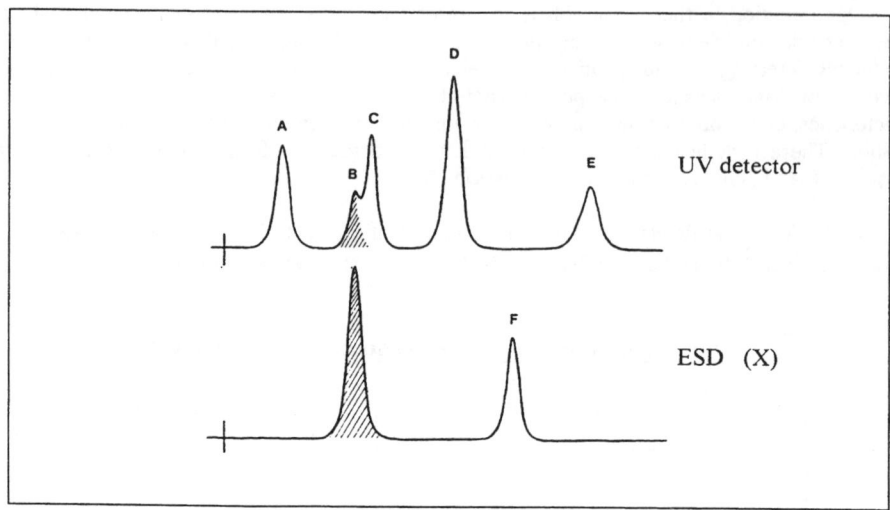

Fig.1. Comparison between two hypothetical chromatograms of a solution containing organic and organometallic components.
a) Conventional HPLC-UV analysis. b) HPLC-ESD analysis.

3.1.3. Considerations for coupled Chromatograph - ESD systems

Efficient methods of transporting chromatographically resolved analytes into the operating detector are critically important in the hyphenated procedures. Transportation of candidate analytes may be accomplished by employing a direct feed of all or of a preselected portion of eluent from the chromatographic system to the ESD. Conventional nonelement-specific detectors and element-specific ones such FAAS, inductively coupled plasma (ICP) and direct current plasma (DCP) detectors provide good examples of directly coupled chromatograph - detector supporting systems (On-line interfaces).

Where interfacing of the chromatograph to the ESD has not been achieved, off-line or indirect characterization remains a viable alternative. In HPLC the general approach to such indirect characterizations involves the use of a fraction collectors which sequentially accumulate predetermined volumes of eluent. These fractions are then individually carried out into the detector for determination by manual or automated methods (Off-line interfaces).

In general the choice of the interface is depending from the atomization mechanism of the ESD and shows several advantages and disadvantages. In Table 2 are summarized these aspects.

A serious problem in the on-line interfacing is related to the dispersion. In fact the use of chromatographic separation prior to the introduction of a sample in a spectrometer causes a loss of sensitivity compared with conventional, continuous, sample introduction. The poorer sensitivity arises from the use of discrete samples and from the dispersion that take place in the system [4]. The main sources of dispersion are the chromatographic column, the connecting tubes, the injection system and the detector. For example, in the use of an ICP spectrometer as an EDS coupled to a chromatographic system, we can considere the observed dispersion as variance of the peak (σ^2_{obs}) and resultant from different sources of dispersion considered as independent [5]:

$$\sigma^2_{obs} = \sum \sigma^2_i = \sigma^2_{injector} + \sigma^2_{tube} + \sigma^2_{column} + \sigma^2_{plasma}$$

Table 2. On - and Off- line interfaces chromatograph - ESD

On - line	Off - line	Advantages	Disadvantages
Flame/plasma atomization FAAS; DCP and ICP-AES; ICP-AES-MS		*Continuous chromatograms * Low analysis time * No contamination	* Monoelemental analysis * Peak broadening and zone spreading * Flows compatibility
	Electrothermal atomization GFAAS	*Total analysis of collected fractions * Multielemental analysis	* Fraction collection * Discontinuous chromatogram * High analysis time

From a theoretical and experimental study of the relative contribution of the various dispersion sources the following results are shown: a) with a commun flow rate of 1 ml/min, the contributions to the dispersion from the plasma, σ^2_{plasma}, and from the tubing, σ^2_{tubing}, are both equal to about 1s; b) the contribution of extra column broadening is insignificant up to 25 μl injection volume.

4. ESD - Atomic Absorption Spectroscopic Detectors

Atomic absorption obeys the same general rules as molecular absorption, but for the former process to take place free atoms of the element in the gaseous phase must be produced, and so sample preparation requires volatilisation followed by dissociation of the molecules. The conversion of metal ions present in a liquid sample to their atomic form in the vapor phase is usually accomplished by heat energy and both flame and electrothermal mothods have been used in chromatography - AAS hyphenated systems.

In GC - FAAS usually a transfer tube is used to deliver the gas-chromatography effluent to the detector. A 1/16 in diameter stainless - tube has been found to be universally suitable for many organometals (e.g., those of Pb, Sn, As and Se), except under special circumstances. For example, in the study of organomercury compounds, a teflon tube of similar diameter was used to avoid the catalytic decomposition of the organometal on a heated metal surface [6]. There are other designs of the transfer line between the gas chromatograph and atomic spectrophotometer but basically it is a narrow - bore tube between the two instruments [7]. One end is connected to the chromatographic column outlet while the other end is connected to an electrically heated silica furnace where the sample is atomized. It is essential that the tube should be as short as possible to avoid peak broadening. It is necessary that the transfer tube is heated by heating tapes to the highest temperature of the oven temperature programme. A detailed description of the construction of the system has been given by Chau and Wong [8].

Atomic absorption spectrometry operated in the flame mode does not have the sensitivity required for environmental analysis. Its sensitivity is at the microgram level. For this purpose, as noted, a silica tube furnace electrically heated is generally used [10-12]. This device is economical, simple to construct and lasts longer than any commercial graphite furnace tubes. A feature of the commercial graphite furnaces however is their high operating temperature ($\cong$ 2000°C) which obviously enhances atomization. Several studies have been published on the use of this commercial graphite furnaces [13-15].

The silica tube atomizer placed in line must be continuously heated to the atomizing temperature, a feature that commercial furnace do not have.

In the classical FAAS the sample solution is aspirated into the flame via a pneumatic nebuliser but for most nebulisers the uptake rate of 2 - 10 ml/min is greater than the HPLC flow - rate which is tipically 0.5 - 2.0 ml/min. After the first work carried out in 1973 by Manahan and Jones [9], several groups of workers have tried to optimise these conditions and now the problems are generally solved. The simplest interface involve direct coupling of the sample stream to the analytical chamber of the detector with a small - bore teflon tubing which satisfactorily minimize peak broadening and zone spreading.

Which HPLC - FAAS hyphenated systems there is no restriction on solvents and in some instances improved sensitivity can be obtained by changing the organic part of the solvent. Gradient elution can also be used and therefore a wide range of samples can be analysed by the method.

The sensitivity of the technique is dependent on the species that is being determined but is generally in the nanogram region.

Electrothermal heating methods are used in HPLC - AAS studies to provide gains in sensitivity for several elements. This increase in sensitivity arises from the ability to select and control the atomization temperature more precisely and a graphite furnace has been generally used in all electrothermal AAS work with HPLC.

The major problem with the graphite furnace AAS detection system is that it cannot used as an on - line monitor. Unlike flame methods, several time consuming stages are required to produce free atoms of an element in a furnace and hence samples must be collected and stored prior the analysis. Thus various indirect couplings have been used to overcome this problem. Typical examples will be described in the next part. Brinckman and coworkers [16] developed two such indirect couplings. The first utilised a PTFE flow-through cell from which the eluent was periodically sampled and injected into a graphite furnace, the so - called pulsed - mode operation. In the second, termed survey - mode, the eluent was collected by an auto - sampler and each collected fraction (10 - 50 µl) analysed by ETA-AAS. Vickerey and coworkers used Zeeman - effect background correction in their interface device which consisted of a sampling valve, timing circuit and automatic co - analyte addition. This interface was controlled by a microprocessor and sample segments were injected into the furnace [17]. Another interface, involving the use of a hydride generator for post - column hydride generation, has resulted in continuous, real - time signals [18]. On - line interfacing of HPLC to a continuosly heated graphite furnace atomic absorption spectrophotometer are also proposed by several research groups. In one method [19] the HPLC effluent is volatilized to an aerosol in a heated silica capillary and enters the furnace through a vitreous graphite tube; limitations are in the use of solvents and buffer solutions in the eluent. Blais and coworkers [20] developed a HPLC - furnace interface that include a thermospray micro - atomizer operated at 700 - 1000 °C where the HPLC effluent is flash evaporated to an aerosol before entering the furnace.

A detailed bibliography about these problems is proposed by Van Loon [21] and Chau [22].

5. ESD - Atomic Emission Spectroscopy

In Atomic Emission spectroscopy (AES) the original emission excitation technique was either d.c. arc or electric sparks; this old emission technique was recently resurrected by the arrival of plasma excitation source technology (PES)[6]. Its sensitivity has beeen enhanced by order of magnitude and its multielement capability feature has been fully exploited. There is now an increasing trend in the use of plasma - excited atomic emission spectrometry as EDS in HPLC speciation studies [22].

Basically three distinct types of plasma sources are being widely used in the areas of interfacing chromatography with PES:

> microwave induced plasma - MIP
> direct current plasma - DCP
> inductively coupled plasma - ICP

A no commercial MIP detector has been used with GC for a number of years; recently a commercial GC - MIP system is available. The characteristic emission spectra of the element can be established during the emission for comparison purposes in the verification of the identity of the element [23].

Significant advantages are obtained by performing on - line speciation analysis by combined HPLC interfaced directly to one or more types of PES.

MIP is virtually incompatible with HPLC at least with the conventionally practiced HPLC using normal mobile phase flow rate. DCP and ICP represents multi ESD which are capable of in situ monitoring of several elements contained within a single analysis. Thus, practically all exisiting literature reports discuss either DCP or ICP but it is the latter which appears to receive the greater amount of attention.

Obviously, a crucial area for successful HPLC - PES work is in the design and utilization of a siutable interface between HPLC column and the PES detector [24]. The interface involves a connection of HPLC to the ICP nebulizer. Nebulizer sample introduction, however, is responsible for some limitations of HPLC - ICP coupling. These include the limitations with various solvents which would be desirable as mobile phases. In fact the frequently used eluents in HPLC, i.e. methanol, acetonitrile and tetrahydrofuran cannot routinely be introduced in the plasma torch because they reduce its excitation properties. There are two limiting factors for the ICP as an ESD. The transport efficiency of the analyte is low for conventional sample introduction with a pneumatic nebulizer. Moreover the tolerance of the ICP for common HPLC solvents is low. The solvent load of the plasma can be reduced by aerosol thermostating [25, 26], cooling and spray chamber [27], application of a condenser [28] and electrothermal carbon cup vaporization [24]. Recently, aerosol generation for ICP - AES with a thermospray vaporizer has been described and the results ar still sufficient to make the application of this device as an interface very promising [5, 29, 30]. For the forming volatile hydride elements the insertion of a hydride generator (gas-liquid separator) between the HPLC volumn and the ICP - AES detector has been suggested [31].

6. ESD - Inductively Coupled Plasma Mass Spectrometry

The inductively coupled plasma mass spectrometry (ICP-MS) system is currently the most reliable and selective detector for identification work. When coupled with HPLC it forms a highly sensitive and element - selective analytical system for speciation. The ICP - MS combines several advantages [32] of ICP - AES:

- broad linear dynamic range (up to 6 orders of magnitude)
- capability of multielemental analysis
- advantage of working with samples in solution
- speed of analysis

to those of MS:
- rapid acquisition of mass spectra
- low detection limits (0.01 - 0.1 μg/l for most of the periodic table)
- simple spectrum (1 - 10 lines/element, corresponding to the number of isotopes of each element)
- capability of isotopic analysis

The behaviour of ICP-MS as a detection unit in the hyphenated techniques with HPLC is dominated by the eluent flow rate and composition. Varying the column diameter from 8 mm i. d. to narrow and microbore columns with 2 or 1 mm i.d., allows a change in flow rate from 10 to 0.1 ml/min with the same packing material. The most critical part of the LC - ICP-MS coupling is the nebulization unit. Water based eluents are often deleterious because of their salt contents, eluents containin organic modifiers or fully organic eluents tend to affect the plasma stability. ICP-MS needs in general more dilute buffers or a lower concentration of organic solvents than ICP-AES. One disadvantage of coupling the techniques in contrast to standard ICP-MS applications is the increased analysis time.

In the last years the HPLC-ICP-MS was proposed in the resolution of many speciation problems but it is a sophisticated and expensive system which only a very few privileged laboratories could acquire[33].

7. Examples of studies with hyphenated techniques

The fast growing of the hyphenated techniques in the detection of metals in marine environment yelds impossible to give an exhaustive review in relation to the aims of this paper. The intention of this paragraph is to give only one idea of the very large possibilities of these techniques and to show a research horizon that represents the achievements of analytical chemistry, biology and geology providing interdisciplinary stimulation and challenge. In this viewpoint in Table 3 some title and relatives abstracts are shown.

Table 3. Examples of studies with hyphenated techniques	
Cromate ion determination in water [34]	*Absorption of CrO_4^{2-} on anionic exchanger from 1 l of water. Elution with 1 M NaCl with FAAS direct connection. Detection limit 0.1 ppb, precision $\pm 20\%$*
Cr speciation [35]	*Partisil 10scx column, mobile phase 8:2 water - 0.2M phosphate buffer at pH 5. Direct connection HPLC-FAAS.*

HPLC - cold vapour AAS in the determination of organic Hg compounds [36]	A liquid chromatographic method with on-line UV irradiation was developed for the determination of organic Hg compounds by cold-vapour AAS. Methyl-, ethyl-, phenyl- and inorganic Hg were separated inon RP C_{18} columns. An UV-irradiation lampe was used for on-line destruction of the organomercury compounds. Sample and $NaBH_4$ solution were continuosly fed to the reaction vessel where Hg was reduced. The volatilized Hg was swept into the absorption cell of a cold-vapour AAS system by N_2. The d.l. for methylHg is 80 pg absolute.
IC coupled with ICP-AES for the determination of Cr(III) and Cr(VI) [37]	The ion chromatography ICP-AEs coupling gives reliable and reproducible results rather quickly. A measurement requires 3 min and a 50 µl sample. Two IC techniques were compared for separation using 7.5 nM potassium hydrogenphthalate and water/ 1M HNO_3..Cr (VI) is retained in the anion exchange column while the Cr(III) passes through without any retention. The detection limits of Cr(III) and Cr(VI) are 0.25 and 0.27 µg/g respectively.
As speciation using HPLC-HG-ICP-AES [38]	In this work the optimization condition are presented for As(III), As(V), monomethylarsonic acid (MMA) and dimethylarsinic acid (DMA) determination using (IC) HPLC coupled to HG with gas-liquid separator and using ICP-AES as ESD.
Determination of organoarsenic compounds by HPLC-HG-AAS with on-line microwave oxidation [39]	An on-line HPLC - microwave oxidation - HG - AAS coupled system has been developed for the determination of arsenite, arsenate, dimethylarsinate (DMA), monomethylarsonate (MMA), arsenobetaine and arsenocholine in environmental samples. An anionic cartridge placed before the HPLC anionic column quantitatively retains anionic species as arsenite, arsenate, MMA and DMA but not cationic species which are separated and quantitatively determined after microwave-$K_2S_2O_8$ decomposition. D.l. of between 0.3 and 0.9 ng are achieved for all species.
Pb speciation by gradient HPLC with ICP-MS detection [40]	Speciation of inorganic Pb, triethyllead chloride (TEL), triphenyllead (TPhL) and tetraethyllead (TTEL) was investigated using gradient HPLC with detection by ICP-MS. The detection limits were 0.37, 0.14, 0.17 and 3.9 ng of lead for Pb, TEL, TPhL and TTEL respectively.
ICP-MS as an ESD for supercritical fluid chromatography [41]	SFC coupled with ICP-MS shows an high potential for the determination of ultratrace levels of organometallic compounds. In this work a SFC/ICP-MS interface is developed.Separation of tetraalkyltin compounds shows detection levels in the subpicogram range: 0.034 and 0.047 pg for tetrabutyltin and tetraphenyltin respectively.The linear ranges are over 3 orders magnitude (1-1000 pg) and the reproducibility of sample injections are better than 5% RSD.

Determination of metalloproteins by HPLC with ICP-AES detection [42]	The HPLC has been directly coupled to ICP-AES for trace organometallic studies. The eluates from the gel-permeation columns pass through an UV detector and then are directly aspirated into the plasma torch. Data are presented concerning the evaluation of the Zn, Cu and Cd thionein contents of different tissues from marine mussels. A comparison of the S and metallic chromatograms shows the relationships between the whole group of S-rich proteins and the metallothioneins.
Organotin compounds in marine mussel samples determined by using HPLC-HG-ICP-AES [43]	Analysis of butyltin, dibutyltin and tributyltin chloride with HPLC-HG-ICP-AES is described. A d.l. of 7 ng for Sn is obtained and the organotin species were completely separated within 7 min. The system has been used for the determination of organometallic compounds in marine mussel samples.

Bibliography

1. Lund,W. (1990) *Fresenius' J. Anal. Chem.* **337**, 557
2. White, P.C. (1984) *Analyst* **109**, 67
3. Jewett, K.L., and Brinkman, F.E. (1983) in T.M. Vickrey (ed), *Liquid Chromatography Detectors*,
   Marcel Dekker, New York and Basel, 205
4. Katz,E.B., and Scott, R.P.W. (1985) *Analyst* **110**, 253
5. Laborda, F., De Loos-Vollebregt, M.T.C., and De Galan, L. (1991) *Spectrochimica Acta* **46B**, 1089
6. Chau, Y.K (1992) *Analyst* **117**, 571
7. Forsyth, D.S., and Marshall, W.D. (1983) *Anal. Chem.* **55**, 2132
8. Chau, Y.K., and Wong, P.T.S., (1989) in G.E. Batley (ed), *Trace Element Speciation:Analytical Methods and Problems*, CRC PRESS; Boca Raton, Fl, 220
9. Manahan, S.E., and Jones, D.R. (1973) *Analyst* **6**, 595
10. Andreae, M.O. (1977) *Anal. Chem.* **49**, 820
11. Chau, Y.K., Wong, P.T.S., and Goulden, P.D. (1975) *Anal. Chem.* **47**, 2279
12. Radziuk,B., and Van Loon, J. (1976*) J. Sci.Total Environ.* **6**, 251
13. Robinson,J.W., Kiesel, E.L., Goodbread, J.P., Bliss, R., and Marshall, R. (1977) *Anal.Chim..Acta* **92**, 321
14. Radziuk, B., Thomassen, Y., Van Loon, J.C., and Chau, Y.K. (1979) *Anal.Chim.Acta* **105**, 255
15. Parris, G.E., Blair, W.R., and Brinckman, F.E. (1977) *Anal. Chem.* **49**, 378
16. Brinckman,F.E., Blair, W.R., Jewett, H.L., and Iverson, W.P., (1977) *J. Chromatogr.* **15**, 493
17. Vickrey,T.M., Howell, H.E., and Paradise, M.T. (1979)*Anal.Chem.* **51**, 1880
18. Burns, D.T., Glockling, F., and Harriott, M. (1981) *Analyst* **106**, 921
19. Nygren,O., Nilsson, C.A., and Frech, W. (1988) *Anal. Chem.* **60**, 2204
20. Blais, J.S., and Marshall, W.D. (1989) *J. Anal. At. Spectrom.* **4**, 271
21. Van Loon, J.C., and Barefoot, R.R. (1992) *Analyst* **117**, 563
22. Chau Y.K., and Wong, P.T.S. (1991) *Fresenius J. Anal. Chem.* **339**, 640
23. Scott, B.F., Chau, Y.K., and Rais-Fifouz, A. (1991) *Appl. Organomet. Chem.* **5**, 151
24. Nisamaneepong, W., Caruso, J.A., and Ng, K.C. (1985) *J. Chromatogr. Sci.* **23**, 465
25. Maessen, F.J.M.J., Seeverens, P.J.H., and Kreuning, G. (1984) *Spectrochim. Acta* **39B**, 1171
26. Maessen, F.J.M.J., Kreuning, G., and Balke, J. (1986) *Spectrochim. Acta* **41B**, 3
27. Hausler, D.W., and Taylor, L.T. (1981) *Anal. Chem.* **53**, 1223
28. Boorn, A.W., and Browner, R.F. (1982) *Anal.Chem.* **54**, 1402

29. Koropchak, J.A., Aryamanya-Mugisha, H., and Winn, D.H. (1988) *J. Anal. At. Spectrom.* **3**, 799
30. Roychowdhury, S.B., and Koropchak, J.A. (1990) *Anal. Chem.* **62**, 484
31. Rivaro,P., Zaratin, L., Frache, R., and Mazzucotelli, A. (1995) *Analyst* **120**, 1937
32. Beauchemin, D. (1991) *Trends Anal.Chem.* **10**, 71
33. Seubert, A. (1994) *Fresenius J. Anal.Chem.* **350**, 210
34. Janauer, K., and Pankow, J. (1974) *Anal.Chim.Acta* **69**, 97
35. Van Loon, J.C. (1977) *Pittsburg Conference*
36. Falter,R.,and Scholer,H.F. (1994) *J. Chromatog.* **A 675**, 253
37. Prokisch,J., Kovacs,B., Gyori,Z., and Loch,J (1994) *J. Chromatog.* **A 683**, 253
38. Rauret, G., Rubio, R., and A. Padrò (1991) *Fresenius J. Anal.Chem.* **340**, 157
39. Angeles Lopez-Gonzalves,M., Milagros Gomez,M., Camara, C. and Palacios, M.A. (1994) *J. Anal.At.Spectrom.* **9**, 291
40. Al-Rashdan,A., Vela, N.P., Caruso, J.A. and Heitkemper, D.T. (1992) *J. Anal.At.Spectrom.* **7**, 551
41. Shen, W.L., Vela, N.P., Sheppard, B.S., and Caruso, J.A. (1991) *Anal.Chem.*, **63**, 1491
42. Mazzucotelli, A., Viarengo, A., Canesi, L., Ponzano, E., and Rivaro,P. (1991) *Analyst*, **116**, 605
43. Rivaro, P., Zaratin,L., Frache,R., and Mazzucotelli,A. (1995)*Analyst* **120**, 1937

ANALYSIS AND SPECIATION OF ORGANOMETALLIC COMPOUNDS IN THE MARINE ENVIRONMENT - GENERAL CONSIDERATIONS

P. J. CRAIG and D. MILLER
Department of Chemistry, De Montfort University
Leicester LE1 9BH, UK

1. Speciation Analysis

It is by now a commonplace that it is not sufficient to identify or quantify the total metal content of a sample, but that it is really necessary also to determine the actual metal compound that is present; ie the chemical formula. The reason for this is that the toxicity and other environmentally important properties will differ between compounds derived from the same metal (see below). For organometallic compounds this reduces in many cases into trying to identify and distinguish the immediate organic surroundings of the metal (eg Bu_3Sn^+ in the presence of Bu_2Sn^{2+}, where Bu is the n-butyl group). Where there is a counter ion present it is often inorganic (eg X in Bu_3SnX where X may be F etc) and not usually so important in toxicity terms, or it may be a complex organic group, not necessarily linked to the metal by a metal-carbon bond (eg the riboside groups found in marine organo arsenic compounds - here there is a metal carbon bond). Often such "counter ions" may be long chain, mostly unknown proteinaceous systems, which are felt not to influence much the toxicity properties of the eg Bu_3Sn^+ moiety. This assumption may not always be correct. However, the basic idea that counter ions do not matter as much has led to a form of speciation analysis for organometallic compounds in the environment that in most cases is satisfied with determining the organometallic moiety only (eg Bu_3Sn^+). This means that much of the procedure of organometallic analysis in the environment is concerned with releasing (extracting) the organometallic moiety (moieties) from the usually intractible or involatile "counter ion" or general environmental matrix (unless HPLC is being used, when this process is usually simpler). This is not usually the end of the matter, however, as the released organometallic(s) are likely to be involatile and not separable by GC and similar procedures. When several organometallic species of the same element are released by the extraction, they then need to be analysed separately because of different toxicity properties. The extract will usually then need to be derivatized in order that volatile separable compounds be produced. The result of all of this is that the organometallic species in the environment has been split in half by the extraction process and that we are achieving, in effect, a half speciation analysis. In principle, the ideal of speciation analysis is full molecular identification, following a non destructive extraction, a separation process and a molecular identification of each separated compound (ie by MS). It is instructive to review briefly the toxicity properties of organometallics that have resulted in this attitude, ie that each compound needs to be identified.

2. Toxicity Properties of Organometallics

These can be summarized for our purposes by a few general observations, viz:

(1) Organometallic compounds are usually more toxic than inorganic compounds of the same metal (arsenic being the main exception). This is because the organic groups confer lipid solubility and thereby "improve" delivery of the metal to the site of action (often the central nervous system), where the normal toxicity action of the particular metal comes in to play. This usually involves coordination to the metal by sulfur, oxo or nitrogen ligand from enzymes etc. The most general problem with organometallics in higher animals is attack on the myelin coating of neuron fibres and degradation of the central nervous system. There can also be attack on non-enzyme sites on proteins, on hemoglobin, on cytochrome P450, and on bone marrow. Interference with ATP synthesis and leaching out of essential trace metals may also occur[1,2].

(2) Usually toxic effects are at a maximum for R_nM^+ species derived from the neutral or saturated $R_{n+1}M$, *ie* the loss of one R group from the fully saturated species. This gives the right balance of aqueous and lipid/membrane solubility for transport around the organism and consequent binding as in (1) above.

(3) Toxicity to a given animal species varies widely between different organometallics of the same element. Using this as an example, towards humans, in R_3Sn^+, the most toxic are methyl, ethyl and propyl (Me, Et, and Pr respectively). The octyls are not very toxic. The di-octyls are used in food contact materials. Table 1 illustrates this property. Table 2 shows the effects of the degree of saturation.

TABLE 1. Toxicity of Different R_3Sn^+ Species

Compound	Toxicity*
Me_3Sn^+	0.07
Et_3Sn^+	0.04
Pr_3Sn^+	0.3
Bu_3Sn^+	0.7
Pr_3Sn^+	0.3
Oct_3Sn^+	>8.0

*LD50 towards the rat (mmol kg^{-1}); adapted from Ref 1. (reproduced with Author's permission)

TABLE 2. Effect of Degree of Saturation

Compound	Toxicity*
Bu_3Sn^+	0.7
Bu_2Sn^{2+}	0.5
$BuSn^{3+}$	8.0
Oct_3Sn^+	>8.0
Oct_2Sn^{2+}	>12.0

*LD50 towards the rat in mmol kg^{-1}; from Ref 1. (reproduced with Author's permission)

(4) The toxic effects of a single organometallic compound vary considerably between different biological species. Tables 3 and 4 illustrate this for tin.

TABLE 3. Toxicity Effects of the Same Compound on Different Biological Species

Biological Species	R in Most Active R_3SnX Compound
Insects	CH_3
Mammals	C_2H_5
Gram +ve bacteria, fish, fungi, molluscs, plants	$n-C_3H_7$
Fish, fungi, molluscs	$n-C_4H_9, C_6H_s$
Fish, mites	$c-C_6H_{11}$

TABLE 4. Toxicity of Bu_3Sn^+ Towards Different Organisms*

Rat	0.7 mmol kg^{-1}
E Coli [†]	1.4 mmol dm^{-3}
Botrytis (fungi) [†]	1.0×10^{-3} mmol dm^{-3}
Crab larvae [†]	5.5×10^{-6} mmol dm^{-3}

*LD$_{50}$ adapted from Ref 1; [†] In water (reproduced with Author's permission)

It is clear from the above that the identity of the actual compound needs to be confirmed. Also, of course, several compounds of interest may coexist at the same site; in principle, as can be seen from a previous Chapter, up to thirteen organo or hydrido tin compounds could be analysed from a single environmental site, although in practice only two or three are usually detected. The following sections of this Chapter discuss the processes of extraction and separation, and the final section summarises the methods of detection themselves.

3. Sample Treatment

3.1 EXTRACTION OF ORGANOMETALLIC FROM THE MATRIX

Numerous extraction techniques, often acid based, have been used and Figures 1 and 2 give typical examples of methodology (for tin and mercury respectively). Other extractants used include, for example, acetic acid and methanol.

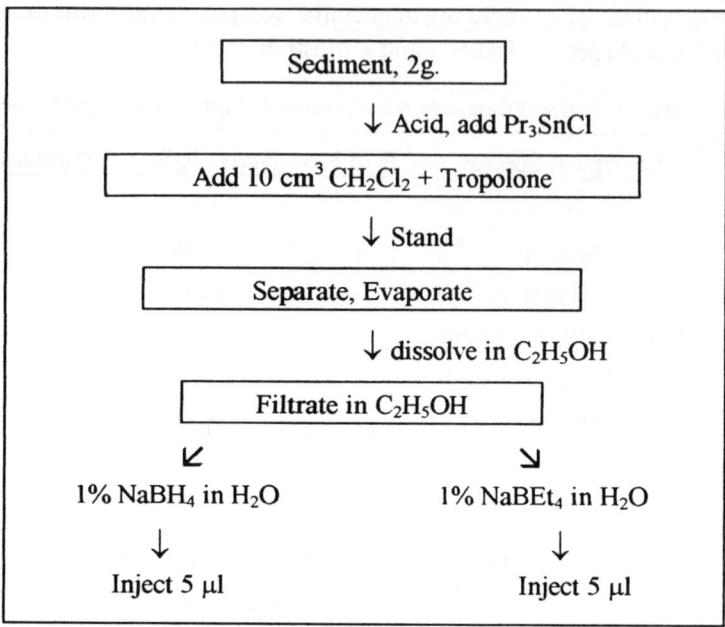

Figure 1. *Extraction scheme of Organotin Species* (adapted from Ref. 3)

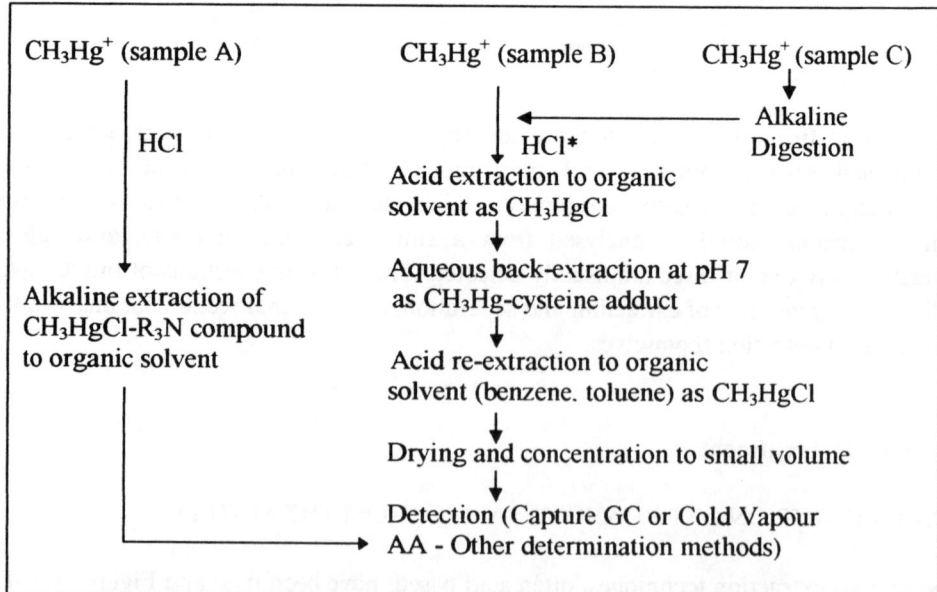

Figure 2. *Extraction scheme of CH_3Hg^+* (adapted from Ref. 4, with Author's permission)

3.2 DERIVATIZATION OF THE EXTRACT

Extraction will generally produce a mixture of the compounds of interest (if indeed more than one is present). If the available or chosen method of separation is other than HPLC, then the detector will require that it receives the analyte in volatile form. Generally then, the mixture of compounds for analysis will have to be separated (see below) and this is normally done by chromatography type methods. For this the extract has to undergo a process that will make it volatile (ie derivatization). This is normally done by direct reaction of the extract with a *quasi* Grignard reagent; the following examples illustrate:

1. $Bu_3Sn^+ + H^-$ (from $NaBH_4$) $\rightarrow$ Bu_3SnH
2. $MeHg^+ + C_2H_5^-$ (from $NaBEt_4$) $\rightarrow$ $MeHgC_2H5$
3. $Et_3Pb^+ + C_5H_{11}^-$ (from $C_5H_{11}MgBr$) $\rightarrow$ $Et_3PbC_5H_{11}$
4. $MeHg^+ + C_4H_9^-$ (from C_4H_9Li) $\rightarrow$ $MeHgC_4H_9$

The above reagents can be considered to carry out a SN_2 type substitution on the environmental ligand binding to the metal, replacing it with an alkyl group, not the same as of that suspected to be environmentally present. This produces a volatile fully substituted organometallic (or organometallics), which is (are) then detected. The reagents in brackets are in fact the more commonly used derivatizing reagents. The first two have the advantage that the reaction can take place in water, whereas the last two require dry conditions (although an excess can be used). All of them, then, convert an extract from a non volatile form into one which is capable of passing through a GC column or which can undergo a purge and trap method (see below). A prime requirement of the derivatizing reagent is that it does not alter in any way the type or distribution of the organometallic groups around the metal (dismutation). It should also convert as much as possible of the organometallic in the extract to the product form, although this is not essential as internal standards can be added in the extraction phase prior to derivatization. In the case of butyl tin compounds, propyl tin is often added as the internal standard (as Pr_3SnCl). Propyl tin species have not been detected in the environment and so are suitable as standards. Limitations owing to chemical stability sometimes occur; in the case of methyl mercury, only recently has $NaBH_4$ been used as it was thought that the organo mercury hydrides were not stable. Ethyl derivatizing agents are of less use for organo leads as the compounds of environmental interest may be ethyl leads; use of these reagents would preclude discrimination between Pb^{2+} and Et_4Pb, Et_3Pb^+ and Et_2Pb^{2+}. In some cases differences in the effectiveness of the derivatization under different conditions may be used as part of the speciation process. Inorganic arsenic (III) species will react with $NaBH_4$ to give AsH_3 at a pH range from 0 to 10; arsenic (V) as arsenate will not derivatize with this reagent at pH values greater than 5. Hence the arsenic (III) and (V) species may be selectively generated and analysed (by subtraction).

The above derivatizing reagents will often convert <u>inorganic</u> compounds to fully saturated and volatile organometallics. This provides the opportunity of analysing both inorganic and organometallic species of the same element which may be present together in the same environmental matrix. Methyl mercury in the presence of inorganic mercury (the normal environmental situation in fact) may be analysed with a single extraction and a single derivatization and detection, in contrast with previous methods where inorganic and methyl mercury had to be analysed separately, viz.:[5]

$$HgCl_2 + 2NaB(C_2H_5)_4 \rightarrow (C_2H_5)_2Hg + 2B(C_2H_5)_3 + 2NaCl$$
$$CH_3HgCl + NaB(C_2H_5)_4 \rightarrow CH_3HgC_2H_5 + B(C_2H_5)_3 + NaCl$$

Inorganic lead compounds may also be derivatized by these reagents

$$2PbCl_2 + 4NaBEt_4 \rightarrow PbEt_4 + 4BEt_3 + 4NaCl + Pb°$$

Recent work has also produced analysible systems for mercury using $NaBH_4$, viz:

$$RHgCl + NaBH_4 \rightarrow 2RHgH + \tfrac{1}{2}B_2H_6 + NaCl.\ [6\text{-}8]$$

$NaBEt_3H$ has also been used as a derivatizing reagent, producing the alkyl rather than the hydride species.[9]

4. Separation of Environmental Organometallic Species

Following the derivatization process, the final extract may contain several organometallics (not necessarily of the same element, although usually only compounds of a single element are being studied). These require to be separated. If the extract has been derivatized the separation methods envisaged will be GC or a Purge and Trap variant, if the analyte has been simply extracted (ie not derivatized), a HPLC method of separation is planned. This latter technique is applied when the analyte may be non volatile or unstable to heat. In general the coupling of HPLC systems to detector systems is more complex than eg GC to AA couplings; there is as mentioned above usually no need for a sample derivatization stage (although one may be incorporated after the separation in order to prepare the analyte for a particular detector). Initial work with coupled HPLC systems tended to produce poor detection limits. However, interfacing of HPLC with plasma systems (eg ICP-MS) overcomes this problem, although this is inevitably more expensive than coupling of GC and AA, where the interfacing of two quite modest (or even old) instruments can produce a synergistically enhanced analytical system cheaply - and one which often gives equivalent information.

TABLE 5. Interfaced Systems Including Separation and Speciation Detection*

Gas Chromatography (GC)	High Performance Liquid Chromatography (HPLC)
Flame AAS	Flame AAS
Electrically heated (quartz) furnace AAS	
Graphite furnace AAS	Graphite furnace AAS
Microwave plasma AES	Microwave plasma AES
Direct current plasma AES	Direct current plasma AES
Inductively coupled plasma (ICP) AES	Inductively coupled plasma AES
Atomic fluorescence spectrometry (AFS)	Atomic fluorescence spectrometry (AFS)
ICP-MS	ICP-MS
Flame photometric detection (FPD)	

*Adapted from Ref 11

Many interfaced systems have by now been devised and used. In a work of this length it is only possible to summarize and note the method used, and to refer the reader to some of the more specialized and lengthy review sources. Table 5 gives an indication of some common interfaced systems, and Figures 3 and 4 give stylised outlines of two of the most frequently used methods, viz interfaced GC-AA and interfaced Purge and Trap with AA. Both of these types of system can be "laboratory home made".

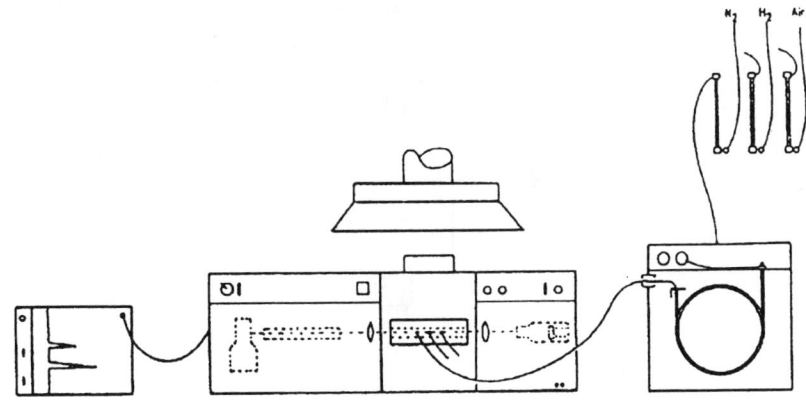

Figure 3. Interfaced GC-AAS Apparatus (adapted from ref.10)

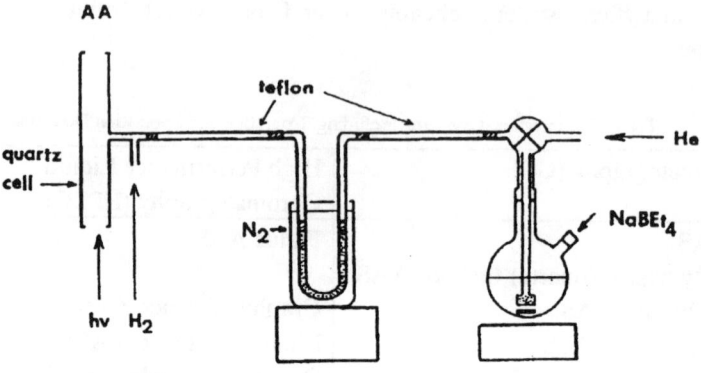

Figure 4. Purge and Trap System.

The AA detection is normally carried out using a quartz constructed open ended cell (Figure 5) via a tube leading from the GC or Purge and Trap system. Various carrier gases are used (eg helium, nitrogen or argon) and, just prior to the tube, combustion gases (hydrogen, oxygen etc) may be introduced. The transfer line may be lagged and electrically heated. The quartz furnace is also electrically heated (usually to several hundred degrees C) and the analyte there breaks down into the atomic form where it is detected by an appropriate lamp.

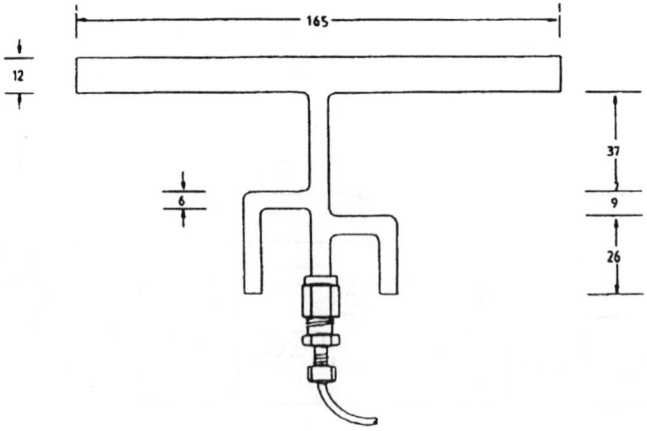

Figure 5. Quartz Atomisation Cell (adapted fron ref. 10)

Clearly this kind of system is element, not compound, specific, *ie* it will detect any compound, *eg* of tin, that passes, but it can only differentiate between the forms on the basis of retention times, thereby necessitating the chromatographic or quasi chromatographic phase. In effect, all of the systems shown in Table 5 are of this type, ie element specific, except for full MS interfaced systems. In that sense, interfaced HPLC-ICP-MS instruments can seem extraordinarily complex and expensive for what they appear to do. Here the justification has to be detection limits and the necessity of HPLC delivery. Apart from MS, the systems in Table 5 can be considered as alternative detectors for the chromatograph. However, as element specific systems, they do have the merit of detecting compounds only of the metal of interest (although there are sometimes interferents). In more general GC detectors (eg FID or EC), compounds of several or many elements will be recorded, including the generality of organic compounds present.

5. Detection of Organometallic Species

5.1. SOME KEY POINTS ON SPECIFIC DETECTOR SYSTEMS

5.1.1 Atomic Absorption (AA). General considerations.
AA spectroscopy depends on the absorption of electromagnetic radiation by the atom, leading to promotion of an electron to a higher energy level. Such absorptions are specific to the atom and transition concerned and can be quantified by the Planck relationship $E = h\nu$. The principle is simple but the detailed theory and practical aspects are complex. As an atomic phenomenon, clearly AA is element specific, not compound specific. In general there are two types of detection cell, one employing a heated silica (quartz) furnace in which the molecules are atomized and one using a heated graphite tube. The analyte can be delivered to the detector by an interfacing transfer line or by direct incorporation of a GC capillary column from the GC into the AA. With liquid chromatography (LC) it may be possible to couple the two instruments directly by a tube as the flow rates can be similar. Flow rate differences can be overcome using various reservoir type devices if necessary. The normal types of interferences which can hinder AA detection may also exist here: *viz* chemical, physical, spectral and non-specific.

Methods for the interfacing and use of GC or LC AA systems, together with discussions on the elimination of interferents are given in Reference 13.

5.1.2. Atomic Emission (AES). General considerations
Here the radiation emitted when an electron falls from a higher to a lower energy level in the atom is recorded. As an atomic phenomenon, AES detection methods are, like AA, element specific. Numerous excitation techniques are used to first promote the

electron, and these different excitation processes give rise to several analytical techniques which exist under the blanket title of AES. Where the excitation is achieved by UV or Visible light (usually by using a flame) the technique is Atomic Fluorescence Spectroscopy (AFS). Where the excitation is thermal, the resulting technique is Flame Emission Spectroscopy (FES), or Flame Photometry. Such thermal excitation may also take place in an Inductively Coupled Plasma (ICP). Where atoms are excited using an electric arc or a high voltage arc, we have Optical Emission Spectrometry.

5.1.3. *Flame Photometric Detection (FPD)*

As mentioned above, this methodology is based on emission in the near UV and Visible regions resulting from the creation of excited molecular species in a hydrogen rich flame. It was first developed for detection of sulphur and phosphorous species, indeed its sensitivity to sulphur makes this element a potential interferent when organometallics are being analysed from environmental samples. The method has been frequently used in recent years for detection of organotin, organogermanium, organoselenium and organotellurium samples. In general FPD is used interfaced or as a detector with GC. Detection limits are generally poorer when FPD is interfaced with LC separation.

5.1.4. *Inductively Coupled Plasma (ICP) Methods*

Here the compounds of interest are excited into a very hot plasma consisting almost completely of atoms. This almost eliminates interferences from molecular species and the method is very sensitive for metals. The ICP is created by the interaction of a radiofrequency field with a flowing gas (usually argon). The gas flows through a silica tube in a solenoid to create a varying magnetic field in the argon. This then generates an eddy current in the gaseous analyte to produce a very hot plasma which discharges at 10^4 °C. As ICP systems are normally designed with liquid inlet flows, ICP has been much used coupled with HPLC.

5.1.5. *Mass Spectroscopy*

Clearly the instrument can be used in the total Ion Current (Selected Ion Monitoring) mode, which of course is much more sensitive than in the GC MS mode. Every effort should be made to obtain the full Mass Spectrum (MS) as only this mode achieves the full speciation analysis required. With organometallic compounds a problem with the MS mode is that the parent ion (giving the Relative Molar Mass) may appear only weakly, but it is usually detectable. With metals, the use of MS allows the full analysis security offered by observations, including the isotopic fingerprint of the metal concerned.

6. Conclusions

As a result of the vastly enhanced ability to carry out speciation analysis over the past twenty years, a large range of compounds (or compound-moieties) has been shown to exist in the marine and terrestrial environments. The discovery, *eg* of the marine natural product chemistry of arsenic, would not have been possible without some of the techniques mentioned above. Scheme 4 in Chapter 1, using arsenic as an example, illustrates and summarises the generality of the knowledge that has been accumulated on the presence and role of environmental organometallics in the natural environment.

7. General Sources of Reference

It is clear that the above account is of a general nature, giving an overview of the possibilities and strategies for the analysis of organometallic compounds in the environment. References. 1,2,4,11,13 give some useful detailed reference sources in this area.

8. References

1. Craig, P.J. and Mennie, D., (1993), in *Metal Ions in Biological Systems*, (ed. Sigel H. and A.), **29**, 37-77, Marcel Dekker.
2. For other Toxicity and Biological Details for Organotin Compounds, see Selwyn, M.J. (1989) in *Chemistry of Tin* (ed. Harrison, P.G), Blackie, Glasgow and London, 359-397.
3. Ashby, J.R. and Craig P.J., (1991), *Applied Organometallic Chemistry*, **5**, 173-191
4. Craig P.J., (1986) in *Organometallic Compounds in the Environment*, (ed. Craig, P.J.) Longman. London, 65-111.
5. Rapsomanikis S., and Craig P.J. (1991), *Anal Chim Acta*, **248**, 563-567.
6. Weber J.H. and Puk, R. (1994), *Applied Organometallic Chemistry* **8**, 709-713.
7. Filipelli ,M., Baldi, F., Brinckman, F.E., and Olson, GJ, (1992). *Environ Sci Technol.* **26**, 1457-1460. 1
8. Craig P.J., Garraud H., Laurie S.H., Mennie D. and Stojak GH. (1994), *J. Organometallic Chemistry*, **468**, 7-11.
9. Mennie, D., and Craig P.J., (1994), *Main Group Metal Chemistry*, **19**, 453-445.
10. Clark, S., and Craig, P.J., (1988). *Applied Organometallic Chemistry* **2**, 33-46.
11. Hill, S.J., Brown, A., Rivas, C., Parkes, S. and Ebdon, L., (1995) in *Quality Assurance For Environmental Analysis* (eds. Quevauviller, Ph., Maier, EA, and Gripink, B) Ellsevier, Amsterdam, 412-437.
12. Menan, E., (ed.) (1991), *Metals and Their Compounds in the Environment*, VCH Weinheim, Germany.
13. Harrison, M. M., and Rapsomanikis, S., (eds.) (1989), *Environmental Analysis Using Chromatography Interfaced with Atomic Spectroscopy*, Ellis Horwood, Chichester, UK.

CHROMATOGRAPHIC ANALYSIS OF ORGANIC MICROPOLLUTANTS IN MARINE ENVIRONMENTS

R. FUOCO and M.P. COLOMBINI
Dipartimento di Chimica e Chimica Industriale
Via Risorgimento, 35 - 56125 Pisa (Italy)

1. Introduction

Organic micropollutants include a large number of classes of both natural and anthropogenic chemical compounds, which have a different impact on the environment depending on their toxicity and concentration level. Many of them are very harmful for living organisms, particularly those which bioaccumulate in organs and tissues, and pass through the cellular membrane. Inside the cell they may interact with the DNA and cause genetic damage and metabolic activation leading to tumour initiation or other diseases. They may also undergo chemical transformations thus generating other pollutants even more toxic than the parent ones. Monitoring the most dangerous ones (polychlorobiphenyls (PCBs), polycyclic aromatic hydrocarbons (PAHs), dioxins, chlorinated pesticides, etc.) in the environment has thus become of prime importance. Organic micropollutants generally occur in "real-life" samples as complex mixtures that vary greatly in terms of the concentration and toxicity of each individual component. Very often only some of them need to be determined at very low concentration levels (pg/g or less). For the analytical chemist this is a formidable challenge, which can only be tackled if all the steps of an analytical procedure have been suitably optimised to ensure the reliability of the results in terms of accuracy and precision.

In this paper, the most significant aspects relating to the determination of PCBs and PAHs in marine matrices (namely, sea water and sediment samples) are considered, particularly:
- the environmental significance of PCBs and PAHs;
- sample collection and storage;
- sample preparation (extraction of the analytes, and cleanup of the extract);
- instrumental analysis;
- data evaluation, including analytical quality control.

2. Environmental significance of PCBs and PAHs

2.1. POLYCHLOROBIPHENYLS (PCBs)

Polychlorobiphenyls are a class of non-polar semivolatile organic compounds which includes 209 congeners. They were synthesised at the end of the last century, and have been widely used in many industrial activities for about 40 years [1, 2]. Their discharge without any concern for the environment, has led to a wide diffusion of these contaminants, even in remote areas such as Antarctica [3,4].

TABLE 1. Occurrence of PCBs in the environment.

Matrix	Location	Concentration	References
Air (particulate)	Remote Urban	20 - 50 pg/m^3 0.5 - 20 ng/m^3	4,8 8
Water	Ocean (remote) Mediterranean coast	50 - 200 pg/l 1 - 15 ng/l	3,4 9,10
Sediment	Ocean (remote) Mediterranean Sea	50 - 150 pg/g 0.2 - 1.5 µg/g	3 11
Soil	Remote Industrial area	50 - 100 pg/g 0.1 - 20 µg/g	3 12
Organisms (fish)	Ocean	0.02 - 150 µg/g	13

Humans
  adipose tissue
  liver
	Industrial area	
		10 - 60 pg/g
		10 - 30 pg/g
			
			13
			13

Food (milk)	Industrial area	0.01 - 0.4 µg/g	13

PCBs are chemically very stable, and are thus one of the most persistent environmental pollutants, with half-life times of ten to twenty years for higher chlorinated congeners [1,5]. The lipophilicity of these compounds is responsible for their ability to bioaccumulate, particularly in adipose tissues, and their presence in living organisms has been associated with carcinogenesis [1,2]. For all these reasons the monitoring of PCBs in the environment has gained more and more attention [6], with a special emphasis on the determination of the most toxic congeners [7]. Table 1 shows typical PCB concentrations in different environmental matrices.

2.2. POLYCYCLIC AROMATIC HYDROCARBONS (PAHs)

PAHs are a class of organic compounds which are included in the wider family of polycyclic aromatic compounds (PACs) [14]. These compounds have shown carcinogenic and/or mutagenic activity in laboratory experiments with animals [14-16]. PACs are generally formed during incomplete combustion or pyrolysis of organic matter occurring in a variety of natural processes or human activities [14,17-20]. In fact, the recombination of radicals generated at temperatures between 650°C and 900°C promotes the formation of more stable aromatic compounds [14,21]. Consequently, PAHs are ubiquitous pollutants which are present in all environmental components. Sixteen of them are included in the list of priority pollutants by US EPA [22]. Along with the parent PAHs, there are hundreds of substituted PAHs, such as hydroxy, thio, chloro, amino, nitro, etc. In particular, nitro-PAHs have been intensively studied since they are considered as one of the most biologically active classes [14]. Nitro-PAHs can be generated by the photochemical transformation of parent PAHs in the atmosphere, and have been found in the exhaust emission of diesel and gasoline engines, and in fly-ash [14]. The half-life time of these compounds in the atmosphere is about 20-50 hours [23], which allows them to be transported even long distances by atmospheric circulation. Benzo[a]pyrene (BaP) shows the highest biological activity [14], and has thus been intensively studied [24,25]. Fossil fuels are the major sources of PAHs; while diesel exhaust is a very important source of these compounds in urban environments [24,25]. Table 2 shows typical PAH concentrations in different environmental matrices.

3. Sample collection and storage

Whatever chemical species have to be monitored in a given marine ecosystem, the first step is to correctly define the information needed, then the analytical procedures to obtain it should be chosen and tested. Before planning the sampling program, all the information available on the studied area should be collected, thus enabling different chemical, physical and biological parameters which may affect the concentration level of the analytes to be taken into account. When characterising an aqueous system the effects due to vertical and horizontal movements of watermasses should also be considered. Based on this preliminary study, the minimum number of sampling stations, their spatial position, and their time frequency can be suitably defined. Generally, sea water samples are collected by a Teflon or stainless steel pumping system; while sediment samples are collected by a stainless steel grab or by a box-corer system, if depth profile is required. Samples are generally stored in stainless steel containers below 0°C.

TABLE 2. Occurrence of PAHs in the environment.

Matrix	Location	Concentration		References
		BaP(*)	Total PAHs	
Air	Urban	0.5 - 2 ng/m^3	10 - 20 ng/m^3	26
(particulate)	Urban (polluted)	10 - 20 ng/m^3	<5,000 ng/m^3	27
Water	River	0.5 - 1 ng/l	5 - 10 ng/l	28
	Mediterranean coast	0.5 - 5 ng/l	15 - 150 ng/l	29
Sediment	Mediterranean Sea	30 - 500 ng/g	2 - 50 µg/g	30
	River	0.2 - 1.5 µg/g	10 - 35 µg/g	28
Soil	Industrial area	15 - 30 ng/g	20 - 100 ng/g	31
	Industrial area (polluted)	0.06 - 0.3 µg/g	0.5 - 1.2 µg/g	31
Organisms				
mussels	Harbour	6 - 20 ng/g	50 - 500 ng/g	32
fish	Ocean	0.05 - 1.1 µg/g	---	33
Humans	Industrial area			
adipose tissue		10 - 60 pg/g	---	34
liver		10 - 30 pg/g	---	34
Food	Industrial area			
olive oil		30 - 70 ng/g	200 - 700 ng/g	33
smoked food		0.1 - 7 ng/g	---	33

(*) BaP = Benzo[a]Pyrene

4. Sample preparation

4.1. EXTRACTION

4.1.1. Sea water

Non-polar or low polar organic pollutants from water samples can be extracted using any water-immiscible solvent since their solubility in water is generally very low. N-hexane and dichloromethane, or a mixture of them, are the most widely used solvents in liquid-liquid extraction techniques [1,14,22,35]. Toluene and benzene have also been used for PAH extraction [22]. Solid phase extraction (SPE) by using different sorbents, i.e. : XAD-2, XAD-4, Tenax, C_{18}, C_{18}-NH_2, etc., and elution with different mixtures of acetone/hexane/ dichloromethane, is a very attractive alternative to liquid-liquid one [22,36,37]. It can be performed either on column or by fixing the sorbent onto a suitable membrane disk [38]. It has several advantages, for example, low solvent consumption, use in field applications and easy automation [22,39]. Solid-phase microextraction (SPME) is a modified SPE procedure based on the use of a coated fibre, usually made by fused silica, which in many cases eliminates the use of organic solvents [37,39]. Chromatographic stationary phases, such as poly (methylsiloxane), are generally used as chemically bonded coatings of the fibre [41]. SPME can be directly coupled with gas chromatography (GC) [39] and high performance liquid chromatography (HPLC) [40]. However, solid phase extraction has some drawbacks that limit its application to environmental samples, such as low recoveries due to matrix effects, low capacity for samples which have a high content of organic matter, and the need of critical calibration procedures for quantitative determinations [39]. Finally, purge and trap on activated carbon (elution with pentane, CCl_4 or diethyl ether) have also been used for the extraction of volatile organic compounds (VOCs), though they do not allow exhaustive extraction to be performed, and quantitative determination is also critical [42].

3.1.2. Sediments

Solvent extraction of organic pollutants from sediment samples can be performed by several techniques, such as manual shaking at room temperature, soxlet apparatus, sonication at room or higher temperatures, microwave assisted, and supercritical fluid extraction (SFE). Exhaustive comparisons of all these techniques have been reported in the literature [43-45]. In all cases a wetting agent should be added to the solvent mixture. Sonication at 40°- 50°C using acetone/hexane/dichloromethane mixtures and wet samples seems to be the fastest and cheapest procedure, though microwave assisted extraction has higher recoveries [46]. SFE can be considered as an alternative to the classic methods, and has recently been used in many applications [47-50].

4.2. CLEANUP OF THE EXTRACT

The cleanup procedure of organic extracts of sea water and sediment samples should be carried out very carefully since it is decisive for the performance of instrumental analysis. The main aim of this procedure is to eliminate or to reduce to as low a level as possible any other compounds which may interfere with the determination of specific analytes. Cleanup is generally performed on microcolumns packed with different sorbents which have been suitably activated (e.g.: silica/allumina, Florisil, C_{18}-CN, etc.) [1,22]. Standard solutions are generally used to select the most suitable solvent or solvent mixture and its optimum volume for selectively eluting the analytes. N-hexane and dichloromethane are the most widely used eluents [1,10,22].

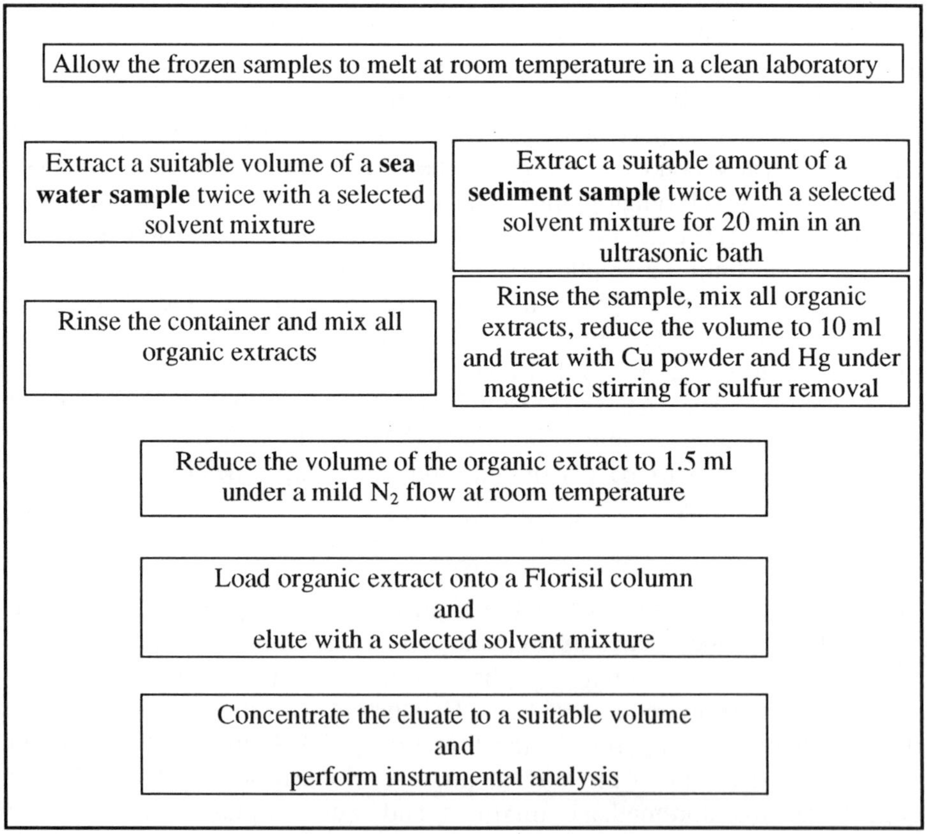

Figure 1. Typical sample preparation procedure for the determination of organic pollutants in sea water and sediment samples.

Other more efficient chromatographic techniques (HPLC, gel permeation chromatography, etc.), along with special treatments to eliminate specific interferences (e.g.: mercury and copper for sulfur removal in sediment extracts) have also been used [10]. Typical sample preparation procedures for the determination of PCBs and PAHs in sea water and sediment samples are schematically described in Figure 1.

5. Instrumental analysis

The choice of the most suitable instrumental technique depends on several factors such as the physical-chemical characteristics of analytes, the detection limit required, the level and type of interferences, the resolution needed, identification power, quantitative accuracy and precision, the availability of instrumentation, analysis time, and cost. Moreover, extraction and cleanup procedures have to be suitably matched with instrumental analysis. GC, HPLC, and supercritical fluid chromatography (SFC) coupled with a variety of detectors have been widely applied for the determination of PCBs and PAHs in organic extracts of environmental samples. These separation techniques should be considered as complementary, not competitive, since each of them allows specific and unique information to be obtained. In the following sections some of the most significant applications are briefly described for each of them.

Finally, a variety of instrumental methodologies for solving specific problems have been developed. The determination of PAHs in sediment samples by SFE coupled on-line with gas chromatography-mass spectrometry (GC-MS) [50], and the determination of non-ortho PCBs in organic extracts of sediment and sea water samples by HPLC separation and gas chromatography-electron capture detector (GC-ECD) or GC-MS analysis [1,7,51], are two examples which will be discussed in more detail.

5.1. GAS CHROMATOGRAPHY

GC on a fused silica capillary column with a mass spectrometric detector should be used whenever possible for the analysis of organic compounds with suitable volatility and thermal stability at trace level in complex mixtures. In fact, it allows the extremely high resolution of GC to be combined with the very high sensitivity and identification power of MS, which makes it possible to determine an analyte at low pg/µl levels in the final organic extract. However, GC-ECD is very common for PCB determination [1,10], since it is the most sensitive technique for chlorinated compounds. It is the most widely used for the analysis of samples with a very low PCB concentration level, such as sea water and sediment samples from Antarctica [3]. PCBs and PAHs up to 24 carbons can be separated on a 30-50 meter fused silica capillary column with 5% phenyl-methylpolysiloxane chemically bonded stationary phase [1,14]. In some cases capillary GC with conventional stationary phases is unable to separate PAH isomers whose chemical affinities and boiling points are very similar. Liquid-crystalline stationary phases can be very useful in such cases since they are able to separate isomers on the basis of their shapes and sizes [14,52]. In fact, different molecular geometries may produce different interaction with these stationary phases. A typical application is the separation of benzo[a]pyrene and benzo[e]pyrene [52].

5.2. HIGH PERFORMANCE LIQUID CHROMATOGRAPHY

HPLC has been very rarely used for PCB analysis, while it is very popular for the separation of PAHs, and it is preferred to GC in many cases [14,52]. It enables both volatile and non-volatile PAHs to be analysed, along with thermally-unstable ones, and it has been proposed in official methods by the US EPA [52]. Reversed-phase HPLC on chemically bonded C_{18} stationary phases either monomeric (separation based on the number of carbons in the molecules) or polymeric (separation based on the stereochemical structure of the molecules) has been widely used for the determination of PAHs in environmental samples [14,22]. Planar and non-planar PAHs have also been separated with this technique, and a pyrenyl-silica column has been used to separate aliphatic hydrocarbons from PAHs, and to isolate some specific PAHs which were difficult to separate by capillary GC [53]. Spectrofluorimetry is the most commonly used detection technique for PAH determination since it has greater selectivity and higher sensitivity than UV detection [14]. The improvement of the last generation interfaces between HPLC and MS, recently reported [54], has made this instrumentation very attractive, though it is very expensive.

5.3. SUPERCRITICAL FLUID CHROMATOGRAPHY

SFC is based on the use of a supercritical fluid as a mobile phase, and is a very useful alternative to HPLC since it has several advantages. In fact, the higher diffusivity of a solute in supercritical fluids than in liquids, enables the use of fused silica capillary columns with a consistent improvement in chromatographic resolution [14,55]. Moreover, supercritical fluids which are gaseous at ambient temperature and pressure facilitate the coupling of SFC with both infrared [56] and mass spectrometry [55] in respect to HPLC. Generally, capillary columns with an internal diameter of less than 100 µm, and 10-30 meter in length are employed [14]. The same variety of stationary phases that are used for the separation of PAHs and PCBs by GC can also be used in SFC, though cross-linked polysiloxane, methylpolysiloxane, and 5% phenyl-methylpolysiloxane are the most widely used [14,55]. The most popular supercritical fluid is carbon dioxide, whose major limitation is the low solubility of high molecular weight PAHs [14]. SFC has been used for PCB analysis in very few cases [49], while it has been very extensively used for the determination of volatile, non-volatile and thermally-unstable PAHs, with several detectors, including both those specific for GC and HPLC [14], e.g.: flame ionisation detector (FID) [57,58], UV [59,60] and fluorescence detectors [61,62], and MS [55,63].

5.4. SUPERCRITICAL FLUID EXTRACTION ON-LINE WITH GC-MS.

Supercritical fluids (SFs) allow the extraction of analytes from solid samples, i.e. marine sediments, to be performed faster and more efficiently since they have lower viscosity and higher diffusivity than liquid solvents [64]. CO_2 is the most widely used supercritical fluid with or without a modifier, e.g. methanol and toluene. A very exhaustive discussion on the role of a modifier in the enhancement of the extraction efficiency was recently published [47]. Few procedures have been described in the literature based on the supercritical fluid extraction of organic pollutants from environmental samples, including PCBs and PAHs [46,47,64,65,67,68]. Generally, the extraction is performed off-line of the chromatographic-detection system, i.e. GC-MS, and only a few examples of on-line SFE-GC coupling are reported in the literature [46,50,65,66]. Recoveries of 70-100% and 30-40% for PAHs with low and high molecular weights, respectively, have been reported for off-line static extraction of PAHs from certified marine sediments [47]. In this case the modifier was directly added to the sample in the extraction cell. This approach showed higher recoveries than adding the modifier to the supercritical fluid prior to extraction, and also required lower quantities of modifier itself.

The most simple way to link on-line an SFE system to a GC is based on the use of either a split-splitless [46] or an on-column injection port [65,66]. In the former, supercritical CO_2 is allowed to expand inside it, and only a small percentage of analytes is transferred into the column, according to the split ratio used. If a mass spectrometric detector (MS) is used, a split ratio as low as 1:100 should be applied. In the second case, the supercritical CO_2 expands directly into the column. The high gas flow and the accumulation of high boiling compounds may cause serious problems for the MS detector and the chromatographic column, respectively. SFE coupled on-line with a GC-MS by a suitably shaped home-made accumulation cell has also been described [50]. In particular, this system allows the effect of CO_2 flow on the MS detector, and the effect of high boiling compounds on chromatographic efficiency to be significantly reduced. It also allows the CO_2 modifier to be added directly into the extraction cell. Specifically, a three-step extraction of PAHs from harbour certified marine sediment reference material was performed at 70°C and 20 MPa, adding 20 µl of methanol as a modifier each time. After a total extraction time of 15 min (5 min for each extraction step), recoveries between 80 and 100% for all PAHs with an r.s.d. of about 10% were obtained [50].

5.5. DETERMINATION OF NONORTHO PCBs

Biological assays have shown that the most toxic PCB congeners are PCB15, PCB37, PCB77, PCB81, PCB126, and PCB169 [69,70]. These PCBs are named nonortho-PCBs, and have a planar configuration since they don't have any chlorines in the four ortho positions available on the biphenyl structure. In fact, planar configuration is mainly responsible for their very high toxicity, since it facilitates the penetration of the cellular membrane in living organisms [1,7]. The concentration of nonortho-PCBs in final extracts of environmental samples is generally lower than the detection limit of GC-MS, and GC-ECD becomes the only technique useful, but it suffers from peak overlapping between ortho and nonortho-PCBs. This makes the determination of the latter impossible without a pre-separation step. In particular, HPLC on specific stationary phases (e.g.: 2-(1-pyrenyl)ethyldimethylsilylated silica gel (CosmoSil 5-PYE) [69], and porous graphitised carbon [1,7]) has been used for this purpose. Several procedures for nonortho-PCB determination in organic extracts based on liquid-solid chromatographic separation followed by GC-ECD or GC-MS analysis have been proposed [1,7,71], and recoveries better than 80% at 0.1 ng/g concentration level in spiked sediment samples have been reported [7].

6. Data evaluation

6.1. IDENTIFICATION AND QUANTITATION

Analytical procedures are generally tested and optimised by using standard solutions and certified reference materials [72], which are also used for signal quantitation, and analytical quality control (see sect. 6.2). The relative retention time (RRT) and the relative response factor (RRF) for each analyte is calculated by using one or more internal standards (ISs), and applied for peak assignment on chromatograms of real samples. If an MS detector in the Selected Ion Monitoring mode is used, at least three ions should be selected: one is used as target and two as qualifiers. RRTs and RRFs reported in literature have also been used [73].

As far as sediment samples are concerned, it can be assumed that every particle is coated with a thin layer of organic matter, mainly humic acid, on which organic pollutants are adsorbed. This means that the total amount of organic pollutants is much more likely to be related to the particle surface area per volume unit than to the mass unit of sample [74]. The calculated specific surface area (CS) can be obtained by particle size analysis, and it is expressed in square meters per cubic centimetre of sample (m^2/cm^3). Comparisons among concentration values of organic pollutants relevant to samples with different particle size distribution, may lead to erroneous conclusions if they are expressed in a conventional way, i.e. ng/g dry weight. For example, a difference up to a factor of four was observed among the total PCB concentrations (ng/g dry weight) of sediment samples from the Ross Sea (Antarctica), which may indicate a non-uniformity in the pollution level [3]. These differences became less than 20-30% when PCB concentrations were normalised by dividing them for the calculated specific surface area of each sample. Thus, the normalized concentrations expressed in (ng/g dry weight)/(m^2/cm^3) showed that the pollution level was uniform in the observed area, as expected [3].

6.2. ANALYTICAL QUALITY CONTROL

The main goal of every procedure for analytical quality control is to allow data within assigned values of accuracy and precision to be obtained. Analytical quality control is primarily achieved by the use of certified reference materials (CRMs), and participation in intercomparison exercises, though calibration solutions, and spiked samples can also be used.

6.2.1. *Certified reference materials (CRMs)*
Certified reference materials are the most useful tool for analytical quality control [75,76]. However spiked samples might be an alternative in many cases, though it must be remembered that spiked analytes generally behave differently from native ones [77]. For a correct use of CRMs, their analysis should be scheduled within the time sequence of the analysis of real samples, and the results should be reported, for example, on a working analytical control chart [78].

6.2.2. *Intercomparison exercises*
Participation in intercomparison exercises is a unique opportunity for a laboratory to assess the quality of its analytical capability. Also, they are very useful for estimating the interlaboratory coefficient of variation for that specific analysis. These exercises are conducted using homogeneous and stable materials [78].

6.2.3. *Calibration solutions*
Calibration solutions are very useful to test an analytical procedure, but their preparation and storage are still one of the main sources of error in these analyses [78]. Certified analytes in neat form (purity higher than 99%) should be preferred for preparing calibration solutions following suitable procedures [79], although in fact commercial standard solutions in various solvents are often used [78].

7. References

1. Erickson, M.D. (1986) *Analytical Chemistry of PCBs*, Butterworth Publishers, Stoneham, MA

2. Hutzinger, O., Safe, S., and Zitko, V. (1974) *The Chemistry of PCBs*, CRC Press, Cleveland, OH
3. Fuoco, R., Colombini, M.P., Abete, C., and Carignani, S. (1995) *Intern. J. Environ. Anal. Chem.* **61**, 309
4. Tanabe, S., Hidaka, H., and Tatsukawa, R. (1983) *Chemosphere*, **12**, 277
5. Moolenaar, R.J. (1983) in R.J. Davenport and B.K. Bernard (eds.), *Advances in Exposure, Health and Environmental Effects Studies of PCBs:Symposium Proceedings*, U.S.EPA, Report No. LSI-TR-507-137B, NTIS PB84-135771, Washington D.C., p.67
6. Albaiges, J. (1993) *Environmental Analytical Chemistry of PCBs*, Gordon and Breach, Reading, UK
7. Fuoco, R., Colombini, M.P., and Samcova, E. (1993) *Chromatographia*, **36**, 65
8. Eisenreich, S.J., Looney, B.B., and Hollod, G.J. (1983) in D. Mackay, S. Paterson, S.J. Eisenreich, and M.S. Simmons (eds.), *Physical Behaviour of PCBs in the Great Lakes*, Ann Arbor Science Publishers Inc., Ann Arborn, MI, p. 442
9. Elder, D. (1976) *Marine Pollution Bullettin* **7**, 63
10. Fuoco, R. and Colombini, M.P. (1995) *Microchem. J.,* **51**, 106
11. Fowler, S.W. (1986) in J.S. Waid (ed.), *PCBs and the environment*, CRC Press, Boca Baton, Fla., vol III, p.209
12. World Health Organization (1976) *Polychlorinated Biphenyls and Terphenyls. Environmental Health Criteria 2.*, World Health Organization, Geneva
13. Wassermann, M., Wassermann, D., Cucos, S., and Miller, H.J. (1979) *Ann. N.Y. Acad. Sci.* **320**, 69
14. Vo-Dinh, T. (1989) in T. Vo-Dinh (ed.) *Chemical Analysis of Polycyclic Aromatic Compounds*, John Wiley & Sons, Inc., N.Y., p. 1
15. Harvey, R.G. (ed.) (1985) *Polycyclic Hydrocarbons and Carcinogenesis*, ACS Symposium Series 283, American Chemical Society, Washington, D.C.
16. National Academy of Science (1983) *Polycyclic Aromatic Hydrocarbons: Evaluation of Sources and Effects*, Washington, D.C.
17. Bjorseth, A. (ed.) (1983) *Handbook of Polycyclic Aromatic Hydrocarbons*, Marcel Dekker, New York
18. Grimmer, G. (ed.) (1983) *Environmental Carcinogenesis: Polycyclic Aromatic Hydrocarbons*, CRC Press, Boca Raton, Fla.
19. White, C.M. (ed.) (1985) *Nitrated Polycyclic Aromatic Hydrocarbons*, Huething Verlag, New York
20. Futoma D.J., Smith S.R., Smith T.E., and Tanaka, J. (1981) *Polycyclic Aromatic Hydrocarbons in Water System*, CRC Press, Boca Baton, Fla.
21. Badger, G., Kimber, R., and Novotny, J. (1964) *Austr. J. Chem.*, **17**, 778
22. Peltonen, K. and Kuljukka, T. (1995) *J. Chromat.*, **710**, 93
23. Murray, J. and Pottie, R. (1974) *Can. J. Chem.*, **52**, 557

24. Baum, E. (1978) *Polycyclic Hydrocarbons and Cancer*, Academic Press, New York
25. U.S. Environmental Protection Agency (1974) *Preferred Standard Path Report for Polycyclic Organic Matter*, Strategies Air Standard Division, Durham, N.C.
26. Moschandreas, D. and Zambransky J. (1980) *Environ. Intern.*, **4**, 413
27. Jacob, J. and Grimmer, G. (1991) *Fresenius J. Anal. Chem.*, **339**, 730
28. Perry, R. (1975) *Water Pol. Contr.*, **77**, 887
29. Bouloubassi, I. (1991) *Marine Pol. Bull.*, **22**, 558
30. Bouloubassi, I. (1991) *Oceanologica Actal*, **16**, 145
31. Ciusa, W. and Morgante, A. (1988) *Riv. Merceol,*, **27**, 3
32. Wise, S.A. (1993) *Fresenius J. Anal. Chem.*, **345**, 325
33. Swiffer, J. (1994) *Mutat. Res.*, **323**, 169
34. Reddy, M. (1986) *Carcinogenesis*, **7**, 1543
35. Millar, J.D., Thomas, R.E., and Schattemberg, H.J. (1981) *Anal. Chem.*, **53**, 214
36. Lintelmann,J., Waadt,K., Sauerbrey,R., and Kettrup,A. (1995) *Fresenius J. Anal. Chem.*, **352**, 735
37. Poole,S.K., Poole,C.F. (1996) *Anal. Commun.*, **33**, H15
38. Chee,K.K., Wong,M.K., Lee,H.K. (1996) *Anal. Chim. Acta*, **330**, 217
39. Zhang, Z., Yang, M.J., and Pawliszyn, J. (1994) *Anal. Chem.*, **66**, 844A
40. Chen, J. and Pawliszyn, J. (1993) *Anal. Chem.*, **67**, 2530
41. Zhang, Z. and Pawliszyn, J. (1993) *Anal. Chem.*, **65**, 1843
42. Colenutt, B.A. and Thorburn, S. (1980) *Inter. J. Environ. Anal. Chem.*, **7**, 231
43. Majors, R.E. (1996) *LC-GC Inter.*, **9**, 638
44. Lopez-Avila, V., Young, R., Benedicto, J., Ho, P., Kim, R., and Beckert, W. (1995) *Anal. Chem.*, **67**, 2096
45. Dean, J.R., Barnabas, I.J., and Fowlis, I.A. (1995) *Anal. Proc.*, **32**, 305
46. Hawthorne, S.B., Miller, D.J., and Langenfeld, J.J. (1990) *J. Chromat. Sci.*, **28**, 2
47. Langenfeld, J.J., Hawthorne, S.B., and Miller, D.J. (1994) *Anal. Chem.*, **66**, 909
48. Luque de Castro, M.D., Valcarsel, M., and Tena, M. (1994) *Analytical Supercritical Fluid Extraction*, Springer-Verlag, Berlino
49. Fuoco, R. and Griffiths, P.R. (1992) *Ann. Chim. (Rome)*, **82**, 23
50. Fuoco, R., Ceccarini, A., Onor, M., and Lottici, S. (1997) *Anal. Chim. Acta*, in press
51. Schwartz, T.R., Tillitt, D.E., Feltz, K.P., and Peterman, P.H. (1993) *Chemosphere*, **26**, 1443
52. Lee, H.K. (1995) *J. Chromatogr.*, **710**, 79
53. Wells, D.E., Echarri, I., and McKenzie, C. (1995) *J. Chromatogr.*, **718**, 107

54. Voress,L. (1994) *Anal Chem*, **66**, 481A
55. Laude, D.A., Pentoney, S.L., Griffiths, P.R., and Wilkins, C.L. (1987) *Anal. Chem.*, **59**, 2283
56. Fuoco, R., Pentoney, S.L., and Griffiths, P.R., (1989) *Anal. Chem.*, **61**, 2212
57. Fjeldstend, J.C., Kong, R.C., and Lee, M.L. (1983) *J. Chromatogr. Sci.*, **279**, 449
58. Wright, B.W., Udseth, H.R., Smith, R.D., and Nazlett, R.N. (1984) *J. Chromatogr.*, **314**, 253
59. Fields, S.M., Markides, K.E., and Lee, M.L. (1988) *Anal. Chem.*, **60**, 802
60. Yonker, C.R. and Smith, R.D. (1987) *Anal. Chem.*, **59**, 727
61. Wright, B.W., Kalinosky, H.T., Udseth, H.R., and Smith, R.D. (1986) in L.L. Stavinoha (ed.) *2nd International Conference on Long Term Storage Stabilities of Liquid Fuels*, Southwest Research Institute, San Antonio, Texas, p. 526
62. Wright, B.W., Udseth, H.R., Chess, E.K., and Smith, R.D. (1988) *J. Chromatogr. Sc.*, **26**, 228
63. Smith, R.D., Kalinosky, H.T., and Udseth, H.R. (1987) *Mass Spectrom. Rev.*, **6**, 445
64. Chester, T.L., Pinkston, J.D. and Raynie D.E. (1994) *Anal.Chem.*, **66**, 106
65. Hawthorne S.B. and Miller D.J. (1987) *J. Chromatogr.*, **403**, 63
66. Lohleit, M. and Bächmann, K. (1990) *J. Chromatogr.*, **505**, 227
67. Reindl, S. and Höfler, F. (1994) *Anal. Chem.*, **66**, 1808
68. Langenfeld, J.J., Hawthorne S.B., Miller D.J., and Pawliszyn, J. (1993) *Anal. Chem.*, **65**, 338
69. Haglund, P., Aspund, L., Jarnberg, U., and Jansonn, B. (1990) *Chemosphere*, **20**, 887
70. Jones, K.C. (1988) *Sci. Tot. Environ.*, **68**, 141
71. Lang, V. (1992) *J. Chromatogr.*, **595**, 1
72. Jacob,J. (1995) in P. Quevauviller, E.A. Maier, and B. Griepink (eds.) *Quality Assurance for Environmental Analysis*, Elsevier Science Publ., Amsterdam, Vol.17, p.563
73. Mullin, M.D., Pochini, C.M., McCrindle, S., Romkes, M., Safe, S.H., and Safe, L.M. (1984) *Environ. Sci. Technol.*, **18**, 468
74. Fuoco, R., Colombini, M.P., and Abete, C. (1994) *Intern. J. Environ. Anal. Chem.*, **55**, 15
75. Wells, D.E., Maier, E.A., and Griepink, B. (1992) *Inter. J. Environ. Anal. Chem.*, **46**, 265
76. Porte, C., Barcelo, D., and Albaiges, J. (1988) *J. Chromatogr.*, **442**, 386

77. Langenfeld, J.J., Hawthorne S.B., and Miller D.J. (1995) *Anal. Chem.* **67**, 1727
78. Fuoco, R., Colombini, M.P., Ceccarini, A. (1996) *Microkim. Acta*, **123**, 175
79. Wells, D.E., Maier, E.A., and Griepink, B. (1992) *Inter. J. Environ. Anal. Chem.*, **46**, 255

APPLICATION OF MASS SPECTROMETRIC TECHNIQUES TO THE DETECTION OF NATURAL AND ANTHROPOGENIC SUBSTANCES IN THE SEA

MARCO VINCENTI
*Dipartimento di Chimica Analitica, Università di Torino,
Via Pietro Giuria, 5 - 10125 Torino, Italy*

1. Introduction

Oceans and seas form the biggest environmental compartment on earth, which raises concern and interest at the same time for its preservation and utilization (fishing, extraction of useful materials, etc.). The marine environment itself comprises extremely different elements such as water, sediment, marine animals, aquatic plants and microorganisms. Each of these elements can be considered in turn as an environmental matrix or as a source of chemical products and/or contaminants. The comprehension of this complex system requires, besides other important tools, the execution of extensive chemical analysis, using sophisticated instrumentation such as for example mass spectrometry. For these reasons, application of mass spectrometry to marine chemistry ranges in a very broad set of analytical problems including anthropogenic pollution control, comprehension of transport and transformation phenomena for organic pollutants, structural identification of substances with potential pharmacological or toxic properties (drugs and toxins) produced by aquatic microorganisms, identification and quantification of biomarkers in marine sediments for geochemical evaluation of climate and environmental variations, characterization of seaweeds composition as well as natural polymers and biopolymers for industial exploitation.

    Mass spectrometry is a powerful analytical and spectroscopic technique that is utilized to (i) identify unknown compounds, (ii) to quantify target substances, (iii) to investigate the chemical and structural properties of specific molecules. All these applications can be carried out with extremely small amounts of material (occasionally, few femtograms) and low concentrations (one part-per-trillion) in very complex chemical mixtures.

From the point of view of marine environmental analysis, mass spectrometry combines two peculiar features, which make it an unique investigating tool, namely its extreme *sensitivity* (essential for trace and ultratrace analysis) and *specificity* (which provides information on the molecular structure of the analytes).

Further strength and analytical performance to these qualities are added by the opportunity of interfacing mass spectrometry with high performance separation techniques such as gas and liquid chromatography (GC-MS and HPLC-MS), which allow one to isolate the components of a chemical mixture from one another and to introduce them sequentially into the mass spectrometer for identification and quantification. Thus, a single GC-MS or HPLC-MS experiment produces tridimensional data (see Figure 1) and is potentially capable to identify and quantify all the components of a complex mixture of organic compounds even those present at trace level.

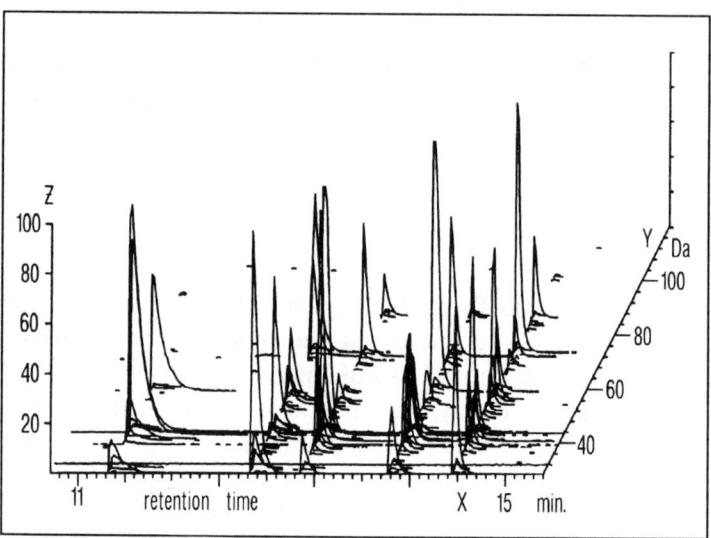

Figure 1. Tridimensional plot of a GC-MS run. The X axis of the diagram reports the chromatographic retention time, the Y axis the m/z values, which are cyclicly scanned every 0.5÷1 s, and the Z axis the abundance of each ion.

The large variety of analytical inquiries concerning marine chemistry, where mass spectrometry is widely utilized, could be arranged into two main categories, namely (i) the studies aimed to identify the chemical structure of large natural molecules (biomarkers, toxins, antibiotics, potential drugs) and materials (polysaccharides, lignin, humic acids, proteins, lipids), of which the marine biosphere supplies plentiful production and (ii) the research devoted to pollution control, mostly from products of anthropogenic origin or from mining and oil industries. The mass spectrometric approach to these two categories of analytical problems is different in many respects, as (i) applies to unknown molecules of generally high molecular weight and takes advantage of the spectroscopic properties of the technique, whereas problems (ii) are mainly devoted to analytes of known structure and low molecular weight, while they stress the analytical performance of mass spectrometry in terms of detection limits, selectivity and linear response. The purely environmental issues of marine analytical chemisty will be considered first in this presentation, since they require the basic elements of mass spectrometric instrumentation and procedures, whereas structural analysis of large biomolecules most often requires specific devices emerging from a rapidly evolving technology.

The present overview is far from being exhaustive on the topic of marine-sample analysis and is only intended to present a few selected examples taken from the scientific literature, useful to highlight the various contributions that mass spectrometry can provides in the elucidation of the marine chemistry.

2. Control of marine pollution

As already cited, the marine environment includes various compartments such as the water itself, sediments, microorganisms and a variety of superior animals and plants. In each of these compartments, specific anthropogenic pollutants are likely to be captured, so each is investigated as a matrix where the pollutant may accumulate or evolve through metabolic or abiotic processes. Analytical chemistry treats each matrix of marine origin similarly to the equivalent terrestrial matrix, except for some important differences and some specific applications. Among the latter, it is worth recalling that some shellfish are utilized in the environmental monitoring of fixed sites as they filter large volumes of water and accumulate the pollutants contained therein. On the other hand, marine water and sediment samples are different from those collected in rivers and lakes in that the dilution factor is much larger and the content of inorganic salts higher. Both elements contribute to make these matrices more difficult to be analyzed than the equivalent terrestrial samples, and emphasize the need of large volume sampling, appropriate clean-up procedures and sensitive analytical techniques. A recent work highlights the difficulties of seawater sample treatment for the analysis of polychlorobiphenyls and organochlorine pesticides at the ppq (parts-per-quadrillion) level.[1]

The subject of marine sample collection and preparation is presented in detail elsewhere in this book. In general, the analytes are exhaustively extracted from the matrix into an organic solvent, purified and concentrated to small volume, a small portion of which (i.e. 1÷25 µL) is introduced into the mass spectrometer. This sample preparation yields analytical solutions still containing a large number of compounds other than the analytes and often interfering with them. Therefore, an on-line separation device (mostly a GC or an HPLC) is requested before mass spectrometric analysis. The main components of a mass spectrometer are depicted in Figure 2.

The mass spectrometer components are kept under high vacuum conditions ($10^{-11} \div 10^{-8}$ Bar) by an appropriate pumping system (rotary plus diffusion or turbomolecular pumps), so as to prevent the ions to collide with air molecules during their flight from the ion-source to the detector. Under high vacuum conditions, a large number of organic molecules can be easily sublimed, possibly by the aid of heating. Also the substances eluted from a GC column within the mass spectrometer ion-source are maintained in the gas phase. For all the compounds that can be vaporized, the most widely ionization methods are the so-called (1) electron impact (EI) and (2) chemical ionization (CI). The highly polar or thermolabile organic compound and those with high molecular weight, which can not be vaporized without inducing their thermal decomposition, are not amenable to ionization by methods (1) and (2). These compounds can nevertheless be ionized by the so-called "desorption methods", among which (3) laser desorption (LD), (4) fast atom bombardment (FAB) and (5) electrospray (ESI) are the most frequently encountered. Electrospray ionization can also be used as an efficient interface to chromatographic techniques such as HPLC and capillary electrophoresis.

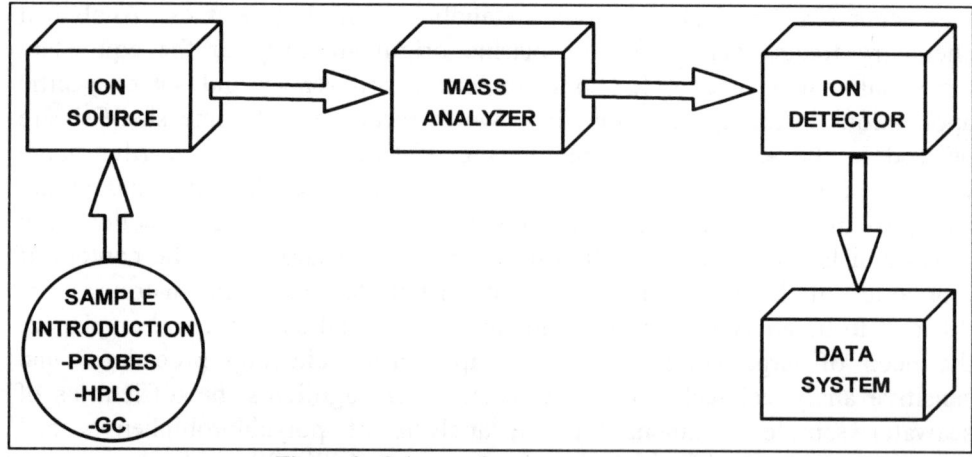

Figure 2. Scheme of a mass spectrometer.

In EI, ions are generated by a beam of electrons accelerated to 70 eV, which collide with gaseous molecules [M]. This bombardment extracts an electron from the molecule producing vibrationally excited "molecular ions" $[M^\bullet]^+$, which in turn can decompose into fragment ions and neutral species, provided that the internal energy of the molecular ion is large enough to exceed the binding energy of some chemical bond. Fragmentation processes for each compound are typically multiple, competitive and consecutive, yet perfectly reproducible even on different mass spectrometers. All the ions generated are extracted from the ion source, focused to form a beam by a series of electric lenses and then driven into the mass analyzer, that separate them on the basis of the ratio between their mass (m) and their charge (z, usually unitary). The separated ions are lastly driven in sequence to the detector which measures their abundance. The whole set of ions generated for each compound (molecular ion and fragments), together with their relative abundances, forms its mass spectrum which constitutes a fingerprint of the molecular structure.

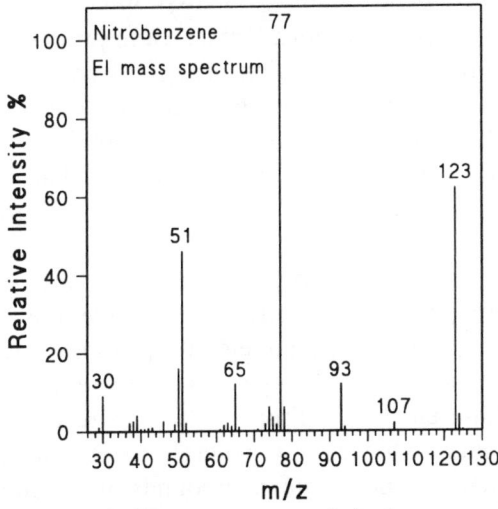

Figure 3. EI mass spectrum of nitrobenzene

For example, the EI mass spectrum of nitrobenzene (Figure 3) contains the molecular ion at m/z 123, corresponding to its molecular weight, the fragments 93 and 77, relative to the loss of NO and NO_2, respectively, and the fragment ion at m/z 30, corresponding to $[NO]^+$. Note that m/z 77 represents the phenyl ion and m/z 51, 65 are also fragments of the aromatic ring. By interpreting the mass spectrum the corresponding molecular structure can be desumed.

An interesting application of EI mass spectral interpretation to marine chemistry is provided by a paper of Kristiansen and coworkers.[2] They investigated the formation of byproducts during the potabilization of seawater. Disinfection of seawater by chlorination is often effected before driving it to the desalting process, so that the growth of microalgae in the desalting plant is prevented. However, in such conditions oxidation of bromide and iodide ions occurs, yielding products (i.e. hypobromous acid) which in turn react with the organic compounds accidentally present in the seawater. The whole process results in the formation of large amounts of a variety of bromine- and iodide-substituted hydrocarbons, both aliphatic and aromatic. All these products could by identified (and their potential risk assessed) only by interpreting the corresponding EI mass spectra, which turned out generally rich of fragments and structurally informative.

EI is the sole ionization technique that provide extensive fragmentation. For the remaining techniques (2)-(5), the molecular ion is almost uniquely produced from each analyte, yielding the molecular weight information, but not structural details. In such cases, fragmentation (and, consequently, structural information) can be induced by collisional activation in experiments of tandem mass spectrometry (see below), which requires specific instrumentation. Despite the cited limitation, CI turns out extremely useful in environmental analysis, including marine applications. CI differs from EI in that a "reagent gas" is introduced into the ion source in large excess with respect to the analytes. Upon electron bombardment, the reagent gas is primarily ionized and the ions generated thereupon undergo a series of ion-molecule collisions, due the relatively high pressure in the ion-source. These reactive collisions lastly ionize the analytes by means of proton transfer, yielding positive ions $[MH]^+$, or by thermal electron capture, resulting in negative ions $[M^\bullet]^-$. Both processes are scarcely exothermic and do not surpass the activation energy of chemical bonds, so that extensive fragmentation is generally not observed. CI positive ions are useful, as a complement to EI, when the analyte structure contain very loose bonds, that are extensively dissociated upon electron impact, so that no molecular ion appears in the EI mass spectrum and the molecular weight information is lost.

Negative ions generated by CI play a crucial role in several branches of pollution analysis. The electron capture process that generates negative ions is similar in many respects to the one occurring in an electron capture GC detector. A common feature of these two techniques is their extreme selectivity. The analytes exhibiting high electron affinity yield high response factors in electron capture experiments, whereas, in general, most matrix components do not provide any signal. Among the compounds with high electron affinity there are many common pollutants, including polyhalogenated hydrocarbons and pesticides, polycyclic aromatic hydrocarbons and their nitro-, keto- and quinone derivatives. Response factors vary by several orders-of-magnitude, depending on the chemical structure of the analyte, but intense signals are observed for the cited classes of substances even from picograms of material. Moreover, suppression of the chemical noise arising from the matrix contributes considerably to the enhancement of the signal-to-noise ratio.

To illustrate these aspects further, the case of two important classes of pollutant can be discussed, namely polychlorobiphenyls (PCB) and polychlorodibenzodioxins (PCDD). Both originate from industrial processes as undesired byproducts, are extremely toxic, persistent in the environment and rather resistent to chemical and microbial purification treatments of wastewater. Since they are lastly released into seas and oceans in high percertage, tremendous effort has been devoted in the last 20 years to their determination in marine samples. Besides water and marine sediment, PCB and PCDD were frequently determined in fish and shellfish samples, for their primary importance in the nutritional cycle.

Biphenyls have an extensive system of 6 conjugated double bonds conferring large electron affinity to the structure, which is increased further by the chlorine substituents. Therefore, polychlorobiphenyl can be determined by electron capture CI (negative ions) obtaining high sensitivity (extremely low detection limits) and selectivity (matrix suppression). In contrast, in the dioxin structure double bond conjugation is broken by the two ether bridges, so that

2,3,7,8-Tetrachlorodibenzodioxin   3,3',4,4'-Tetrachlorobiphenyl

the electron affinity of tetrachlorodibenzodioxins is approximately the same as the corresponding dichlorophenol. In practice, PCDD are not determined by electron capture techniques, but rather in EI conditions. The selectivity needed to cut the chemical noise arising from the matrix is achieved by operating the mass analyzer in the high resolution mode, i.e. by selecting the exact mass of the analyte corresponding to the dioxin chemical formula, while all the ions with different exact mass are filtered out of the detector.

High mass resolution is commonly obtained by *magnetic* mass analyzers, which, however, are by far the most expensive ones. From this fact stems the continuous research for alternative mass analyzers, capable of high resolution, as well as alternative procedures for selective detection of target analytes. High sensitivity and adequate selectivity is often obtained by running the mass analyzer (any kind) in the selective ion monitoring (SIM) mode. The principle of SIM operation is evident from the GC-MS tridimensional plot depicted in Figure 1. While the total ion chromatogram is built by adding up the signals from all the mass-scanned ions, resulting in a complex trace where all the sample components produce a peak, the profile of a single mass-value, given by the variation of a single-ion abundance with retention time, contains few peaks commonly without interferences. Instead of continuously scanning the mass analyzer over a large mass range, mass-hopping among specific ions of interest is accomplished in SIM. In this way, *selective* detection of the molecules whose mass spectra contain the chosen ions is obtained. Concurrently, the mass analyzer rests on the selected ions for a much longer time than it does when it is continuously scanned, yielding an improvement in *sensitivity* of about two orders-of-magnitude. Quite obviously, the selected analytes and their mass spectra should be known in advance, since the spectroscopic information is lost.

A wide range of mass spectrometric techniques has been exploited by Albaigés and coworkers,[3] to discriminate anthropogenic hydrocarbon pollution from natural inputs (plant waxes, marine phytoplankton, biomass combustion, etc.) in a variety of marine sediments from the Mediterranean basin. The chemical structure of each individual hydrocarbon component depends on its origin, viz. on the chemical, biological, geological and physical processes (including the industrial ones) that have contributed to produce it. Thus, these components have to be structurally identified and individually quantified, in spite of their concomitant presence in the sample extracts. The most abundant aliphatic fraction (mostly of natural and fossil origin) was determined by GC using the non-selective flame ionization detection. Polycyclic aromatic hydrocarbons (PAH) and the corresponding alkylated species were selectively detected in GC-MS by coupling the unselective EI ionization mode with the selective SIM mass-analysis, which provided suppression of the aliphatic hydrocarbon interference. Most PAH have preminent pyrolytic origin, with the exception of perylene and retene, that originate from geochemical sources, while alkyl-PAH arise predominantly from fossil sources (for example, from accidental oil spills). Polycyclic aromatic ketones and quinones are good markers for combustion processes of fossil fuels and originate mainly from urban emissions. Since they are commonly much less concentrated than the corresponding PAH, they had to be analyzed by GC-MS using a highly selective ionization technique such as electron capture CI. The mass fragmentograms for a series of polycyclic aromatic ketones are reproduced in Figure 4. Multi-component analytical determinations in a variety of marine samples (sediments, invertebrates and fish) have also been reported in several papers studying the Antarctic environment. In a typical case,[4] several laboratories have collaborated to correlate PAH, PCB and pesticide pollution to local activities, geological causes and transport phenomena. Approximately the same pollutants were monitored using the same analytical and mass spectrometric techniques as in Ref. 4, in an inventory study[5] aimed to define the sources of pollution in coastal Maine and the potential for biological impact.

Another interesting application of GC-MS to marine contamination is provided by the analysis of sediment cores at different depths from the surface to determine historical trends in anthropogenic pollution of the sea. In fact, deposited organic matter reflects the evolutional record of the sea, including the contribution of human activity. These studies are favored in the cases of seas with limited water exchange, so that the pollutants are not transported to large distances, as in lagoons, protected coves and small water depths. For example, Palm and Lammi have determined a variety of chlorophenols, chloroguaiacols and chlorocatechols in the gulf of Bothnia, to ascertain the the environmental impact of bleached pulp mill effluents in the last 35 years.[6] The selective determination of that many analytes in the complex sediment matrix was accomplished by working in SIM after EI ionization.

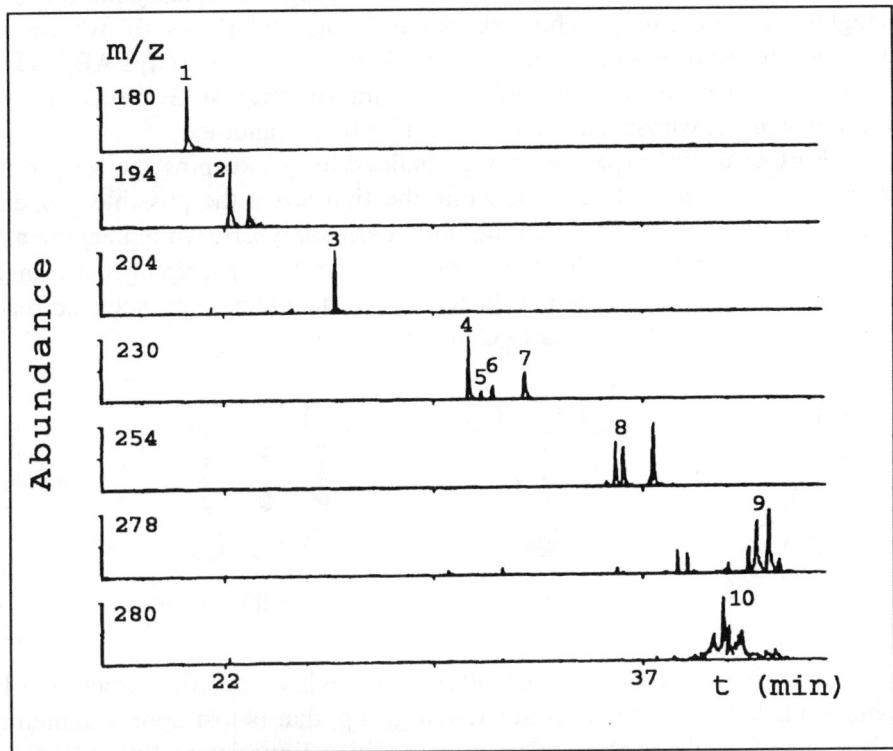

Figure 4. Mass fragmentograms of the molecular ions of polycyclic aromatic ketones by GC-MS (electron capture CI). (1) 9H-fluoren-9-one; (2) anthrone or phenanthrone; (3) 4H-cyclopenta[def]phenanthren-4-one; (4) benzo[a]fluoren-9-one; (5) benzo[c]fluoren-9-one; (6) benzo[b]fluoren-7-one; (7) 7H-benz[de]anthracen-7-one; (8) benzopyrenones; (9) indenopyrenones; (10) dibenzofluorenones. Reprinted with permission from *Environmental Science & Technology*, **1996**, *30*. 2495, by I. Tolosa, J. M. Bayona and J. Albaiges. © 1996 American Chemical Society.

Extreme selectivity can be achieved by experiments of tandem mass spectrometry (MS/MS). This technique usually requires instruments carrying two consecutive mass analyzers between which a collision cell is interposed. This configuration do not introduce any limitation in the choice of the introduction device (i.e. HPLC-MS/MS) nor in the ionization method. In the easiest MS/MS experiment, the first mass analyzer remains fixed and selects a unique ionic species, for example the molecular ion of a certain analyte of interest. This species is driven in a volume (the collision cell) with a relatively high pressure of an inert gas, upon which the selected ionic species collides one or more times. Owing to these collisions, part of the impinging ion kinetic energy is converted into internal energy, resulting in activation of various fragmentation processes. The resultant fragment ions are separated by the second mass-analyzed and lastly detected, yielding a secondary mass spectrum (daughter ion spectrum). This experiment is particularly useful whenever the ionization technique needed do not provide fragmentation (CI, FAB, LD, ESI) and when it is necessary to eliminate the interference of GC peaks coeluting with the analyte, whose mass spectra overlap to one another.

Further useful experiments are obtained by (ii) keeping the second mass-analyzer on a selected fragment, while the first scans the possible progenitors (parent ion spectrum); (iii) scanning both mass-analyzers, while keeping a fixed mass-difference between them (neutral loss scan); (iv) keeping both analyzer fixed or mass-hopping them on selected parent-daughter pairs (selected reaction monitoring, SRM). According to Cooks notation:

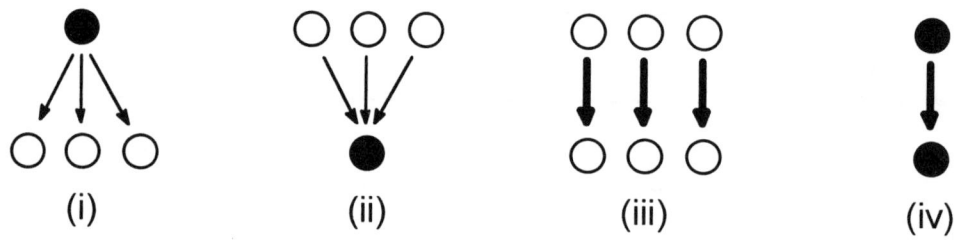

Experiments (ii) and (iii) allow the highly specific selection of the compounds having a common functional group, that is lost upon fragmentation either as an ion (ii) or as a neutral species (iii). Experiment (iv) (SRM) is the MS/MS analogue of SIM and provide high sensitivity and unsurpassed selectivity at the same time.

The most widely used application of MS/MS to environmental monitoring is devoted to the determination of PCDD in extremely complex matrices, such as for example marine sediments. The parent-daughter transition followed in this case is the consecutive loss of a Cl• radical and a CO molecule from the molecular ion, resulting in a net mass loss of 63 Da. PCDD with any chlorine content, as well as positional isomers, can all be selectively detected by a single neutral loss scan (iii), where the mass difference between the analyzers is fixed at 63 Da. Higher sensitivity can be obtained by mass-hopping the analyzers over the expected PCDD signals, instead of continuously scanning them, while keeping the same mass difference between the first and the second analyzer (experiment iv). In the absence of major interferences, MS/MS analysis of PCDD offers similar performace as high resolution MS, but MS/MS proved to provide much more accurate quantitative data than HRMS whenever greater selectivity is needed.[7] Application of MS/MS to the chemical structure elucidation of molecules of marine origin will be discussed herein after.

Skilful use of several GC and MS techniques, including chiral chromatography, electron capture CI and EI-MS/MS (multiple reaction monitoring), was accomplished by Buser and Müller[8] to determine a variety of toxaphene components in tissue extracts of aquatic vertebrates (herring, salmon, seal and penguin).

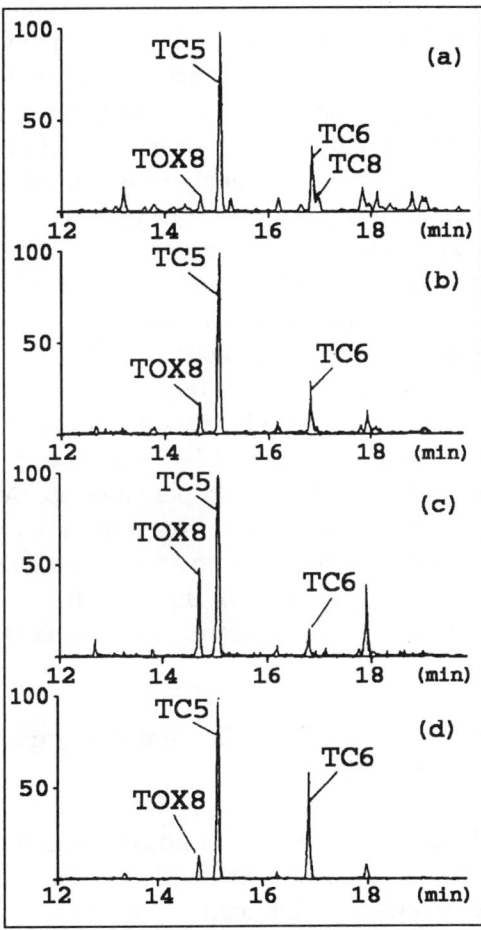

Toxaphene is an organochlorine insecticide produced by the chlorination of camphene and consists of a complex mixture of products such as polychlorobornanes (see above). By using GC-MS/MS, instead of GC/MS, higher selectivity toward individual isomers was obtained, provided that isomer-specific parent-daughter transitions could be singled out. Thus, much simpler chromatograms could be observed, from which isomer recognition was easily derived (Figure 5).

Figure 5. SRM chromatograms of the fragmentation from 376^+ to 280^+ of octachlorobornanes in various samples. (a) technical toxaphene; (b) Baltic herring; (c) Arctic seal; (d) Antarctic penguin. Reprinted with permission from *Environmental Science & Technology*, **1994**, *28*. 119, by H.-R. Buser and M. D. Müller. © 1994 American Chemical Society.

Another class of pollutants frequently determined in seawater are the "volatile organic compounds" (VOC), a definition usually referred to a list of 59 EPA-regulated substances of suspected cancerogenic risk. They are usually present at ppb level in seawater samples and, consequently, require preconcentration before the analysis by conventional techniques, such as GC-MS in the EI mode. This preconcentration is most effectively performed by the purge-and-trap technique, where VOCs are purged from quite large volumes of water by an inert gas and trapped into a cartridge containing an appropriate sorbent. Subsequently, the cartridge is heated and backflushed by the gas, releasing the VOCs, which are then cryofocused in the GC injector and analyzed. Even though the whole process assures very low detection limits and accurate quantitation, it is nevertheless time-consuming. A new and promising approach, that considerably shorten the analysis time by cutting out the GC separation stage, is membrane introduction mass spectrometry combined with on-line flow injection analysis. The method proved not to be affected by high saline content and particulate matter typical of marine water samples.[9] The technique makes use of a semipermeable membrane, acting as an interface between the flowing sample solution and the vacuum system of the mass spectrometer. The analytes dissolved in seawater selectively pass through the non-porous membrane by a process of pervaporation. This implies the analyte adsorption to the external side of the membrane, diffusion through the membrane structure, and vaporization of the analyte from its internal surface into a low-pressure area of the mass spectrometer such as the ion-source region. For most VOCs, detection limits are in the low part-per-trillion range, much below the threshold requested by regulatory laws. The combination of microporous membranes and flow injection analysis allowed the VOC determination on 12-15 seawater samples per hour, without any preconcentration or sample work up. Mass spectrometric analysis is generally performed in the electron impact mode under SIM conditions. The mass analyzer is cyclicly set over selected m/z values, corresponding to abundant and characteristic ions of the analytes of interest. In the cases when matrix interferences are present, MS/MS analysis may possibly be required.

3. Structural identification of organic and biological compounds in marine matrices

As mentioned above, besides pollution control, organic mass spectrometry can also be utilized strictly for the structural analysis of more or less purified substances. Applications of this sort to marine chemistry ranges from genuinely environmental (mostly aimed to the comprehension of environmental chemistry, biochemistry and geology) to applied biochemistry, in which seas and oceans are merely regarded as source of useful biochemical products. The two domains cannot be separated neatly, as any practical utilization of a biomolecule implies the comprehension of its biological response and the mechanism of action. However, a large number of mass spectrometric investigations approach these biomolecules making no difference on whether they originate from marine organisms or from the terrestrial biosphere. Thus, the latter applications will be considered only briefly in the present overview.

At the intersection of chemistry, paleontology and geology lies the so-called molecular stratigraphy. The basics is that molecular organic components of a marine sediment are strictly related to the bio-environment existing at the time the sediment was laid down. Thus, stratified sediments preserve a chemical record of the variations in the biological species distribution and their evolution during geological periods.[10] The conceptual approach is similar to stratigraphic analysis conducted for evaluating the historical fluctuation of marine pollution,[6] except for the much larger time scale and the different analytes, which are biomarkers instead of pollutants. Biomarkers are generally hydrocarbons whose structure depends not only from the organism that generated them, but also from the environmental conditions under which they were produced (oxic vs. anoxic conditions, redox potential and acidity of the sediment, temperature).

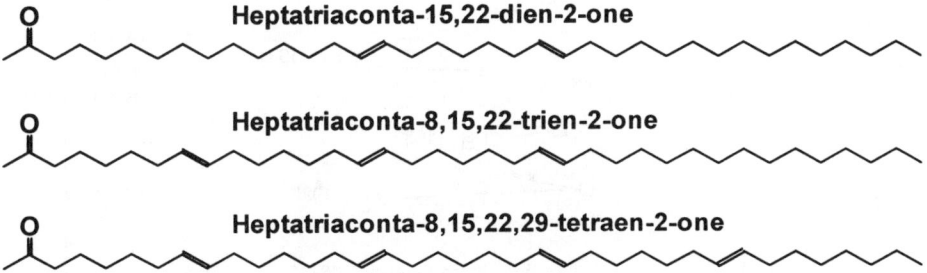

For example, several alkenones (see above) were biosynthesized by an ubiquitous unicellular marine alga. It has been found that the relative abundance of the various alkenones, differring from one another in the number and position of double bonds, depended on the habitat temperature at which the algae lived and produced their biosynthetic metabolites. Thus, the biomarker distribution in the stratified sediment represent a record of the ocean temperature during geological periods. Correlation of climatological perturbation of the ocean temperature (for example, the well-known El Niño events) with the relative abundance of the most unsaturated alkenones has been established (Figure 6).

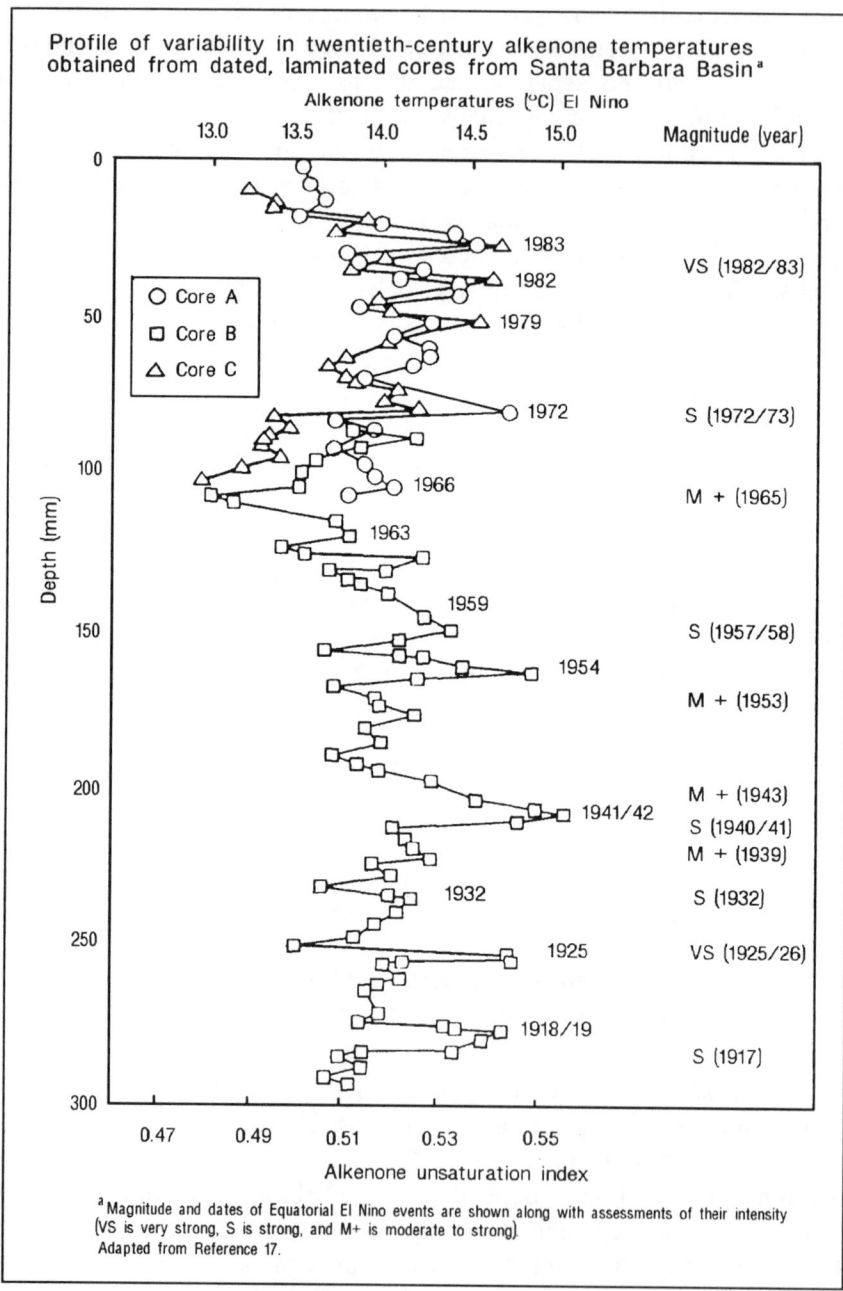

Figure 6. Caption inside the figure. Reprinted with permission from *Environmental Science & Technology*, **1993**, 27. 29, by S. G. Wakeham. © 1993 American Chemical Society.

From an analytical point of view, the various alkenones have to be identified, separated and quantified in a complex matrix. All these tasks can be accomplished by the use of mass spectrometric techniques. In a GC-MS instrument, effective product separation is performed by capillary GC, while EI mass spectra allow one to determine univocally the presence of the keto-group and its position, as well as the number of double bonds present in the alkenyl chain. Although EI mass spectra do not provide unequivocal information on the double bond position, the location of multiple double bonds on any hydrocarbon chain can either be determined by MS/MS by activating "remote-site fragmentations"[11] or even by EI-MS after appropriate derivatization of the double bonds.[12] For example, Curtis and coworkers used two different methods to determine the double bond position in the fatty acids liberated from a complex mixture of sulfoquinovosyl diacylglycerols produced by a marine microalga.[13] They first completed a reductive cleavage of the sulfoquinovosyl diacylglycerols, obtaining the corresponding alcohols, that were derivatized to form nicotinate esters. From EI mass spectra of these derivatives the double bond position could be determined. In the alternative method, the starting material was directly analyzed in a MS/MS intrument. The first analyzer was used to select the molecular ion of each mixture component, which was collisionally activated, so as to produce daughter ion spectra containing structurally informative remote-site fragments.

Biomarkers are also extremely important in petroleum geochemistry, because they indicate the ancient biological origin of organic matter that has been converted into fossil oil. This is of course of vital significance for oil industries, but biomarker characterization has practical interest also for the analysis of marine environment. For example, Wang and coworkers were able to detect, and inequivocally attribute, residues of contamination from an oil spill (Arrow spill, February 1970) occurred 22 years before sediment sampling.[14] This attribution was made possible by comparing the specific biomarkers (triterpanes and stearanes) present in the oil spilled with those recovered from marine sediment samples. Biomarker patterns are resistent to weathering and degradation, and are very specific for each oilfield. This mixture of complex hydrocarbons has to be characterized by GC-MS under EI conditions and collecting complete mass spectra.

Marine materials of wide interest are humic and fulvic acids. Analytical and environmental chemists have started investigating their structure, but this tasks appears to be singularly arduous, due to both the polymeric nature of humic acid framework and the extreme variability of its "monomeric" constituents. Thus, it is impossible to isolate individual humic acid components, unless its complex structure is extensively broken up by chemical, thermic or biological treatments, so as to form small molecular units. Mass spectrometric investigations on humic and fulvic substances follow the same criteria. Either molecular weight distributions are assessed on the untreated materials, or individual building blocks are identified following some form of degradation. Rice and coworkers were able to determine the molecular weight distribution of five different fulvic acid materials by using laser-desorption MS.[15] As mass spectrometry provide a direct measurement of molecular masses, it is relatively safe from bias (Figure 7).

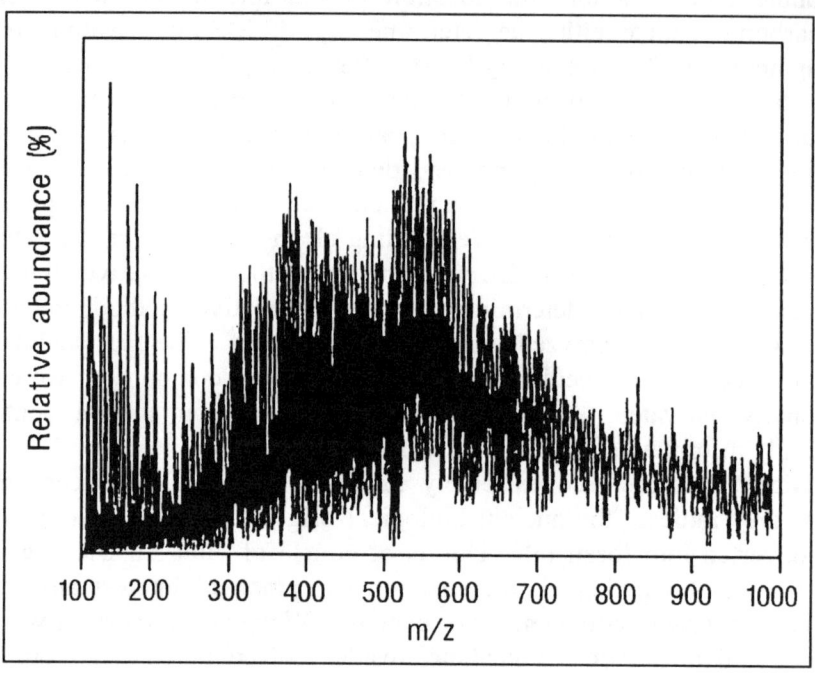

Figure 7. Positive-ion laser desorption mass spectrum of fulvic acid sample. Only the portion of the mass spectrum from 100 to 1000 Da is presented. Reprinted with permission from *Environmental Science & Technology*, **1995**, *29*. 2464, by F. J. Novotny, J. A. Rice and D. A. Weil. © 1995 American Chemical Society.

Thus, the Authors could demonstrate that the alternative analytical methods, namely gel filtration chromatography and vapor pressure osmometry, largely overestimated the average molecular weight of fulvic acids, as a consequence of the lack of suitable calibration standards, as well as the polydispersity factor and the non-ideal behavior of fulvic acids. In LD-MS, a layer of sample is deposited onto an appropriate matrix, capable of absorbing the photons emitted by a laser pulse and to transfer part of the corresponding energy to the analyte. Alternatively, the sample itself absorbs the photons. As a consequence of this energy absorption, intermolecular bonds are broken and the analyte is desorbed into the high vacuum of the mass spectrometer, partly in its protonated (charged) form. These ions can then be mass-analyzed and detected. Given the pulsed nature of the laser shot, the fast transient analytical signal has to be intercepted by a mass-analyzer provided for the purpose, either a "time-of-flight" or an "ion-cyclotron resonance Fourier transform" analyzer, not a traditional magnet or quadrupole.

Another way to examine humic substances is by pyrolysis-GC/MS.[16] The subject has been recently reviewed.[17] Analytical pyrolysis is a controlled thermal degradation method that endeavours to infer the chemical structure of a material from its pyrolysis products. Since these products are generally multiple, in low amount and rather volatile, they are typically identified by GC/MS in the EI ionization mode. Thus, while the analytical characterization of pyrolysis products presents little problems, major interpretation puzzles remains unanswered about the mechanism of product formation during the thermal degradation process. As a matter of fact, pyrolysis often induces dramatic modification of the original building blocks present in the humic substance, which lead to false deduction on its original structure.

Further information on the chemical structure of humic substances may be gained by studying the products formed from chemical degradation (for example, hydrolysis) of the original material, but also from this approach analogous interpretation problems arise as from analytical pyrolysis. A type of study with a different perspective, and direct consequences on public health, is aimed to find out the products that are possibly released by humic substances as the result of a chemical treatment of waste or drinkable water.[18]

A variety of other substances and materials from marine biota are susceptible of refined mass spectrometric analysis. These include polysaccharides, glycopeptides, sphingo- and phospholipids, glycosides, oligonucleosides, proteins, polyether macrocycles and steroids. The complexity of their structure can range from quite simple to very complicated, as is the case for maitotoxin (depicted below), one of the toxins responsible for seafood poisoning produced by the dinoflagellate *Gambierdiscus toxicus*.[19,20]

MAITOTOXIN

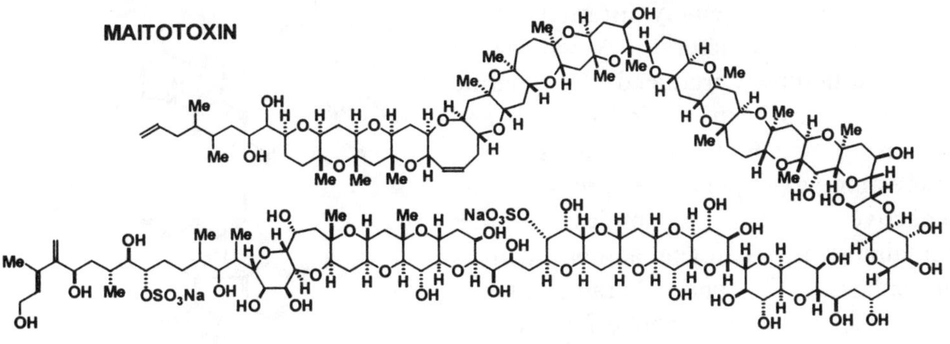

In general, the structure of a large number of polyether marine toxins are rapidly being elucidated.[21] To determine the structure of these molecules a combination of several analytical techniques has to be employed, including mass spectrometry,[22] though not in the most prominent role. In contrast, mass spectrometric analysis is decisive whenever the availability of the sample is extremely limited and the toxin structure is simpler, as is the case for DTX-4 and DTX-5, water-soluble diarrhetic shellfish poisoning toxins, produced by *Prorocentrum* dinoflagellates (Figure 8).[23]

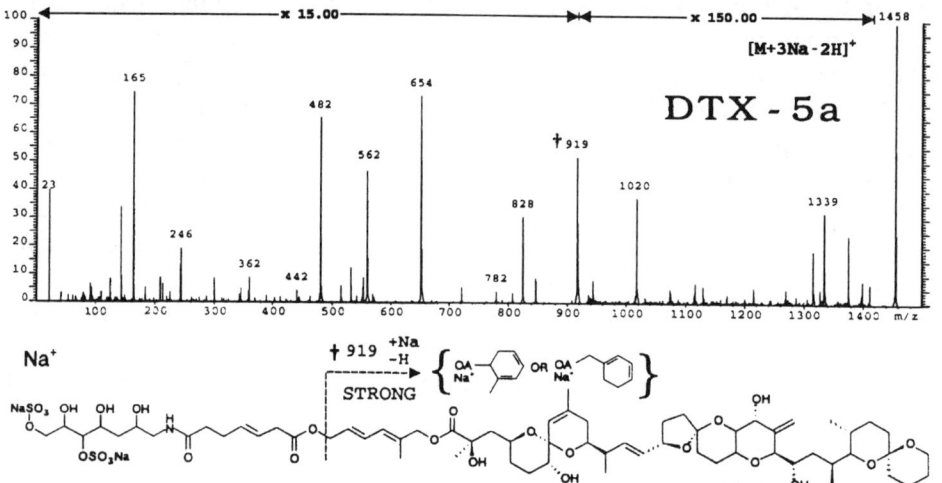

Figure 8. LSIMS MS/MS daughter ion spectrum of [M+3Na–2H]$^+$ of DTX-5a. Reprinted by kind permission of the Authors. *Proc. 44th ASMS Conference on Mass Spectrom. & Allied Topics*, **1996**, p. 884, by J. M. Curtis, T. Hu and J. L. C. Wright.

Recently,[24] fast-atom bombardment (FAB or LSIMS) was used to induce the desorption of intact cationized molecular ions from the surface of a DTX solution, and fragmentation was produced by collisional activation in tandem mass spectrometry. The resulting MS/MS spectrum (Figure 8) contains several fragment ions extremely informative on the molecular structure. A variety of other structures produced by marine organisms has been studied by low and high resolution FAB/MS. Among these, interesting examples are provided by bioactive cyclic peptides, often containing also halogen atoms, such as orbiculamides[25] and keramamides[26], produced by a marine sponge *Theonella*.

Despite the capability of FAB to produce intact molecular ions from quite large biomolecules, the most powerful tool to accomplish this job has become, without doubt, electrospray mass spectrometry (ESI-MS). Although this technique has been developed very recently, already thousands of applications has been published, mainly dealing with the structural characterization of biomolecules with molecular weights up to 100,000 Da. Among these, quite rare are the ones taking into account products of marine origin. This stems from reasons of budgets and delayed availability of the technique. However, also this sector of life science is likely to be strongly influenced by the expansion of ESI/MS in the near future, since the biological substances arising from marine microorganisms, animals and plants are structurally analogous to the ones from the terrestrial biosphere, that have been subjected to extensive ESI-MS analysis. Recent examples of application of ESI/MS (and the related ion-spray and atmospheric pressure chemical ionization techniques) to marine chemistry were addressed to the determination of shellfish toxins,[27-29] the identification of chlorophyll metabolites in zooplankton[30] and the detection of chitin in algae.[31]

ESI can either be used to directly introduce analyte solutions into the mass spectrometer or it is used as an efficient interface between chromatographic systems such as HPLC or capillary electrophoresis and the mass spectrometer. In ESI, small volumes of the sample solution (mostly aqueous) pass through a stainless steel hypodermic needle while a strong electrical field charges the surface of the emerging liquid to form a fine spray of multiply charged droplets. These droplets migrate to a capillary tube through a desolvating chamber. A countercurrent of dry nitrogen often helps evaporate the solvent from the droplets and remove the neutral species. Solvent evaporation also increases the charge density on the droplets, which experience explosion into smaller droplets. Free expansion of the gas drives the multicharged droplets though the capillary tube, a series of vacuum stages and lastly into the mass spectrometer. At this stage, most droplets have been dried up to multi-charged molecular ions that can be mass-analyzed and detected. The efficiency of the whole process is so high that useful mass spectra can be obtained from as little as few femtograms (10^{-15} g) of analyte, quite a unique feature for the analysis of marine products (i.e. mixtures of toxins) of very limited availability.

Since ESI produces essentially multicharged molecular ions, its analytical performance is enormously enhanced when it is connected to an instrument with MS/MS capabilities. In fact, the molecular weight information alone is of little help in the structural characterization of a complex biological molecule. On the other hand, fragmentation of the molecular ion upon collisional activation is favored by its multiple charge, as the dissociation process leads to charge separation and the production of two complementary fragment ions, instead of one fragment ion and one undetectable neutral species.

4. Conclusions

A variety of studies regarding the marine chemistry have been presented, with different aims and adopting different procedures, but having in common the use of a mass spectrometer as a mean to provide accurate and reliable analytical answers to the questions raised by each study. Despite the apparent multiplicity and large variety of the investigations showed, these represent a small fraction of the inquiries concerning marine chemistry and susceptible of mass spectrometric investigation. In general, the marine environment is by far the less extensively studied among the environmental compartments and very little is actually known on the complex chemistry of the oceans. Much larger investments have been addressed to the comprehension of compartments with supposed more relevant impact on the human health, as, for example, groundwater and atmosphere. In this context, an "expensive" technique such as mass spectrometry has found relatively little application in the investigation of the marine environment.

Even if the oceans are not part of any single nation, it is emerging in several countries an increased awareness of the importance of their preservation. The need of analytical control of marine pollution, together with the generalized price cut of mass spectrometric instrumentation and the development of ionization methods capable of analysing molecules of increasing molecular weight and polarity, opens the way to a progressive diffusion of mass spectrometric instrumentation in schools of marine science and in oceanographic laboratories.

The recent introduction of compact and portable[32] (tandem) mass spectrometers makes this instrumentation compatible with field analysis and laboratories installed on ships. In this perspective, analysis of increasing sensitivity, specificity and structural information can be obtained from instruments requiring progressively smaller investment, space and power supply. Thus, mass spectrometry is likely to become, in a very short time, a leader analytical technique also in marine science.

5. References

1) Kelly, A.G., Cruz, I. and Wells, D.E. (1993), *Anal. Chim. Acta* **276**, 3.
2) Kristiansen, N.K., Frøshaug, M., Aune, K.T., Becher, G. and Lundanes, E. (1994), *Envir. Sci. Technol.* **28**, 1669.
3) Tolosa, I., Bayona, J.M. and Albaiges, J. (1996), *Envir. Sci. Technol.* **30**, 2495.
4) Kennicutt II, M.C. et al. (1995), *Envir. Sci. Technol.* **29**, 1279.
5) Kennicutt II, M.C. et al. (1994), *Envir. Sci. Technol.* **28**, 1.
6) Palm, H. and Lammi, R. (1995), *Envir. Sci. Technol.* **29**, 1722.
7) Charles, M.J., Green, W.C. and Marbury, G.D. (1995), *Envir. Sci. Technol.* **29**, 1741.
8) Buser, H.-R. and Müller, M.D. (1994), *Envir. Sci. Technol.* **28**, 119.
9) Kasthurikrishnan, N. and Cooks, R.G. (1995), *Talanta* **42**, 1325.
10) Wakeham, S.G. (1993), *Envir. Sci. Technol.* **27**, 29.
11) Adams, A. (1990), *Mass Spectrom. Rev.* **9**, 141.
12) Vincenti, M., Guglielmetti, G., Cassani, G. and Tonini, C. (1987), *Anal. Chem.* **59**, 694.
13) Keusgen, M., Curtis, J.M. and Ayer, S. W. (1996), *Lipids* **31**, 231.
14) Wang, Z., Fingas, M. and Sergy, G. (1994), *Envir. Sci. Technol.* **28**, 1733.
15) Novotny, F.J., Rice, J.A. and Weil, D.A. (1995), *Envir. Sci. Technol.* **29**, 2464.
16) Peulve, S., De Leeuw, J.W., Sicre, M.-A., Baas, M. and Saliot, A. (1996), *Geochim. Cosmochim. Acta* **60**, 1239.
17) Saiz-Jimenez, C. (1994), *Envir. Sci. Technol.* **28**, 1773.
18) Kanniganti, R., Johnson, J.D., Ball, L.M. and Charles, M.J. (1992), *Envir. Sci. Technol.* **26**, 1998.
19) Murata, M., Naoki, H., Matsunaga, S., Satake, M. and Yasumoto, T. (1994), *J. Am. Chem. Soc.* **116**, 7098.
20) Sasaki, M., Matsumori, N., Maruyama, T., Nonomura, T., Murata, M., Tachibana, K. and Yasumoto, T. (1996), *Angew. Chem. Int. Ed. Engl.* **35**, 1672.
21) Yasumoto, T. and Murata, M. (1993), *Chem. Rev.* **93**, 1897.
22) Naoki, H., Murata, M. and Yasumoto, T. (1993), *Rapid Commun. Mass Spectrom.* **7**, 179.
23) Hu, T., Curtis, J.M., Walter, J.A. and Wright, J.L.C. (1995), *J. Chem. Soc., Chem. Comm.* 597.
24) Curtis, J.M., Hu, T. and Wright, J.L.C. (1996), *Proc. 44th ASMS Conference on Mass Spectrom. & Allied Topics*, Portland, OR, May 12-16, p. 884.
25) Fusetani, N., Sugawara, T. and Matsunaga, S. (1991), *J. Am. Chem. Soc.* **113**, 7811.
26) Kobayashi, J., Itagaki, F., Shigemori, H., Ishibashi, M., Takahashi, K., Nagasawa, S., Ogura, M., Nakamura, T., Hirota, H., Ohta, T. and Nozoe, S. (1991), *J. Am. Chem. Soc.* **113**, 7812.
27) Quilliam, M.A. and Ross, N.W. (1996), *ACS Symp. Ser.* **619**, 351.
28) Quilliam, M.A. (1996) *J. Chromatogr. Libr.* **59**, 415.
29) Draisci, R., Lucentini, L., Giannetti, L., Boria, P. and Stacchini, A. (1995), *Toxicon* **33**, 1591.
30) Harris, P.G., Carter, J.F., Head, R.N., Harris, R.P., Eglinton, G. and Maxwell, J.R. (1995), *Rapid Commun. Mass Spectrom.* **9**, 1177.
31) Shahgholi, M., Ross, M.M., Callahan, J.H. and Smucker, R.A. (1996), *Anal. Chem.* **68**, 1335.
32) McDonald, W.C., Erickson, M.D., Abraham, B.M. and Robbat, A. Jr. (1994), *Envir. Sci. Technol.* **28**, 336A.

TECHNIQUES OF EXTRACTION AND ANALYTICAL METHODS FOR HUMIC SUBTANCES IN SEA WATER AND SEDIMENTS

B.M. PETRONIO
  Department of Chemistry, University of Rome "La Sapienza"
  Piazzale Aldo Moro 5, 00185 Rome, Italy

Humic substances are a general class of biogenic, refractory, yellow black organic substances with various high-molecular weight that are present in all terrestrial and aquatic environments.

1 Recovery of humic compounds from water

Aquatic humic substances may be found in groundwater, river water, seawater, lakes, marshes, bogs and swamps and are the major component of the hydrophobic acid fraction of dissolved organic carbon, according to the division of Leenheer [1] in six fractions: hydrophobic acids, bases and neutrals, and hydrophylic acids, bases and neutrals.

  Humic compounds are present in concentrations from 20 µg/L in groundwater to 30 mg/L in surface water [2], from 150 µg/L to 200 µg/L in open ocean and from 400 µg/L to 800 µg/L in productive coastal water [3], 90% of which are fulvic acids.

  Because of their low concentration, to obtain sufficient amount of humic compounds, free both of inorganic salts and low-molecular weight organic acids, and in a stable form, it is generally necessary to process large volumes of water and employ in combination different techniques.

  The recovery procedure can be summarized in the following four steps:
    - *filtration*
    - *concentration*
    - *isolation*
    - *preservation*

1. Filtration
The filtration is the slowest step in the recovery procedure of aquatic humic compounds. It is employed to separate aquatic humic substances into dissolved and particulate fractions [4]; the particulate consists of suspended organic compounds and suspended clay minerals. Clay minerals are similar to humic compounds in many properties such as exchange capacity, metal uptake, etc. and their presence hinders both the characterization of humic substances and the determination of the metal complexing ability [5]. Besides, clay mineral colloids can interfere in

the concentration process by sorption method: clay colloids can clogged the pores of the resin, consequently hydrous oxides of aluminum, iron and manganese can be sorbed on.

The amount of particulate organic carbon depends on the type of aquatic sample: in sea water ranges from 0.7% of the total organic carbon (TOC) in the North Central Pacific to 24% in the Artic Ocean [6], in rivers of United States ranges from 7% to 24% of TOC [7].

Generally 0.45 µm pore size filters are used, even if colloidal species are of approximate size range of 0.001-1.0 µm [6]. The employment of 0.1 µm pore size filter, as suggested by Kennedy et al.[8], induces decrease in flow rate, that is disadvantageous when large volumes of water are to be filtered. The 0.45 µm filter is thence the compromise between the flow rate and the clay mineral removal.

It is to point out that flow rate drops when samples with high particulate concentration or large volume of sample are filtered, but the effect on dissolved organic compounds is not significant for water with low dissolved organic carbon [9].

A number of filter types with 0.45 µm pore size varying in chemical composition, uniformity of pore size, flow characteristics are used, and the advantages and disadvantages of the different matrices are reported in Table 1 [10].

TABLE 1 Advantages and disadvantages of different filter matrices

Matrix type	Advantages	Disadvantages
Silver membrane	Uniform pore size Bactericidal properties	Slow flow characteristics Expensive Slight sorption of certain organic compounds
Glass fiber	Good flow characteristics	Particles larger than nominal pore size can pass through filter Slight sorption of certain organic compounds
Organic membrane: cellulose acetate cellulose nitrate	Uniform pore size Economical	Wetting agents contaminate filtrate Sorption of organic compounds to matrix

The considered filters can be subdivided in membrane filters (silver- membrane filters, cellulose-acetate and cellulose-nitrate filters), having uniform pore sizes, and depth filters (glass-fiber filters) with not uniform pore sizes. In this case it is defined a nominal pore size, given to the particle size retained to a predetermined percentage (usually given as 98% retention). Silver membrane filter are recommended for the isolation of particulate organic carbon [11] and for aquatic humic substances, considering that organic membrane filters sorbed 15% of humic compounds present in the water sample, glass-fiber filter 5% and silver-membrane filters 3% [12]. As reported by Cranston et al. [12], filtering time for 47 mm diameter filters increased in the order:

glass-fiber filter < silver-membrane filters < organic-membrane filters

Substantial differences exist between different silver filters due to the number of pores per unit area that are not observed with the other filters.

2 Concentration

Numerous methods, based on different techniques, have been developed to concentrate dissolved humic substances, and different sorbents have been employed in the column chromatographic methods. In any case during the concentration process also the removal of inorganic or organic compounds occurs.

Aiken has evaluated the more used methods[10], that are summarized in Tables 2 and 3 with the respective advantages and disadvantages

TABLE 2 Methods commonly used to concentrate aquatic humic substances

Method	Advantages	Disadvantages
Vacuum distillation	Low temperature	All solutes concentrated
Lyophilization	Mild High concentration factors Sample taken to dryness	Method is slow All solutes with the exception of volatiles are concentrated
Freeze concentration	Mild Inexpensive Simple	Method is slow All solutes concentrated
Reverse osmosis	Ambient conditions, mild Large volumes can be processed	All solutes concentrated Efficiency dependent on concentration
Coprecipitation	Inexpensive Effective for waters high in DOC	Efficiency dependent on initial DOC Inefficient on large volumes of water Isolated organic matter must be separated from inorganic salts
Ultrafiltration	Organic solutes fractionated by molecular size Large volumes can be processed	Interaction with membrane possible Fouling of membrane possible
Solvent extraction	Inorganic salts effectively escluded	Humic insoluble in many solvents Method is slow

TABLE 3 Sorption methods commonly used to concentrate aquatic humic substances

Method	Advantages	Disadvantages
Alumina	Organic acids readily sorbed to basic adsorbent Mild eluents Efficient adsorption	Insufficient desorption Structural alteration of organic matter possible
Nylon and polyamide powder	Efficient adsorption	Irreversible sorption probable
Carbon	Inexpensive Simple procedure Large volume of water can be readily processed Organic blanks are low	Irreversible sorption possible Slow elution rates Slow sorption rates with high-molecular weight species Chemical alteration of organic solutes possible
Anion exange: Strong-base resins	Method is simple Large volumes can be processed High capacities for macroporous resins	Irreversible sorption probable Fouling of resins possible Resin bleed All anions concentrated
Weak-base resins on amphoteric matrix	Method is simple Large volumes can be processed High capacities for macroporous resins Efficient desorption Inorganic salts removed	All organic anions concentrated. Humic substances must be isolated from hydrophilic acids Extensive cleanup of resins required Resin bleed Desorption with NaOH
Non ionic macroporous sorbents	Method is simple Resins easily regenerated Large volumes can be processed High capacities Efficient desorption of acrylic ester resins	Irreversible sorption possible on styrene divynil benzene resins Desorption with NaOH. Precaution required to prevent oxidation of humic substances Resin bleed pH adjustment to pH 2 prior to adsorption

Among these methods, sorption on weak anionic exchange resins, such as phenol-formaldheyde resin (Dualite A-7), that combines weak-base secondary-amine functional groups with the relatively hydrophilic phenol-formaldehyde matrix, or diethylamino ethyl cellulose (DEAE-cellulose) [13,14] and on non ionic macroporous resins, such as Amberlite XAD at low pH values, has beee shown to be the most convenient method for isolation of aquatic humic substances (AHS) from water.

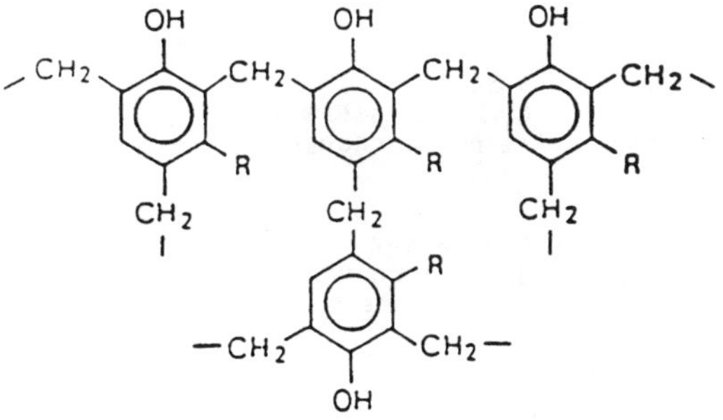

Phenol formaldehyde

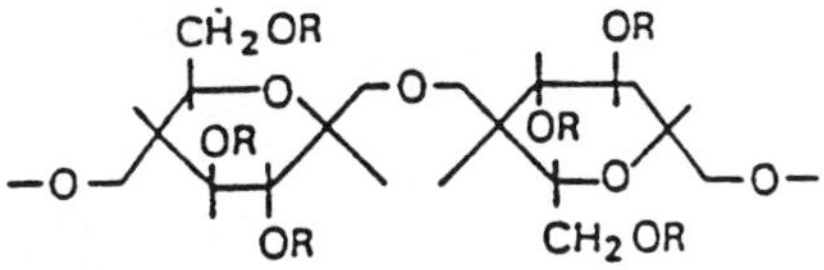

Cellulose

Comparison of DEAE and XAD-2 (divynilbenzene and styrene copolymer) (Table 4) show that both adsorption and desorption from peatbog water are higher on DEAE resin [15]

TABLE 4 Recovery of AHS (as COD) on DEAE-cellulose and XAD-2

	COD (%)	
	DEAE-cellulose	XAD-2
Original	100	100
Adsorbed	94.2	74.0
Desorbed with NaOH	71.6	67.7
Not adsorbed	5.8	26.0
Adsorbed irreversibly	22.6	1.2

Recovery with DEAE-cellulose shows also several advantages, such as rapid sorption, higher flow rates, retention only of molecules which possess acidic groups, not necessity of low pH values, but flow characteristics are poor.

Dualite A-7 shows a high capacity for anionic organic compounds and excellent elution characteristics when the loading is from one-half to two-thirds of the resin capacity [16], but also inorganic anions are retained. They are co-eluted and included in the eluate.

Leenheer and Noyes [17]) have proposed the method shown in Figure 1 for isolation of humic compounds.

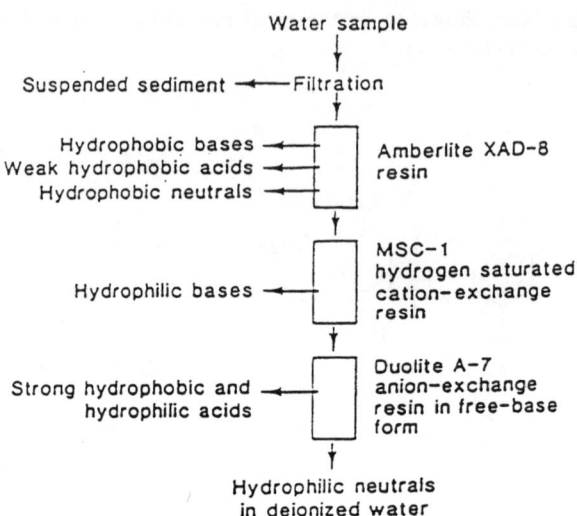

FIGURE 1 Scheme for isolation of humic compounds from water [17]

XAD-8 is not a necessary part in the system if the only compounds of interest are organic acids. On the contrary, cation-exchange resin is essential for the adsorption of organic acids on the Duolite A-7 and has two functions:
- all organic acids are in the hydrogen form passing through the resin
- the pH of the effluent from the resin is low

Duolite A-7 sorbs the most part of organic acids, both by hydrogen bonding and ion exchange and both the mechanisms are pH dependent. An efficient sorption is favored by acidic pH.

DEAE-cellulose is not widely used and XAD resins are commonly employed. They are non ionic, macroporous polymers with large surface areas, as summarized in Table 5[18,19].

TABLE 5 Properties of some XAD resins

Resin	Composition	av pore diameter A	Specific surface area (m^2/g)	Specific pore vol (cm^3/g)	Solvent uptake g/g of dry resin
XAD-1	Styrene divinylbenzene	200	100	0.69	---
XAD-2	Styrene divinylbenzene	90	330	0.69	0.65-0.70
XAD-4	Styrene divinylbenzene	50	750	0.99	0.99-1.10
XAD-7	Acrylic ester	80	450	1.08	1.89-2.13
XAD-8	Acrylic ester	250	140	0.82	1.31-1.36

Styrene divinylbenzene

Acrilic ester

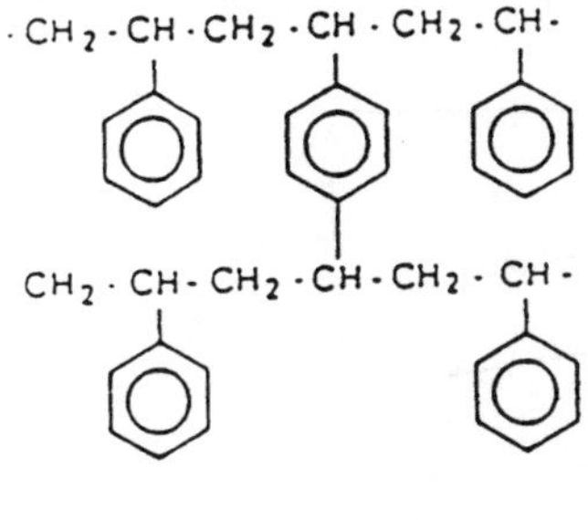

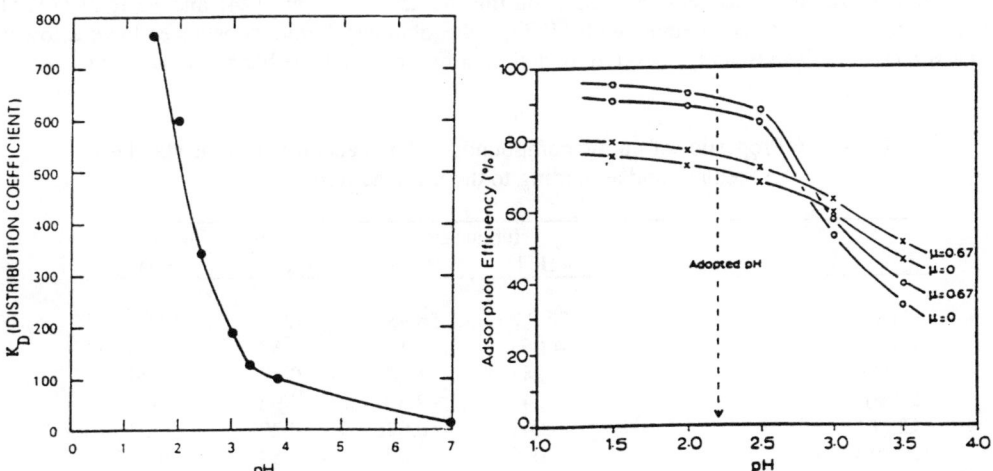

XAD resins adsorb organic acids when they are in the uncharged state. The driving force for sorption of non polar organic solutes is the hydrophobic effect. Desorption is favored when the acids are in the ionic form.

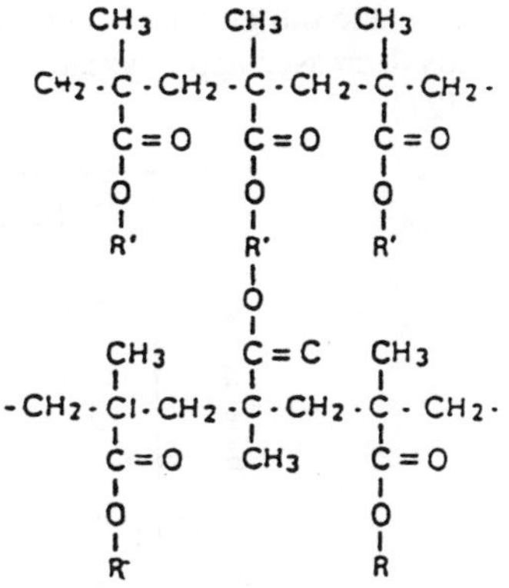

In Figure 2 is reported the relationship between the pH of the solution and the distribution coefficient of fulvic acids on XAD-8 [20]; the efficiency of adsorption of humic (o) and fulvic (x) acids on XAD-2 resin as a function of pH is shown in Figure 3 [21].

FIGURE 2 pH dependence of the distribution coefficient of fulvic acids on XAD-8 [20]

FIGURE 3 pH dependence of the adsorption efficiency of HA (o) and FA (x) on XAD-2 [21]

The most efficient adsorption is obtained at very low pH values; pH 2,2 is selected in order to prevent the possible denaturization process of fulvic acids that occur in strong acidic solution.

At this pH value the effect of flow rate was studied: the adsorption percentage of humic and fulvic acids decreases approxymatively linearly with the increasing of flow rate, as shown in Figure 4 [21]

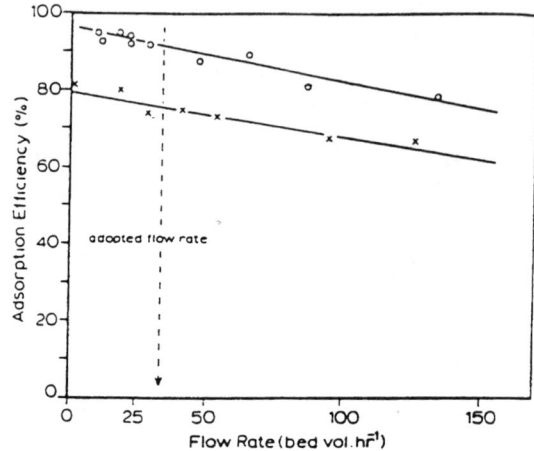

FIGURE 4 Adsorption efficiency of HA (o) and FA (x) on XAD-2 as a function of flow rate (μ 0.005) [21]

Sample acidification is carried out by hydrochloric acid [2, 22,23]. Production of chlorinated structures has been evidenced during the acid treatment [24] and Krog et al.[25] has proposed a modified method with HNO_3 : the obtained humic substances have a lower halogen content (Tab.6) and excessive oxidation or hydrolysis have been not observed.

TABLE 6 Properties of humic compounds isolated according to the standard procedure and according to the nitric acid modification

	Humic acids		Fulvic acids	
	HCl	HNO_3	HCl	HNO_3
C (%)	56.93	56.98	42.73	42.90
H (%)	4.06	4.13	4.37	4.26
N (%)	1.24	1.25	0.63	1.74
S (%)	2.24	2.50	2.96	--
Cl (%)	0.041	0.020	0.20	0.031
Br (%)	0.009	0.008	0.53	0.005
I (%)	0.008	0.007	0.012	0.019
Ash (%)	1.7	0.8	15.7	10.0
Carboxylic acidity (meq/g)	2.1	2.3	--	--
Amino acids (nmol/g)	13	15	--	--

Styrene divinylbenzene resins, aromatic in character, very hydrophobic and lacking in ion-exchange capacity, are the slowest to attain equilibrium, as adsorption-rate curves show (Figure 5) and adsorb a lower amount of humic compounds, as breakthrough curves evidence (Figure 6), in comparison with acrilic ester resins, that are non-aromatic and posses very low ion-exchange capacity [20].

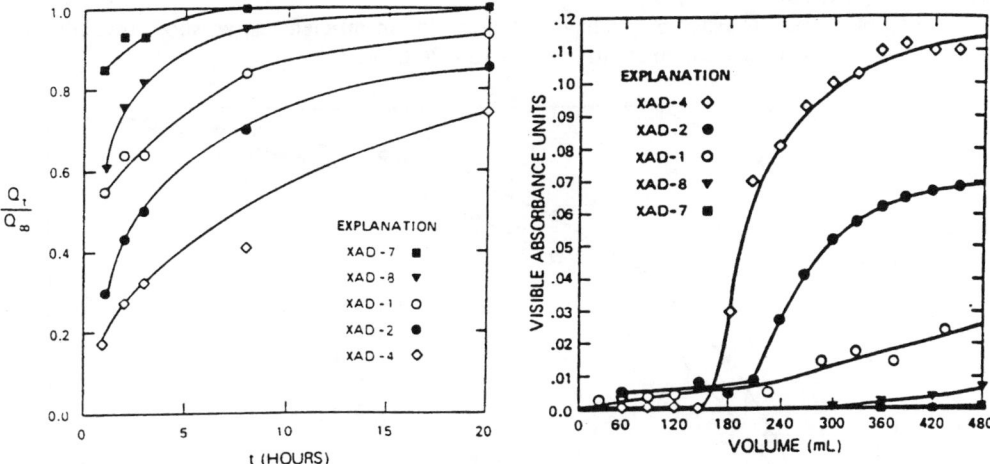

FIGURE 5. Adsorption-rate curves [20]

FIGURE 6 Breakthrough curves of FA at pH 2 based on absorbance measurements at 460 nm [20]

Desorption is carried out with aqueous solutions; organic solvents, such as THF, methanol, ethanol and acetone, are inefficient eluents because of the limited solubility of humic compounds in them [21].

TABLE 7. Elution of adsorbed humic and fulvic acids from amberlite XAD-2

Eluent	Molarity	pH	Percentage desorption	
			Humic acids	Fulvic acids
Potassium hydrogenphtalate	0.1	4.0	57	63
Tris buffer	0.4	7.0	71	75
Ammonium hydroxyde	1.0	10.6	83	89
Sodium hydroxide	0.2	13.3	95	96
Methanol	--	--	35	38
Ethanol	--	--	21	27
Acetone	--	--	33	35
Methanol-ammonia	1.0	--	91	95

With aqueous solution pH is the principal parameter controlling the efficiency of desorption; usually desorption of humic compounds is carried out with 0.1 M NaOH solution. Ammonium hydroxyde can also be used as eluent, but NH_4^+ can strongly interact with humic compounds [26] and it is difficult to eliminate them. Consequently a higher content of nitrogen can be determined in final product. Percentage of desorbed humic compounds from XAD-2 resin are reported in Table 7 [21].

At pH 13 acrylic ester resins have the best elution efficiencies, as show breakthrough curves obtained with fulvic acid solution at pH 13 (FIGURE 7).

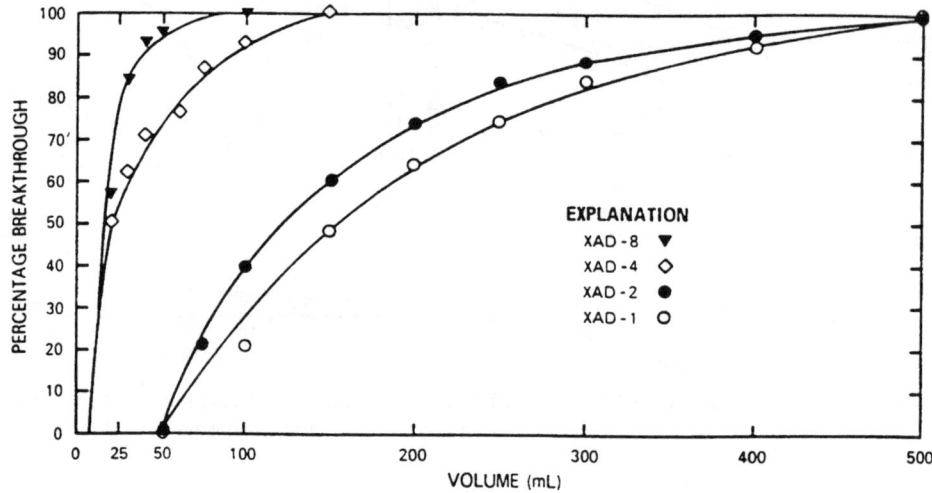

FIGURE 7 Breakthrough curves of FA at pH 13 based on absorbance measurements at 460 nm [20]

In conclusion the macroporous acrylic-ester resins (XAD-7 and XAD-8) are excellent sorbent for humic substances. XAD-8 is preferred because of bleed problems of XAD-7 with NaOH [20].

Incorporation of impurities into isolated humic compounds is possible using syntetic materials as sorbents, such as XAD or Duolite A-7, and this problem is more significant when humic compounds are present at low concentration in water sample. Bleed characteristics of the styrene-diviniylbenzene XAD resins are much better than the acrylic ester resins. XAD-8 bleeds in the order of 3 mg/L in 0.1 N NaOH and the major component is methyl acrylic acid. Duolite A-7 bleeds in the order of 30 mg C/L at pH 11.5; the major components are formic acid, formaldehyde and amino phenols [17, 27].

Elimination of bleed components can be obtained with Soxhlet extraction, but it is not possible to eliminate all the bleed components

Cleaning of XAD-8 resin is carried out by soxhlet sequential extractions for 24 h with methanol, diethylether, acetonitrile and methanol. The resin is stored in methanol. Before column packing, methanol is removed, the resin washed with distilled water and packed in the glass column. Then rinsed three time with 0.1 NaOH and 0.1 N HCl alternatively [20].

The scheme proposed by Thurman and Malcom [2] to concentrate aquatic humic substances using XAD-8 can be summarized:

-Filter sample through 0.45 µm silver-membrane filter and lower pH to 2.0 with HCl
-Pass acidified sample through column of XAD-8; aquatic humic substances adsorb to resin
-Elute XAD-8 resin in reverse direction with 0.1 N NaOH; acidify immediately to avoid oxidation of humic material
-Reconcentrate on smaller XAD-8 column until dissolved organic carbon (DOC) is greater than 500
-Adjust pH to 1.0 with HCl to precipitate humic acid. Separate HA and FA by centrifugation. Rinse humic acid fraction with distilled water until $AgNO_3$ test shows no Cl^-.
-Dissolve HA in 0.1 N NaOH and hydrogen saturate by passing solution through cation-exchange resin in H-form
-Reapply FA fraction at pH 2 to XAD-8 column. Desalt FA by rinsing column with 1-void volume of distilled water to remove HCl and inorganic salts; elute FA in reverse direction with 0.1 N NaOH
-H-saturate FA fraction by immediately passing 0.1 N NaOH eluate through cation-exchange resin in H-form. Continue cation-exchange process until final concentration of Na^+ is less than 0.1 ppm
-Freeze-dry humic acid and fulvic acid fractions

3. Isolation
Isolation of humic compounds from other organic substances with low molecular weight and from inorganic salts can be carried out by dialysis against distilled water or by diafiltration technique , that is particularly employed for the fractionation of humic compounds

2 Recovery of humic compounds from sediments

The amount of humic compounds present in sediments depends on the evolutionary stage of the sediments, on the sedimentation environment (geography and climate) and on the productivity of the continental and marine environments.
Generally, the content of humic compounds and the ratio of fulvic and humic acids in sediments decrease with depth.
Extraction procedures are used for isolating humic compounds from sediments. Stevenson [26] has listed four criteria for an ideal extraction method:

-the method leads to the isolation of unaltered materials
-the extracted humic substances are free of inorganic contaminants, such as clay and polyvalent cations
-extraction is complete, thereby ensuring representation of fractions from the entire molecular weight range
-the method is universally applicable to all samples
and Whitehead [28] has proposed four criteria for solvents that are used for extraction. The solvents should have:
-a high polarity and a high dielectric constant to assist the dispersion of the charged molecules
-a small molecular size to penetrate into the humic structures
-the ability to distrupt the existing hydrogen bonds and to provide alternative groups to form humic-solvent hydrogen bonds
-the ability to immobilize metallic cations

The solvents employed are selected on the basis of some important properties such as the dipole moment, the ability to hydrogen bond, the acidity and basicity. Organic solvents (hydrocarbons and chlorophorm) have been used occasionally, but only moderate amount of humic compounds are extracted; up to about 30% of humic compounds can be isolated with water solution containing sodium and potassium salts of inorganic and organic acids [29], but the better results have been obtained with basic water solutions (NaOH and ethylendiamine, NaOH and $Na_2P_2O_7$, NaOH [30]).

The procedure generally employed is that outlined by King [31]. In this scheme basic solutions (0.1 M NaOH) are alternated with acid solutions (0.1 M HCl) in order to remove humic compounds, present in sediments also as clay-metal-humic complexes in which the main polyvalent cations responsible for the binding of humic and fulvic acids to clays are Ca^{2+}, Fe^{3+} and Al^{3+}. Ca^{2+} ion not forms strong coordination complexes with organic matter [32] and humic compounds can be easily extracted from the matrix; on the contrary, Al^{3+} and Fe^{3+} ions form strong coordination complexes, consequently the displacement of the bound metal is difficult. Only a part of humic compounds present in the sediment can be extracted without using acid solutions.

3. References

1 Leenheer, J.A. (1981) Comprehensive approach to preparative isolation and fractionation of dissolved organic carbon from natural waters and waste waters. *Environ. Sci. Technol.* **15**, 578-587.

2 Thurman, E.M. and Malcolm, R.L. (1981) Preparative isolation of aquatic humic substances. *Environ. Sci. Technol.* **15**, 463-466.

3 Harvey, G.R., Boran, D.A., Chesal, L.A. and Tokar, J.M. (1983) The structure of marine fulvic and humic acids. *Mar. Chem.* **12**, 119-132.

4 Danielson, L.G. (1982) On the use of filters for distinguishing between dissolved and particulate fractions in natural waters. *Water Res.* **16**, 179-182.

5 Malcolm, R.L. (1976) Method and importance of obtaining humic and fulvic acids of high purity. *J. Res. U. S. Geol. Surv.* **4**, 37-40.

6 Sharp, J.H. (1973) Size classes of organic carbon in sea water. *Limnol. Oceanogr.* **18**, 441-447.

7 Leenheer, J.A. (1982) United States Geological Survey Data Information Service , in E. Degens (ed.), *Transport of Carbon and Minerals in World Rivers. Pt. I*. Report of a workshop, Hamburg, Germany, March 8-12, 1982, SCOPE pp. 355-356.

8 Kennedy, V.C., Zellweger, G.W. and Jones, B.F. (1974) Filter pore-size effects on the analysis of Al, Fe, Mn and Ti in water. *Water Resour. Res.* **10**, 785-790.

9 Sheldon, R.W. (1972) Size separation of marine seston by membrane and glass-fiber filters. *Limnol. Oceanogr.* **17**, 494-498.

10 Aiken, G.R. (1985) Isolation and Concentration Techniques for Aquatic Humic Substances, in G.R. Aiken, D.M. Micknight, R.L. Wershaw and P.Maccarthy (eds), *Humic Substances in Soil, Sediment, and Water*, Wiley-Interscience, New York, pp.363-385.

11 Wangersky, P.J. and Hincks, A.V. (1978) The shipboard intercalibration of filters used in the measurement of particulate organic carbon. *Research Council of Canada Publication Number NRCC- 16767*

12 Cranston, R.E. and Buckley D.E. (1972) The application and performance of microfilters analysis of suspended particulate matter. *Bedford Institute of Oceanography Report Series B1-R-72-7*.

13 Sirotkina, I.S., Varshall, G. M., Lur'e, Y.Y. and Stepanova P.N. (1974) Use of cellulose sorbents and Sephadexes in the systematic analysis of organic matter in natural water. *Zh. Anal. Kim.* **29**, 1626-1632.

14 Miles, C.J.,Tushall, J.R. and Brezonic P.L. (1983) Isolation of aquatic humus with diethylaminoethylcellulose. *Anal. Chem.* **55**, 410-411.

15 Hejzlar J., Szpakowska B. and Wershaw R.L. (1994) Comparison of humic substances isolated from peatbog water by sorption on DEAE-cellulose and Amberlite XAD-2.*Water Res.* **28**, 1961-1970.

16 Abrams, I.M. and Breslin R.P. (1965) Recent studies on the removal of organics from water. *Proc. Int. Water Conf.*, Pittsburgh.

17 Leenheer, J.A. and Noyes T.I. (1984) A filtration and column adsorbent system for onsite concentration and fractionation of organic substances from large volumes of water. *U.S. Geol. Surv. Water-Supply Paper No.2230*.

18 Kunin, R. (1977) Polymeric adsorbents for treatment of waste effluents. *Polym. Eng. Sci.* **17**, 58-62.

19 Parish, J.R. (1977) Macroporous resins as supports for a chelating liquid-ion exchanger in extraction chromatography. *Anal. Chem.* **49**, 1189-1192.

20 Aiken G.R., Thurman E.M. and Malcolm R.L. (1979) Comparison of XAD macroporous resins for the concentration of fulvic acid from aqueous solution. *Anal Chem.***51**, 1799-1803.

21 Mantoura R.F.C. and Riley J.P. (1975) The analytical concentration of humic substances from natural waters. *Anal. Chim. Acta* **76**, 97-106.

22 Petterson C., Arsenie I.,Ephraim J., Borèn H. and Allard B. (1989) Properties of fulvic acids from deep groundwater. *Sci. Total Environ.* **81/82**, 287-296.

23 Aiken, G.R., Mc Knight, D.M., Thorn K.A. and Thurman, E.M. (1992) Isolation of hydrophilic organic acids from water using non ionic macroporous resins. *Org. Geochem.* **18**, 567-573.

24 Penravuori, J. (1992) Isolation, fractionation and characterization of aquatic humic substances. Does a distinct humic molecule exist. *Finn. Humus News* **4**, 1-99.

25 Krog, M. and Gron, C. (1995) Isolation of haloorganic groundwater humic substances. *Sci. Total Environ.* **172**, 159-162.

26 Stevenson, F.J. (1982) *Humus Chemistry: Genesys, Composition, Reactions*. Wiley-Interscience, New-York.

27 Aiken, G.R. (1988) A critical Evaluation of the use of macroporous resins for the isolation of aquatic humic substances in F.H. Frimmel and R.F. Christman (eds.), *Humic substances and their role in the environment*. Wiley-Interscience, New-York, pp 15-28.

28 Whitehead, D.C. and Tinsley , J. (1964) Extraction of soil organic matter with dimethylformamide. *Soil Sci.* **97**, 34-42.

29 Bremner, J.M. and Lees, H. (1949) Studies of soil organic matter: II. The extraction of organic matter from soil by neutral reagents. *J. Agric. Sci.* **39**, 274-279.

30 Hayes, M.H.B., Swift, R.S., Wardle, R.E. and Brown, J.K. (1975) Humic materials from an organic soil: A comparison of extractants and of properties of extracts. *Geoderma* **13**, 231-245.

31 King, L.H.Isolation and characterization of organic matter from glacial marine sediments on the Scotian Schelf. *Bedford Inst. of Oceanography* Report 1967-4.

32 Bowen, H.J.M. (1966) Trace elements in biochemistry, the biochemistry of the elements in *Trace elements in biochemistry*. Academic Press, London, New-York , pp 173-210

CERTIFIED REFERENCE MATERIALS FOR CHEMICAL ANALYSIS IN MARINE ECOSYSTEMS

ROBERTO MORABITO
ENEA, Environmental Department
Via Anguillarese, 301, 00060 Rome, ITALY

PHILIPPE QUEVAUVILLER
Standards, Measurements and Testing Programme (SM&T)
DG XII - CEC
Rue Montoyer, 75, 1049 Brussels

1. Introduction

The public concern regarding environmental problems has dramatically increased during the last 40 years, starting more or less from the end of fifties and in particular from the London disaster, in which many elder people died as a consequence of a strong air pollution due to the carbon coke heating systems combined with particular meteorological conditions, and from the Minamata disaster, in which people from the fisherman village of Minamata were poisoned from fish strongly contaminated by mercury.

Accordingly to the increase of the public concern, environmental studies have become one of the main research lines in the past few years. They cover a broad range of disciplines, e.g. analytical chemistry, geology and biology, and include several aims such as monitoring (routine analyses), research (studies of environmental pathways), modelling etc. Chemical analyses are in many cases the basis of these studies and hence have an enormous economic impact. Many analytical methods have been developed for the determination of contaminants in air samples (fumes, steams, fogs, etc.), water samples (freshwater, seawater, rainwater, etc.), solid samples (wastes, soils, sediments, plants, fish tissue, etc.), and biological fluid samples (blood, serum, urine, etc.). However, these methods should have at least three main characteristics to be suitably employed in environmental analysis: (i) they should be able to determine contaminants down to the extremely low environmental concentrations: often µg/kg levels, sometimes ng/kg and even pg/kg levels; (ii) they should be able to determine the analytes of concern in presence of a wide variety of interferents; and, (iii) they should be characterized by good precision and accuracy, avoiding mistakes that would lead to wrong experimental results an consequent wrong environmental considerations followed by wrong political decisions in terms of environmental management, health care, etc.

In few words, analytical methods and, in particular, analytical techniques, should be characterized by high sensitivity, high selectivity and high precision and accuracy. Unfortunately, many of these methods are far from being under control. In general, a lack of quality control in chemical analysis leads to economic losses due to repetition of analysis, destruction of goods wrongly considered of bad quality, doubts on government decision, etc.

In case of environmental analysis a lack of quality control means not only economic losses but also, and above all, risks for environment and for public health.

Consequently, awareness of the need for the quality control of environmental analyses has increased considerably as is illustrated by the multiplication of quality assurance (QA) guidelines, standards and accreditation systems. However, the accuracy of analytical results is, in many cases, far from being achieved. Moreover, in the past too many wrong but highly reproducible results have led to misinterpretation of environmental processes. In recent years the quality of analytical data from environmental samples has improved through regular use of reference materials, together with individual analysts participation in interlaboratory exercises. Both the participation to intercomparisons and the use of reference materials are two key aspects, necessary to ensure the accuracy of the measurements.

2. Proficiency testing schemes

Information from marine monitoring programmes forms the basis for national and international policies, providing insight in the effectiveness of measures which have been taken and basic information for strategic decisions for environmental protection [1]. This information is expensive, e.g. a rapid estimate of the monitoring costs in the North Sea amounted upto 20 millions ECU per year [2]. Economic impact of the measures which may be taken on the basis of this information can even be higher and it is easy to figure out which negative economic and environmental impact can be produced by monitoring data of bad quality.

As a consequence, the need for quality control in marine monitoring programmes became a priority and several proficiency testing schemes, within the frame of single monitoring programmes or not, have been developed.

It is worth to cite at least two examples, NOAA National Status and Trends Project and QUASIMEME Project, representing the state of art in U.S.A. and in Europe, respectively.

Since 1984, the National Oceanic and Atmospheric Administration (NOAA) has conducted the National Status and Trends (NS&T) Program, a long-term monitoring study of the coastal water of the United States [3].

This project aims at the evaluation of the environmental conditions of about 100 sites located on the East, West and Gulf coasts, the Great Lakes and on the Alaska, Hawaii and Puerto Rico coasts.

The program envisaged the evaluation of the concentration levels of inorganic and organic contaminants in water, sediment and biota samples.

Since its start, the quality of the produced data has been evaluated through a quality assurance project aimed to document sampling protocols and analytical procedures, and to improve the analytical performances of the laboratories involved in the measurements. All these laboratories were requested to participate in a series of intercomparison exercises, consisting in different types of analysis on different matrices, for the whole duration of the NS&T project. The intercomparisons for organic measurements were coordinated by the National Institute of Standards and Technology (NIST) and those for inorganic analysis were coordinated by the National Research Council of Canada (NRCC).

The detailed description of this project is in [3].

The proposal on Quality Assurance of Information in Marine Environmental Monitoring in Europe (QUASIMEME) was prepared in 1991-1992 and submitted for funding to the M&T Programme at the end of the 1992.

The proposal envisaged the participation of 89 laboratories from 13 EU and 2 EFTA Countries in a three years proficiency testing scheme dealing with the evaluation of analytical performances in the determination of inorganic and organic contaminants as well as nutrients. Furthermore, more than 60 laboratories involved in marine research were proposed to participate at their own costs.

The economical justification was made clear by the amount of resources lost every year owing to a poor quality control of marine monitoring measurements. It was estimated that, in some cases 80% of the data produced were no exploitable due to a poor quality assurance.

The importance for legislation was highlighted (support to Directives on the Protection of Sea), as well as the strong links existing with the Oslo and Paris Commission (OSPARCOM), the Helsinki Commission (HELCOM), the Barcelona Convention (MEDPOL), and the European Comission.

All exercises and selected parameters had their justification in the improvements needed in support to the OSPARCOM mandatory determinands.

A questionnaire, for the identification of the needs of laboratories, guidelines, to support the participating laboratories, and a regularly published bulletin, to ensure a liaison between the coordinating committee, the participants, and the European Commission, were also envisaged.

The answers to the questionnaire evidenced that the laboratories do not have a true measure of their real capabilities [4]. The priority needs, with respect to the three main exercises proposed in QUASIMEME (nutrients in seawater, trace metals in sediments, CBs in fish oil), resulted to be training, establishment of guidelines on sample storage and sampling, and availability of appropriate reference materials for quality control purposes.

The results of the 1993 Laboratory Performance Studies showed that for nutrients more than 80% of laboratories met the target performance at the high concentration level and that the majority of errors were systematic and readily correctable by strict application of reliable (blank and calibration) procedures [5]; for trace metals, differences in the results were mainly due to the use of partial (e.g. aqua regia) and total (e.g. HF) methods [6]; for CBs in fish oil, 60% of all errors were due to inaccurate calibration, poor GC separation and non-analytical mistakes by 30% of the laboratories [7]. The results of the 1994-1995 Laboratory Performance Studies are in course of publication; beside the three main topics, new matrices such as trace metals in biota, CBs and organochlorine pesticides in biota and sediments, and PAHs in solutions, sediment extracts and sediment samples, have been considered.

The establishment of the QUASIMEME proficiency testing scheme enabled to perform a continous monitoring of the quality of the methods used in routine analysis, to detect sources of pitfalls and remove them, and to train staff responsible for providing analytical data for marine monitoring purposes.

The detailed description of this project is in [1].

On the basis of the achievements of QUASIMEME, a new project was prepared, aiming at evaluating and improving sampling, sample pretreatment and storage methods used in marine monitoring. This project, named QUASH (Quality Assurance of Sample Handling for Marine Environmental Measurements), has been funded by the SM&T Programme and its start is foreseen for October 1996 [1].

3. Definition and use of reference materials

A correct implementation of quality assurance guidelines includes procedures for the control of the quality of measurements and procedures to evaluate their quality .

Keeping under control the measurements means not only to control blanks, calibrants and technical instrumentation but also to improve the capability of the laboratory staff, to adopt, where existing, Standard Operation Procedures and in general to adopt good laboratory practice consisting in a correct sample storage, treatment and analysis trying to reduce the very common contamination and losses of analytes problems.

The evaluation of the quality of measurements implies the use of independent methods of analysis, the statistical treatment of data, the maintanance of long term documentation and, above all, the participation in intercomparison exercises (also called collaborative studies, round robin tests, interlaboratory studies, etc.) and the use of reference materials (RMs) and/or certified reference materials (CRMs) depending on the final aim. According to ISO, a reference material is defined as a material or substance, one or more properties of which are sufficiently well established to be used for the calibration of an apparatus, the assessment of measurement method or for assigning values to materials. A certified reference material is defined as a reference material, one or more of whose property values are certified by a technically valid procedure, accompanied by or traceable to a certificate or other documentation that is issued by a certifying body. The fundamental difference between RMs (also known as "Laboratory Reference Materials") and CRMs (also known as "Standard Reference Materials") is that some parameters in CRMs are known with great accuracy [8].

CRMs can be classified according to the following categories:
(a) pure substances or solutions;

(b) matrix reference materials representative of real samples;
(c) methodological-defined reference materials for parameters such as leachable trace element fractions from sediments (certified values defined by the method applied following a strict protocol).

Pure compounds are generally used for calibration while matrix materials are used for the evaluation of the accuracy: analytical procedures may be verified by analysing a CRM with a matrix similar to that of the unknown sample; disagreement between the certified value and the value determined by the laboratory indicates an error, or errors, in the procedure.

Furthermore, CRMs can also be used to demonstrate the equivalence of methods, enabling laboratories to follow the development of new analytical instrumentation within their laboratory.

CRMs are products of very high added value. Their production and certification are very costly (typically some hundreds of thousands of ECUs) and they should be reserved for the final verification of analytical procedures and method validation only and not for routine analysis or for interlaboratory studies for which RMs, being less costly and sufficiently reliable, are more suitable [9].

In interlaboratory studies, the materials used can be both calibrants or extracts designed to tests steps in the analytical methods (e.g., detection or separation) and matrix RMs for testing the complete procedures [9]. Errors due to a specific procedure (or to a specific laboratory) can be detected by comparing different techniques with different analytical steps (e.g., different extraction, separation, derivatization, detection). Similarly, a laboratory's performance with a given technique can be evaluated by comparing the results obtained by that laboratory with those of other laboratories using the same technique. Furthermore, RMs can be used in statistical control schemes, i.e., for the verification of the long-term reproducibility of a method by setting up control charts [10]. The level of precautions and verifications for the production of a reference material is generally much less than for a CRM. As a consequence, reference values should not be considered as 'absolute', i.e., a RM should not be used to verify the accuracy of an analytical procedure.

4. Selection, preparation and storage of reference materials

The main prerequisite that should be addressed in the selection of the raw material to be processed is that the same sources of error should be encountered in analysing the CRM or the unknown, i.e., the CRM used for verification of method performance should have a composition as much as possible similar to the unknown sample. This means similarity of (i) the matrix composition, (ii) the contents of the analytes, (iii) the means of binding of these analytes, (iv) the fingerprint pattern of possible interferences and (v) the physical status of the material [8]. These requirements should be taken into full consideration when preparing a Reference Material. For practical reasons, however, the similarity cannot always be entirely respected and compromises have to be made in some instances in the choice and in the preparation step.

The material must be collected in an amount sufficient to ensure an adequate, sufficiently long-lasting stock. The amount to be prepared is a function of the analytical sample size, stability, shelf size and frequency of use. For RMs, it is generally smaller than 10-20 kg while for CRMs it is generally much higher. The producer needs to be equipped to treat large amounts of material without substantially changing the analytical representativity. Whereas the treatment of few kilograms of raw material can be achieved with normal laboratory equipment and manual processing, the treatment of larger batches and especially larger volumes of material requires equipment of half industrial size.

Typical operations are crushing, grinding, sieving, filtering, mixing or homogenization of the materials. Furthermore, the material requires to be stabilized in order to minimize the risks of chemical and/or biological processes that can lead to an instability of the material. The stabilization has to be adapted to each particular case and should be studied in detail before processing in order to respect the integrity of the material as much as possible. Usually the materials are dried to avoid chemical or microbiological changes, but γ-irradiation is also applied for this purpose. This step is particularly critical for reference materials to be certified for their contents of chemical species as they are often unstable and a stabilization procedure such as γ-irradiation is not applicable owing to risk of degradation of many compounds (e.g., butyltins). Therefore, this step needs to be carefully studied for each type of material and specific chemical form of elements.

The material must be homogenized and stored in adequate vials. In some instances, particularly for liquids, the choice of containers (e.g., polyethylene, borosilicate glass) has to be investigated to ensure that no losses (adsorption onto the container walls or degradation) occur during storage.

For gases and liquids, homogeneity is not the most difficult problem; the stability, however, causes great concern. Solid materials are difficult to homogenize; for particle sizes less than 125 μm and a sufficiently narrow distribution, however, the homogeneity is generally sufficient for sample intakes of less than 100 mg. A proper grinding procedure and thorough homogenization prior to and during the filling procedure are therefore recommended. It should be noted, however, that fine grinding may lead to some drawbacks, e.g., easier digestion of the matrix because of a larger contact surface in comparison with natural samples. It is worth stressing that this could lead to an overestimation of the extraction efficiency of the method to be validated.

5. Homogeneity and stability testing of reference materials

The homogeneity of the reference materials has to be carefully verified to ensure that within a bottle or ampoule and from one vial to another the contents are the same (within and between-vial homogeneity). Homogeneity tests for CRMs produced by the Standards, Measurements and Testing Programme (SM&T) of the European Union are generally carried out by performing ten replicate determinations of the element(s) concerned in the same bottle (within-bottle homogeneity) and by one determination in an adequate number of bottles (2-5% of the whole batch) set aside at regular intervals during the bottling procedure (between-bottles homogeneity); the relative standard deviation (σr) obtained may be compared with the σr of the final step of the method, usually determined by performing ten replicate determinations on the same final extract obtained from one single sample. After subtraction, the residual σr can be associated with the homogeneity of the material. For certain types of matrices and/or analytes it may be necessary to test the homogeneity at different levels of intakes (e.g., 100 and 500 mg). The minimum (analytical) sample intake recommended to the user corresponds to the smaller intake at which the homogeneity has been verified. As segregation of particles may occur during transport, RMs and CRMs have to be carefully re-homogenized before use. This can simply be done by manual shaking for a few minutes.

The properties of the material and the parameters investigated should remain unchanged over long periods and the long-term stability is another parameter to be studied. The stability can be estimated by evaluating the behaviour of the material stored at different temperatures. Analyses may be performed after 1, 3, 6, 12 and sometimes 24 months of storage at e.g. -20 °C, +4 °C, +20 °C and +40 °C. The results obtained at +20 °C may lead to an assessment of the sample stability at ambient laboratory temperature whereas the results obtained at +40 °C are used to assess the worst case conditions (e.g., during transport) and allow the stability of the material over longer periods of time to be evaluated; it is assumed that a sample that is stable at +40 °C for 1 year may be stable at +20 °C for a longer period (Arrhenius law). Results obtained at -20 °C may be used as reference assuming that no chemical or biological changes occur at this temperature.

6. Certification of reference materials

There are a number of suppliers of CRMs for the various fields of analyses. The most important and reliable ones are the National Research Council of Canada (NRCC), the

National Institute for Environmental Studies (NIES, Japan), the National Institute for Standards and Technology (NIST, USA), the Standards, Measurements and Testing Programme (SM&T (formerly BCR) - European Union) and its associated counterpart the Institute for Reference Materials and Measurements (IRMM), and the International Atomic Energy Agency (IAEA - NATO).

Several valid approaches may be used to obtain assigned values for a reference material. One approach consists of using a method well under control in a single laboratory that has given proof of good performance and to determine the content of the elements/compounds concerned along with its analytical uncertainty. It is stressed that the value obtained cannot be proved to be accurate (e.g., the laboratory may have a bias in using their single method) and, in addition, the use of a single method in one laboratory does not give a fair estimate of the uncertainty achievable by other methods. Another approach more often adopted consists of using several (at least two) independent methods within a single laboratory. The results become more reliable when two or more independent methods give the same results but also in this case the risk of systematic errors linked to the laboratory are present. A third approach consists of using several independent methods applied in different laboratories. In this case, it is necessary to take the following precautions: (i) the participating laboratories should be carefully selected on the basis of their experience in the field and after an evaluation of their performance in that field by intercomparison exercises; (ii) the same analytical methodology should be independently performed in at least two or three laboratories; and (iii) the results of intercomparison and certification should be carefully evaluated and discussed in a technical meeting.

The first two approaches are generally used by NIST and NRCC whereas the third approach is used by SM&T Programme, NIES, and IAEA.

7. Reference materials for marine chemistry

There is a number of certified reference materials for marine chemistry that are available on the market. They include several matrices such as water, sediments (e.g. from harbour and coastal environments), marine organisms (e.g. fish, mussels, oysters, plancton, etc.), and plants (e.g. sea lettuce).

With regard to the certified analytes, major and trace elements are the most available: they have been certified in all the above mentioned matrices at different range of concentrations.

A reduced number of marine reference materials are available for organic contaminants. Most of them are sediments from different marine environments certified for PAHs, PCBs, and organics such as DDE and DDD, and chlorobenzenes. Nevertheless, there is also a small number of biological reference materials such as shrimps and copepod certified for pesticides and mackerel and cod liver oils certified for PCBs [11].

Recently, there has been an increasing attention in the development of analytical techniques for the determination of chemical species in environmental matrices. At the same time there has been a correspondent increase in the number of reference materials certified for chemical species [12]. Even if the available number of these reference materials is very low, it is worth to stress that 90% of them have been certified in the last 3-5 years. Mostly of them concern organotin compounds and methylmercury. In particular, one (tributyltin) or more organotin compounds in sediments, fish tissue and seawater as well as methylmercury in several biological matrices (lobster hepatopancreas, dogfish liver and muscle, tuna fish) and sediments are already certified and available on the market.

Other materials (e.g. butyl and phenyltins in mussel tissue) are at the certification stage.

In table 1 are reported the certified reference materials for organic and inorganic contaminants analysis in marine environment available from the 5 Institutions mentioned above (referred to 1993 catalogues).

In table 2 are reported the certified reference materials for chemical species in marine environment available.

TABLE 1. CRMs for organic and inorganic contaminants analysis in marine environment (referred to 1993 catalogues).

MATERIAL	ANALYTE	MATRIX	PRODUCER
OILS			
CRM 349	PCBs	Cod liver oil	BCR
CRM 350	PCBs	Mackerel oil	BCR
SRM 1588	Organics	Cod liver oil	NIST
SEDIMENTS			
BCSS-1	Elements	Marine sediment	NRCC
BEST-1	Elements	Estuarine sediment	NRCC
CS-1	PCBs	Coastal sediment	NRCC
HS-1	PCBs	Coastal sediment	NRCC
HS-2	PCBs	Coastal sediment	NRCC
HS-3	PCBs	Marine sediment	NRCC
HS-4	PCBs	Marine sediment	NRCC
HS-5	PCBs	Marine sediment	NRCC
HS-6	PCBs	Marine sediment	NRCC
MESS-1	Elements	Marine sediment	NRCC
PACS-1	Elements	Harbour sediment	NRCC
SES-1	PAHs	Estuarine sediment	NRCC
IAEA-357	Organics	Coastal sediment	IAEA
IAEA-367	Isotopes	Pacific ocean sed.	IAEA
IAEA-368	Isotopes	Pacific ocean sed.	IAEA
SD-M-2/TM	Elements	Marine sediment	IAEA
SD-N-2	Isotopes	marine sediment	IAEA
SRM 1646	Elements	Estuarine sediment	NIST
SRM 1941	Organics	marine sediment	NIST
CRM 277	Elements	Estuarine sediment	BCR
TISSUES			
IAEA-307	Isotopes	Sea plant	IAEA
IAEA-308	Isotopes	Mediterranean seaweeds	IAEA
IAEA-350	Elements	Tuna homogenate	IAEA
IAEA-351	Organics	Tuna homogenate	IAEA
IAEA-352	Isotopes	Tuna homogenate	IAEA
MA-A-1/OC	Organics	Copepod homogenate	IAEA
MA-A-1/TM	Elements	Copepod homogenate	IAEA
MA-A-3/OC	Elements	Shrimp homogenate	IAEA
MA-B-3/OC	Organics	Fish	IAEA
MA-B-3/RN	Isotopes	Fish	IAEA
MA-B-3/TM	Elements	Fish	IAEA
CRM 278	Elements	Mussel tissue	BCR
CRM 414	Elements	Plankton	BCR
CRM 422	Elements	Cod muscle	BCR
DOLT-1	Elements	Dogfish liver	NRCC
DORM-1	Elements	Dogfish muscle	NRCC
LUTS-1	Elements	Non-defatted lobster hepatopancreas	NRCC
MUS-1	Organics	Domoic acid	NRCC
TORT-1	Elements	Lobster hepatopancreas	NRCC
SRM 1566a	Elements	Oyster tissue	NIST
SRM 1974	Organics	Mussel tissue	NIST
NIES 9	Elements	Sargasso seaweed	NIES
NIES 11	Elements	Fish tissue	NIES
WATERS			
CASS-2	Elements	Nearshore seawater	NRCC
NASS-4	Elements	Open ocean seawater	NRCC
SLEW-1	Elements	Estuarine water	NRCC
IAEA-298	Isotopes	Pacific ocean water	IAEA
V-SMOW	Isotopes	Ocean water	IAEA
CRM 403	Elements	Seawater	BCR

TABLE 2. CRMs for speciation analysis in marine environment.

MATERIAL	SPECIES	MATRIX	PRODUCER
TISSUES			
LUTS-1	CH_3Hg	Lobster hepatopancreas (non-defatted)	NRCC
TORT-1	CH_3Hg	Lobster hepatopancreas (partially-defatted)	NRCC
DOLT-1	CH_3Hg	Dogfish liver	NRCC
DORM-1	CH_3Hg	Dogfish muscle	NRCC
CRM 463	CH_3Hg	Tuna fish	BCR
CRM 464	CH_3Hg	Tuna fish	BCR
NIES 11	TBT, TPhT	Fish tissue	NIES
SEDIMENTS			
IAEA 356	CH_3Hg	Marine sediment	IAEA
PACS-1	TBT, DBT, MBT	Harbour marine sediment	NRCC
RM 424	TBT	Harbour sediment	BCR
CRM 462	TBT, DBT	Coastal sediment	BCR

8. Preparation of a candidate CRM within the frame of the Tin speciation project of the SM&T Programme: an example

Organotin contamination in marine environment has been well documented, starting from the '70s, when these compounds, and in particular tributyltin (TBT) and, to a less extent, triphenyltin (TPhT), replaced copper salts as active components in antifouling paints.
The high toxicity of these compounds together with their high tendency to be accumulated in marine organisms, and in particular in filter-feeding organisms such as mussels and oysters, can cause environmental damage as observed in the past in the Arcachon Bay.
Several monitoring campaigns have been carried out all over the world during the '80s in order to determine the concentration levels of these compounds in marine environment and to control the effectiveness of legal provisions. At the same time, many analytical methods have been developed for this type of analysis.
In 1987 the BCR (now SM&T) launched the "Tin speciation project" in order to: (i) evaluate the performances of the European laboratories involved in this type of analysis; (ii) to improve the quality of organotin measurements; and (iii) to make available certified reference materials for organotin analysis.
Like other similar projects also the tin speciation project followed the classical BCR step by step approach consisting in a series of exercises of increasing difficulties starting from a very simple intercomparison on standard solutions to evaluate the final determination step, arriving to very complicate exercises on real samples to evaluate the performance of the overall method [13]. Performances of separation methods and evaluation of extraction recoveries are also considered (Figure 1).
The first intercomparison consisted in the analysis of TBT in synthetic solutions in presence or not of interferents and the results were very good.
The second one was carried out on a low polluted lake sediment spiked with 3.3 ppm of TBT. The results were very encouraging: 15 out of 18 laboratories were able to provide results and the difference between the highest and the lowest result was around a factor of 2, 2.5. So, it was decide to organize a certification campaign for TBT in sediment. The material was collected in the high polluted area of the Sado Estuary in Portugal.
Unfortunately, the certification was unsuccesfull. Only 12 out of 19 laboratories were able to provide results and the difference between the highest and the lowest result was around a factor of 30. It was decided that this material could not be certified and now is available as reference material at this concentration (RM 424).

BCR STEP-BY-STEP APPROACH
1   **INTERCOMPARISON ON SIMPLE SOLUTION**   To evaluate the methods of final determination
2   **INTERCOMPARISON ON RAW OR CLEANED EXTRACT**   To evaluate the performance of separation methods
3   **INTERCOMPARISON ON SPIKED SAMPLES**   To evaluate the extraction recoveries
4   **INTERCOMPARISON ON REAL SAMPLES**   To evaluate the performance of the overall method

Figure 1.   BCR approach for the evaluation and improvement of the analytical performances of laboratories

Anyway, the campaign clearly indicated that techniques able to determine high levels (mg/kg) of TBT in simple matrices (as the spiked lake sediment of the intercomparison exercise) may fail in the determination of very low levels (μg/kg) in very difficult matrices (as the high polluted Sado Estuary sediment) and that spectroscopic techniques were not able to provide comparable results in such a matrix probably suffering of strong interferences during the hydridization.
On the basis of this experience further research on organotin determination in sediments by hydride generation techniques was carried out and a new certification was attempted.
The certification was successfully and TBT and DBT were certified in coastal sediment (CRM 462).
Furthermore, this second certification campaign pointed out that further research on extraction and derivatization techniques should have been necessary before the certification of a mussel material.
In the mean time a candidate CRM for the determination of butyl- and phenyltin species in mussel tissue was prepared [14].
1200 Kg of mussels *(Mytilus edulis)* were purchased directly from the responsibles of the La Spezia mussel farm at the beginning of July 1991. The month of July was chosen as in this period the best ratio between the weight of the whole mussel and the weight of its edible part is reached.
200 kg per day of mussels were furnished during a week.
After collection the mussels were largely washed with fresh water, in order to eliminate matrix salts as possible interferents in freeze-drying and analysis, immediately frozen by liquid nitrogen and transported to the close ENEA Research Center of Santa Teresa (La Spezia).
Shelling of mussels candidate reference material was performed in the past by cooking or by vapor stream in order to open the shells. In this case, however, it was not possible to use a similar procedure because it is well known that at high temperature organotin degradation can occur. Thus, it was necessary to use a time consuming and difficult procedure shelling directly the frozen material by special mussel knives. The edible part was then put in thermally sealed polyethylene bags (ca 4 Kg per bag) and immediately stored at -25 °C. The total amount of the material after shelling was ca. 325 Kg.

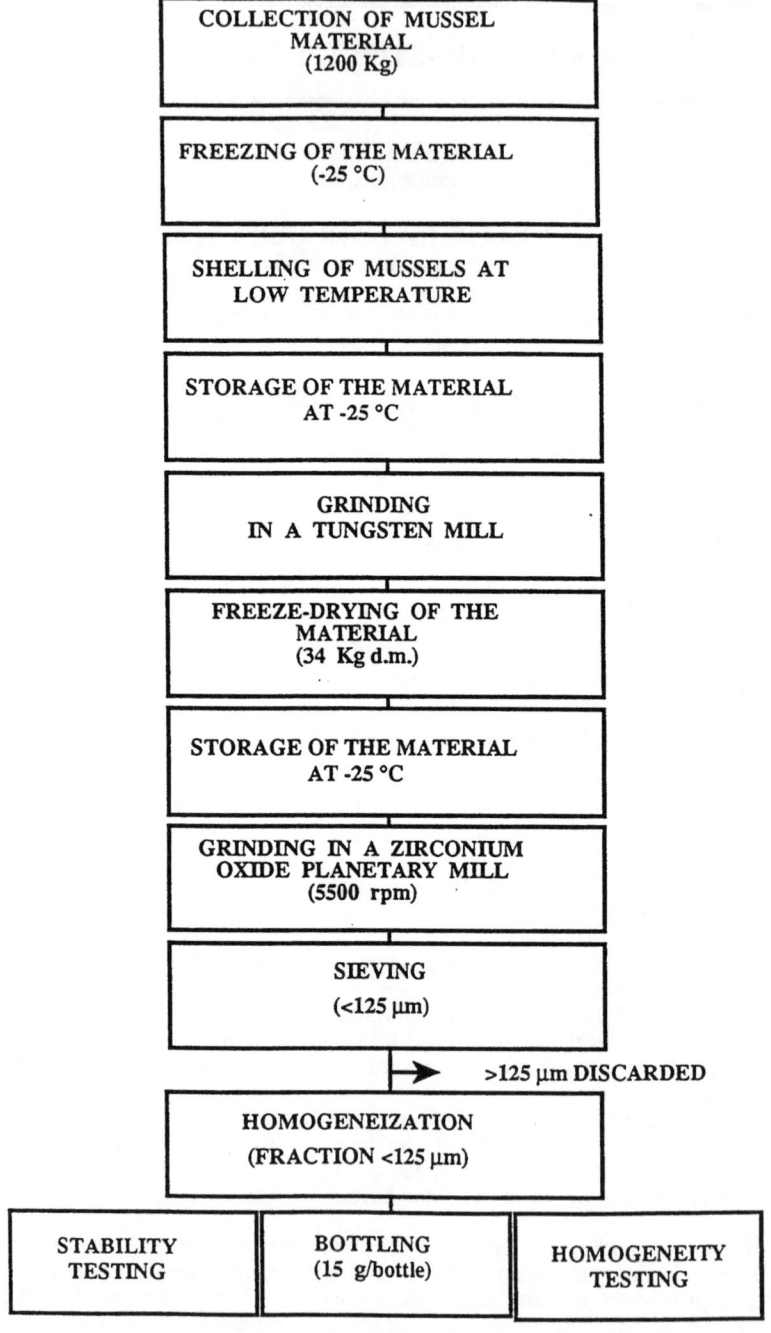

Figure 2   Preparation of candidate CRM 477 for the certification of organotin species

Three freezers containing the material were transported, by a truck equipped with emergency generators, to the Biostarters (Parma) freeze drying facilities.
The frozen material was ground, by a teflon coated grinding mill, and spread on sterilized flat trays (40x30 cm) and then freeze dryed. The process consisted in keeping the material at ca -55 °C in six hours, then applying the vacuum for 48 hours. The freeze drying apparata were Edwards: 1 Mini Fast Mod. D 0.5/NS Capacity 20 Kg, 1 Mini Fast Mod. 3400 Capacity 10 Kg and 1 Mini Fast Mod. D 0.5 Capacity 10 Kg.
Analyses were performed at the end of each freeze drying process on samples collected from the top, the intermediate and the bottom level of the flat trays, to evaluate the quality of the process. Results showed a moisture content of the material below 4% (m/m).
The freeze dried material (ca. 35 Kg) was put in thermally sealed polyethylene bags, stored in freezer at -25 °C and transported, by a truck equipped with emergency generators, to the Joint Research Center of Ispra.
Upon freeze drying, the material lost more than 90% of its mass reducing to ca. 30 Kg.
The material was ground for 15 days in a zirconia ball mill, avoiding material contamination.
The material was then sieved in three days by passing a titanium sieve with apertures of 125 µm and separating the fibrous part.
Successively, the material was mixed for 15 days, under argon atmosphere, in a special polythene lined mixing drum. The argon atmosphere was renewed after 5 and 10 days.
The material was bottled in batches of 40 remixing the material for 30 minutes after collection of each batch. Three bottles were selected randomly out of each batch and set aside for the homogeneity and stability studies.
1000 bottles, with a content of 15 g ± 5%, have been prepared. The content of the single bottle was determined by weighing each bottle before and after the filling.
An accurate control of the potential contamination sources has been performed during the whole collection and preparation procedures.
The flow chart of the whole preparation procedure is reported in Fig. 2.
After preparation, homogeneity and stability tests were carried out. The results have been published elsewhere [14].

9. References

1. Quevauviller, Ph. *Mar. Polut. Bullet.* submitted
2. Quevauviller, Ph., Cofino, W.P., Vijverberg, J., Wells, D.E. and Griepink, B. (1992) *Quality Assurance in Marine Monitoring,* EUR Report, 14297 EN, European Commission, Brussels.
3. Cantillo, A.J. and Lauenstein, G.G. (1993) in M. Parkany (ed.), Quality Assurance for Analytical laboratories, The Royal Society of Chemistry, Cambridge, pp. 34.
4. Bailey, S., Wells, A.S. and Wells, D.E. (1994) *Mar. Pollut. Bullet.* **29**, 187.
5. Aminot, A. And Kirkwood, D. (1994) *Mar. Pollut. Bullet.* **29**, 159.
6. Pedersen, B. and Cofino, W.P. (1994) *Mar. Pollut. Bullet.* **29**, 166.
7. Wells, D.E. and de Boer, J. (1994) *Mar. Pollut. Bullet.* **29**, 174.
8. Quevauviller, Ph., Maier, E.A. and Griepink, B. (1995) in A. Townshend (ed.), *Encyclopedia of Analytical Science,* Academic Press, Vol. 7, pp. 289.
9. Maier, E.A., Quevauviller, Ph. and Griepink, B. (1993) *Anal. Chim. Acta* **283**, 590.
10. Hartley, T.H. (1990) *Computerized Quality Control: Programs for the Analytical Laboratory (2nd ed.),* Ellis Horwood, Chirchester.
11. Cantillo, A.Y. (1993) *Standard and Reference Materials for Marine Science, NOAA National Status and Trend Program for Marine Environmental Quality (4th ed.),* US Department of Commerce, Rockville MD.
12. Quevauviller, Ph. (1996) *Fresenius J. Anal. Chem.* **354**, 515.
13. Quevauviller, Ph., Astruc, M., Ebdon, L., Muntau, H., Cofino, W., Morabito, R. and Griepink, B. (1996) *Mikrochim. Acta* **123**, 163.
14. Quevauviller, Ph., Chiavarini, S., Cremisini, C., Morabito, R., Bianchi, M. and Muntau, H. (1995) *Mikrochim. Acta* **120**, 281.

DISTRIBUTION MODELS OF POLLUTANTS IN THE MARINE ENVIRONMENT

R.CECCHI and G.GHERMANDI
Dept. Scienze dell'Ingegneria, University
via Campi 213/a , 41100 - MODENA (ITALY)

I. Introduction

The pourpose of this work is to outline a simple description of the distribution of pollutant species (trace inorganics) in the marine environment.
In development of a dispersion model, it is necessary to clearly define the system of interest and to identify its boundaries. Both on global scale, and also at a lower one, the most frequent approach is to subdivide the earth into a number of well defined boxes (or reservoirs), each of them corresponding to a part of the system that, respect to the phenomena of interest, may be regarded as a single phase (with which the condition of homogeneous equilibrium deals. Taking the ocean as a single box, this assumption means to consider it as an electrolyte solution).
The bio-geochemical cycle of a substance in the earth surface is generally described by a three box model: land, atmosphere and ocean. Once defined the boxes, the performance of a model requests to identify the transport agents of the substance between the boxes and to evaluate the corresponding fluxes. The transport paths on a global scale follow usually streams, ground water flow, ice, rain, wind.
Obviously the previous knowledge about possible interactions between each box and its surroundings, and the existing data on the chemical-physical and biological characteristics of the studied substances will determine the detail and in some case also the scale at which the model may be approached and the fluxes defined.
The ocean for most elements non involved in biological processes over periods of the order of thousand years may be

considered a geochemically homogeneous system, and be thougth as only one box with a wholesale chemical properties (i.e. a single phase). It is not certainly the case of pollutant behaviour in the marine environment, at first for their presence in the ocean from few hundreds years and also because several of them interact significantly with biota.

Therefore to describe pollutants distribution in the marine environment we assume the ocean to be our system and subdivide it into more sub-boxes (at our extent upper mixed layer, deep layer and bottom sediment), taking into account of the diffusive, advective and biological transport processes inside the ocean in addition to the input and output transport agents to the ocean itself.

A number of pollutants, and trace inorganics among them, can be considered an increase in injection of material into natural cycles: addition to streams and groundwater of toxic minor elements as Zn, As, Cd, Se, Pb and also to ocean and lakes of nutrients, animal wastes, petroleum, radioactive nuclides. Into the ocean the effect of man invasion improves concentration gradients of substances involved in transport processes between ocean sub-boxes, so that pollutant distribution produces also a useful tracer of water motion. On the other hand, standing the lack of data about fluxes to the ocean and the concentration in seawater of the most pollutants, their dispersion model are limited at present. The analytical difficulties in measuring significant trace concentrations in sea water, strictly depending on the ability to avoid sample contaminations, are one of the more relevant problems in approaching pollutant distribution study: multi-element analysis and high sensitivity techniques, performed with ultra-clean procedures, must be applied.

The chemical constituents of sea salt can be notoriously divided in three major categories [1]: the biolimiting, the biointermediate and the biounlimited constituents, depending on their interaction with living organisms. The biolimiting constituents are almost totally depleted in surface waters by plants and animal activities: in addition to the so called nutrients (phospate, nitrate, silicate) also elements like Zn, Cd, Ge follow this behaviour. These constituents fixed into particulate material are carried by gravity back to the deep sea, where their concentration strongly increase because they are largerly redissolved and may become available to organisms when deep water is returned to the surface. A constituent is biounlimited if its ratio to clorinity and salinity is substantially equal in surface and deep water (with measurement errors). Na, K, Mg, B, S, Fl, Cl,

Br and Rb, Cs and U are currently classified in this category. These elements behave as conservative in seawater. The biointermediate elements Ca, Sr, Ni, Cu, Se, C, Ba, Ra are partially depleted in surface water respect to deep water. Their source in deep water may depend on absorption on organic matter both living and dead.

The constituents that fit in this classification are recycled within the sea several time before removal to the sediment. Some elements are so reactive that they pass throught surface water only once, then firmly fixed onto particles fall throught the sea to the sediment: these noncycling constituents are enriched in shallow water and depleted in depth. Mn, Hg, and also ^{3}H, Pb, Al, Bi fall into this category. They reside in the ocean a very short time. It is important to stress that the elements that exhibit a typical recycled profile can be scavenged from solution as well (i.e. Ni, V, Cu, Zn, Fe) and, conversely, scavenged profile does not preclude involvement in metabolism of organisms (Mn, Co, Al).

2. The equation system for the box model of substance distribution

Natural and anthropogenic substances from earth surface enter in the ocean mainly by stream transport. The 98% of the total transport to the ocean follows the agents related to water cycle, in addition to streams, icebergs and ground water, the remaining 2% being dust particles and gas from atmosphere, carried to the ocean by rain, or as dry fall out, or directly dissolving in sea water. The ocean is left by net substance deposition in the bottom sediment (remobilization may occur) and, once buried, the substances follow the processes of soil formation (up to uplift). Output paths from ocean to atmosphere are by water evaporation, ejection by breaking waves, in aerosols or as gas generated in sea water.

The input-output fluxes are generally expressed by the transport agent flux (for example river discharge) times concentration of the studied substance in it (river water), and represent the amount of a constituent transported in a given time (mass per year) across a box boundary, already integrated over the whole boundary surface exposed to the flux.

The recent origin of anthropogenic pollutants restrict the study on a time scale not certainly comparable with the ocean age, so that the substance cycling related to earth crust evolution (uplift) are obviously out of interest.

The mathematical formulation [2] of a box model for a substance is based on the mass conservation equation for the substance in each box (or sub-box) and then on all the equation together to form the mass balance of the investigated system. The mass conservation condition requires the change of mass of a substance i in each box (and then in the whole system) dM_i/dt to be equal to the input flux $F_{in\,i}$ plus production term P_i, minus the output flux $F_{out\,i}$ and the consumption term C_i. (P_i and C_i. having the dimensions of fluxes)

$$dM_i/dt = F_{in\,i} + P_i - F_{out\,i} - C_i \qquad (1)$$

The term in (1) are composed, i.e. expressed with respect to the mean substance concentration in each box, depending on the different transport agents and the numerous phases (generally the more representative in the ocean are dissolved in aqueous phase, suspended in particles, bounded to biota), and the various chemical species (oxidation states) in which the sustance may occur (for conservative, i.e. not reactive, chemical species, $P_i = C_i = 0$; their concentration in the box may be modified only by processes taking place at the boundaries).

In principal for each species and for each phase in a box a differential equation would be written, involving the simultaneous solution of a number of equations pare to the number of phases times the number of species.

If speciation is not taken into account (as for the total concentration of an element in a phase or for conservative substances) an equation for each phase of each box will result. The further semplification of steady state is suitable for substances whose average concentration in the phase does not change over a long period of time; in worldwide models of material dispersion the constant composition of atmosphere, ocean, earth's crust for hundreds of million years is assumed. These semplification correspond to put dM_i/dt equal to zero in (1), all the term in the equation resulting independent of time.

More difficult is the construction of time-dependent concentration models, in which fluxes for transfer of material may be considered a time function and additional time-dependent input to boxes (as in the case of anthropogenic substances) may be evaluated. The approach is particularly hard, standing the lack of knowledge about minor and trace elements: several investigations in the last twenty years have produced data more and more reliable, but for many pollutants it

is not possible to perform a precise model of bio-geochemical cycle.
The mathematical formulation of time-dependent concentration proposed by Albarède [3], starts with the consideration that critical aspects of the evolution of a system, e.g. the ocean internal structure, can often be adequately described with a limited set of geochemical variables. These one are typically concentration of substances, concentration ratios and isotope composition that evolves in funcion of some parameters such as, in the ocean, temperature T, salinity S, dissolved CO_2 and so on. A such variable c being a funcion of time and of the previous mentioned parameters, the rate of change of c per unit time results as:

$$dc/dt = C\ [\ c(t),\ x(t),\ t\ ] \quad (2)$$

where x(t) is a time-dependent external parameter and C (c, x, t) any suitable funcion. Only one external parameter is needed to illustrate a general behaviour of a dynamic system.
Assuming c_0 and C_0 the values of c and C for t = 0, and β = $(\partial C/\partial c)_0$, i.e. the derivative taken at t = 0, the mathematic leads to the solution

$$c - c_0 = (C_0/\beta)\ (e^{\beta t} - 1) \quad (3)$$

Only negative values of β lead to physically bounded values of c. The reciprocal of β has the dimension of a time and is called the relaxation time τ of c in the system (it has the meaning of a residence time) and is a measure of how fast an isolated system adjusts to a change in its condition (the pace of change). C_0 is the forcing constant and produces a systematic drift in the chemical state of the system depending on interaction with the system surroundings: it tells where the system goes.
Generally, most systems have their rate of changes described in the general form

$$\tau\ dc/dt + c = h\ (x,t) \quad (4)$$

where τ is possibly τ(t) and h is a forcing funtion of the time dependent parameters. h can be deterministic if the funcion is exactly known for each t (it is not always the case for trace concentration in interacting reservoirs) elsewise h can be stochastic (only its statistical properties such as mean and variance are known, as in the case of diffusion).

In single-variable analysis, the mean residence time τ for a conservative species i in steady state in a reservoirs results finally

$$\tau = V/Q \qquad (5)$$

where V is the box volume and Q the input, equal to output, volume flux of material (water, gas, sediments....). This parameter (equally defined as the ratio between the total mass of the species i in the reservoir and the mass supplied per year) does not depend on the nature of the species i, as long as it is not reactive, and correspond also to the residence (or renewal or flushing) time τ_H of water (for example, if it is the sea) in the reservoir.

In the case of a reactive species, for example, the species i is added from input Q of material - with i concentration c_{in_i} - to the reservoir ocean of volume V, in which the concentration of i in liquid phase is c_{liq_i}, and the reservoir releases an equivalent output Q, sedimentation take places in the box at a rate S and the species i is entrained by the sediment with a concentration c_{sed_i}. The mass balance equation results

$$dVc_{liq_i}/dt = Qc_{in_i} - Qc_{liq_i} - Sc_{sed_i} \qquad (6)$$

Introducing the solid-liquid partition coefficient D_i, so that

$$c_{sed_i} = D_i c_{liq_i} \qquad (7)$$

the residence time τ_i of the species in the reservoir [4] is

$$\tau_i = V/(Q + SD_i) = \tau_H / \alpha_i \qquad (8)$$

where $\alpha_i = (Q + SD_i)/Q$ is a coefficient that measures the reactivity of an element in the reservoir and is equal to unit for non-reactive species; the forcing terms are inversely proportional to α_i.

Examples of comparative evolution of the concentration for non-reactive and reactive species are given in fig. 1 [3], when mass balance equation is solved in the case of input concentration change at t=0 (i.e. C_{in_i} = constant for $t \geq 0$). The steady-state is established more rapidly for a reactive than for a non-reactive species and the steady-state concentration will be smaller.

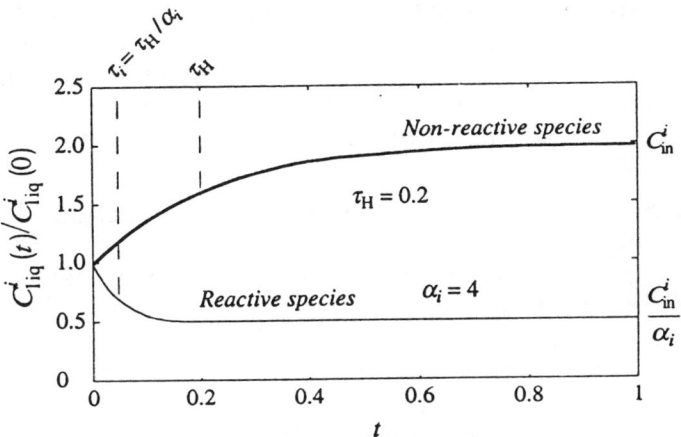

<u>Figure 1</u> - Concentration behaviour for non-reactive and reactive species in the case of doubled input concentration at t = 0. In the example τH = 0.2 (water residence time) and α_i = 4 (reactivity coefficient of the reactive species).
Reproduced from the book "Introduction to Geochemical Modelling" by Francis Albarède, 1995, with kind permission of Cambridge University Press, Cambridge, GB.

3. The box model for the system "ocean"

The ocean is now regarded as subdivided in upper mixed layer, lower deep layer and sediment. The input from atmosphere and streams to the ocean and the net release in marine sediment are expressed as products of each transport agent flux times substance concentration in it. The evaluation of fluxes between the sub-boxes requests the inclusion of advective and diffusive terms in the mass balance equation.
The standard diffusion equation for the stable species i (concentration c_i) in an incompressible medium flow with advection velocity **v** and density ρ, with Ak_i the production (>0) or consumption (<0) rate of the species in the kth reaction (if = 0 for all k, the species is conservative), results from the fundamental transport equation with the only assumption of mass conservation as [3]

$$\delta c_i/\delta t = \kappa_i \nabla^2 c_i - \mathbf{v} \nabla c_i + \sum_k Ak_i/\rho \qquad (9)$$

κ_i being the molecular kinematic diffusivity coefficient (m²/s), typical of each transported species/property, as given by the Fick's law. $\mathbf{v} \nabla c_i$ is the advective term. $\sum_k Ak_i/\rho$ is the in-situ source term, dependent on space coordinates and independent on concentration. This equation applies to an infinitesimal fluid element: it is derived considering the net flux of a species i for a small fluid element and equating it to the total rate of change of i inside the element. For example, the flux of i in z direction is $\kappa_i (\partial c_i/\partial z)$, and the net flux inside an infinitesimal fluid element is $\kappa_i \partial(\partial c_i /\partial z)/\partial z$.

In the most general case the diffusion equation is a partial differential equation of the first order in time and of the second order in x,y,z (space coordinate). Concentration or flux conditions valid at any time >0 must be given along the entire boundary.

Some easily adsorbable metals in the ocean are removed from the water column by falling particles with strong surface reactivity, such as oxi-hydroxides: this is the case of many transition elements, the rare earth, thorium. On the scale of the ocean, molecular diffusion is an inefficient transport process. It becomes significant at the water-sediment interface, induced by concentration gradients between pore water and overlying sea water.

However, the prevalent turbulent transfer in water column is commonly described via the same phenomenological equation (9). For a given local water dynamics in the ocean, the turbulent or eddy diffusivity (viscosity) K_e coefficient describes how fast eddies are transported; it is a property of the flow and not of the fluid. It also measures the efficiency of the transport down the concentration gradients in much the some way of molecular diffusivity. It is larger by several orders of magnitude and, being associated with bulk material transport, dominated by the turbulent flow field, it is common to assume that eddy diffusivity is the same for all scalars. The range of values for eddy vicosity coefficient for each property are similar (probably have the some order of magnitude) to those of the momentum (being the larger one), so that a complete equalization of the properties of the water not necessarily follows the immediate equalization of the momentum; in addition different values of the coefficient attain to vertical and horizontal mixing, and it may vary with space coordinates depending on the particular fluid motion. It is finally common use to assume this coefficient identical for all the elements, sometimes also without distinction between mixing direction, and independent on x,y,z.

More special cases can be derived from this general equation; the physical-oceanographic solutions for a conservative property in stationary condition (i.e. salinity), describe typically transverse mixing (in one direction normal to the flow) or mixing in all directions without average water transport in anyone (advective term = 0). The distribution of a property along a homogeneous turbulent flow results tongue-shaped; it is little affected by the different initial conditions (property and velocity behaviour over the initial transverse cross section) because this distribution shape is largely a consequence of the turbulent mixing [6]. A similar pattern (already described by Sverdrup) results also as the sole effect of pure mixing in horizontal and vertical directions without advections.

The one-dimension case with first-order removal kinetics [5] set the source term in (9) equal to zero and leads to this equation for the stationary state profile :

$$K_e \, d^2c_i(z)/dz^2 - v \, dc_i(z)/dz - Hc_i(z) = 0 \qquad (10)$$

where z (taken as positive upward) is the depth below the ocean surface, v the up-welling velocity, and H the first-order scavenging-rate constant. For conservative reactions H = 0.
Defining the mixing length $l_m = K_e/v$, obtained from the distribution of conservative quantities (salinity, potential temperature) and the scavenging length $l_s = v/H$, the solution is:

$$c(z) = a \exp[(\varepsilon -1)z/2l_m] + b \exp[-(\varepsilon +1)z/2l_m] \qquad (11)$$

where $\varepsilon = \sqrt{(1 + 4 \, HK_e/v^2)}$
and a, b are two constants to be determined from the boundary conditions.

If boundary conditions are given, for example concentrations are known at the top (z=0) and at the bottom (z=Z <0) of the scavenged layer (sub-reservoir), the flux of dissolved element reduced as f(z) may be deduced, and the flux of an element carried downward by sinking particles (the "rain") can be estimated [3] by comparison with it (in steady state, assuming the sum of dissolved and particulate fluxes remains constant). See in fig. 2 the behaviour of concentration and fluxes in the water column for the l_s/Z values labelled on the curves [3].

The inverse calculation (from measured element concentrations at various depth) allows to find the best set of advection-diffusion model coefficients.

The further concept of mixing time is related to stirring processes that levels off heterogeneities, and is sample size dependent: for a given sample size (a water bottle in the ocean),

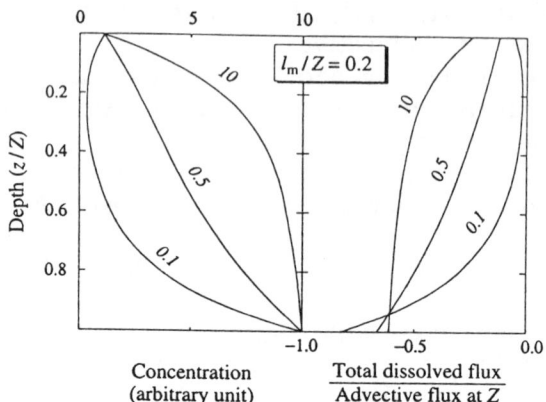

<u>Figure 2</u> - Concentration and fluxes in a water column of depth Z, mixing length l_m and scavenging length l_s, obtained from advection-diffusion model with these boundary condition: concentrations known at the top and at the bottom of the scavenged layer. In this example $l_m/Z = 0.2$, and l_s/Z values are written on the curves.
Reproduced from the book "Introduction to Geochemical Modelling" by Francis Albarède, 1995, with kind permission of Cambridge University Press, Cambridge, GB.

an element is homogeneously distributed in a system if a suitable dispersion parameter, such as the standard deviation of concentration, falls below a critical level.
An appropriate scale for the mixing time is the reciprocal of the local velocity gradient. If residence time is significantly longer than the mixing time, the systems level off changes faster than they are introduced from the surroundings. If mixing time is longer than residence time, stirring is slow relative to external perturbations and the system is heterogeneous (more variable concentrations).
The mass conservation equations (1) of a species i for the three boxes of the ocean system, upper layer, deep layer and sediment (fig. 3), may be derived from (9) integrated onto each box volume and introducing also input and output flux terms as defined in section 2.
If $V^{U,L,S}$ are the whole volumes of upper layer, lower layer and sediment, $A^{U,L}$ and $A^{L,S}$ the interface areas between upper and lower layer and lower layer and sediment respectively, and taking into account only of the terms resulting significant, the equation system for the three boxex finally results (explanation of the other symbols in fig. 3):

upper layer:

$$V^U(dc_i/dt) = F_{in i}{}^A + F_{in i}{}^{R,D} + F_{in i}{}^{R,S} +$$
$$+ (vc_i{}^L + K_e [dc_i(z)/dz]_{z=Z^U})A^{U,L} - F_{in i}{}^S \qquad (12a)$$

lower layer:

$$V^L(dc_i/dt) = \kappa_i \, [dc_i(z)/dz]_{z=z^L} \, A^{L,S} -$$
$$- (vc_i{}^L + K_e \, [dc_i(z)/dz]_{z=z^U}) \, A^{U,L} - F_{in\,i}{}^S \quad (12b)$$

bottom sediment:

$$V^S(dc_i/dt) = F_{in\,i}{}^S - F_{out\,i}{}^S - \kappa_i \, [dc_i(z)/dz]_{z=z^L} \, A^{L,S} \quad (12c)$$

where the first members may be expressed by (2) and the flux $F_{in\,i}{}^S$ throught the whole box interfaces comes from the reaction term in (10), or more simply from (7) when c_i is assumed constant in each box and pare to its three average values $c_i{}^{U,L,S}$.
The solution of this system, in addition to the mathematical approach, mainly depends on availability of flux estimations (it is necessary to distinguish natural inputs from human activity sources) and on reliable concentration data and diffusion coefficients (generally fitted from measured concentrations, as previously mentioned). Examples of the present day cycle of pollutants like mercury and lead are proposed in licterature [2, 7] : the same references give details about type of data and approximations used in the construction of the model.

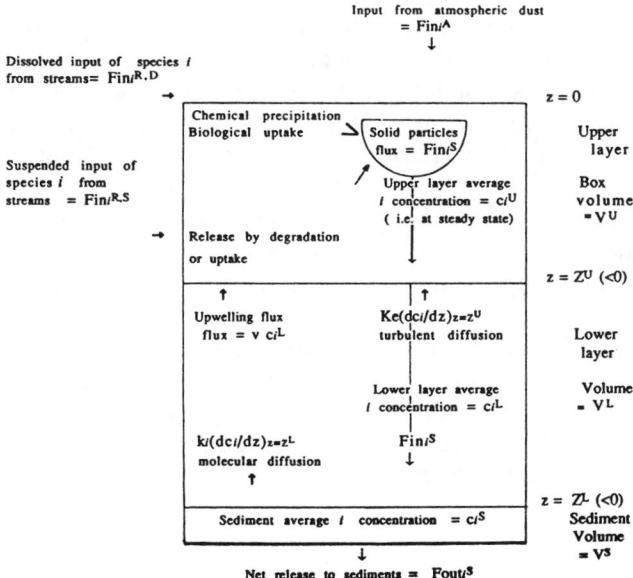

Figure 3 - The system ocean subdivided in three sub-reservoirs: upper mixed layer, lower deep layer and bottom sediment.

The composition of seawater is markedly influenced by growth, distribution and decay of phytoplankton and other organisms. At biologically utilized constituents (biolimiting and biointermediate) better attain a kinetic simple model proposed by Broeker[1].

The more relevant aspect of this model is the consideration of element recycling procesess respect to oceanic mixing cycles. The one process that yields geographic and depth variation in the chemical composition of sea salt components is uptake of dissolved constituents by organisms. The remainss of these organisms sink under the influence of gravity and are gradually destroyed by oxidation. The superimposion of this particular cycle upon the ordinary mixing cycle of the sea accounts for the present distribution of chemical properties.

The Broeker's model assumes the ocean subdivided in only two boxes, corresponding to the two mayor water type, warm surface water (few hundred meters depth) and cold deep reservoir (about 3200-m depth), separated by the oeanic thermocline, the density gradient zone which provides obstruction to vertical mixing.

Further semplification is to take runoff from the continent as the only way of addition of the constituent to the sea. Second, it assumes that biologically utilized constituents are removed from the ocean only by the fall of organism-produced particles to the sea floor, a small fraction of this debris being permanently buried in the sediment. It is also assumed that the ocean (and its two compartments) is at steady state, and the biologically active constituents that pass throught the ocean partecipate to its internal cycle. Warm surface waters receive their supply from river water and deep ocean water which exchanges with surface water: these two inputs are matched by loss with downwelling water and with settling particles. Standing that the concentration of biologically active constituents in warm water is less than in deep cold water, and that the water exported to the cold reservoirs carries away less of each - dissolved - constituent replaced by upwelling, the particulate loss must account for the difference.

Assuming that in present-day the ratio of the flux of up-downwelled water to the flux of river discharged water (both extimated per year) is about 30 [1], it is possible to define "g", the fraction of a given constituent i reaching the surface reservoir removed in particulate form, as

$$g = 1 - 30(c_i^U/c_i^R)/(1+30\ c_i^L/c_i^R) \qquad (12)$$

where $c_i^{U,L,R}$ are the concentrations of the constituent in surface ocean, deep ocean and average river water respectively.
Furthermore it is defined the fraction "f" of a given constituent carried to the deep sea in particles that survive destruction (that must be equal to the river input) as:

$$f = 1/\ [1+30(c_i^L/c_i^R - c_i^U/c_i^R)] \qquad (13)$$

so that the fraction of an element removed per oceanic mixing cycle (i.e. per transfer from the cold to the warm box) will be

$$f\ g = \frac{\text{export into the sediments}}{\text{total flux into surface ocean}} = 1/\ [1+30\ c_i^L/c_i^R] \qquad (14)$$

Finally, to obtain the residence time τ required to remove an amount of a constituent equal to that stored in the sea today, it is necessary to divide the time required for one mixing cycle of the ocean, resulting pare to 1000 years from C^{14} distribution in the ocean, by the product "f g", that gives $\tau = 1000/fg$ (in years).
Typical values of τ results of the order of $10^4 - 10^5$ for the most important biolimiting and biointermediate constituents. For biolimiting elements g is closed to 1 and f<0.1; for biounlimiting elements g must be very small (g<0.01). It follows that for a biolimitig element the ratio c_i^L/c_i^R is about equal to 1/30f, while for a biounlimiting the same ratio result about equal to 1/30fg and the ratio c_i^U/c_i^L is about one [4].

4. Trace elements distribution profiles in the ocean

Several trace pollutants are biologically utilized, but the assumption of steady state and the comparison of their behaviour with oceanic mixing time makes difficult the immediate application of the described model. Nevertheless, for the pollutant constituents that follow this trend a nutrient type profile is currently referred. As an example, the vertical distribution of Cd, Zn, Ni and of Se are similar to those for the nutrient constituents, for Ge and Ba. Cu also interacts with marine organisms [1, 8].
Cd profile has the shape of PO_4 one and Zn of H_2SiO_4, the correlation between Cd and PO_4 beeing also better than between

NO$_3$ and PO$_4$. Furthermore, the difference between the mayor ocean basins (Pacific-Atlantic) observed for nutrients is nearly the same for Cd as well (the Cd/P ratio in major central water masses is about 15% higer for North Pacific respect to North Atlantic). There is not a simple biological explanation of these findings, but the pathway of these elements through the sea are surely determined by living matter formation and destruction; Cd appears to be taken into tissues and regenerate at the same rate as phosphate (and nitrate) during decomposition, for reasons yet unknown. The deep maximum observed for Zn and SiO$_2$ is ascribed to a deep regeneration cycle.

The oceanic chemistry of Ge is partly similar to that of Si, but shows also striking differences. Even if discriminated by some organisms, Ge is taken up into the silica tests and released when they dissolve in the same ratio with Si: Ge/Si about equal to 0.7 10^{-6} both in tests and in the water column, where Ge show a profile exactly like that of Si. The organic compounds of Ge (monomethyl and dimethyl) are present in the sea in greater concentration than the inorganic form, and show perfectly conservative behaviour. Their source is unknown, because the river input of these compounds is very low and they show conservative mixing, just like the salt from the seawater.

Barium follows the same nutrient behaviour of Zn and Ge.

Ni is a biointermediate constituent; its c_i^U/c_i^L ratio is about five (as Ba one). Its curve distribution with depth is similar to that of Cd in the upper 1000 m and that of Zn in deep water. For Ni the correspondence with phosphate is not so close as for Cd, and evidently the release rates of Ni from sinking particles would seem to be rather slower than those of phosphate.

Both the two different oxidation state, Se(IV) and Se(VI), show an interesting nutrient-type profile, and the difference between them is roughly constant, with combination of shallow and deep generation as in the case of Ni. Little is known about the causes of these behaviours [8, 9].

Cu concentration increases regularly with depth, with c_i^U/c_i^L about ten in North Pacific water. It not follows a nutrient type profile, even if its movement from surface water to the sediment is ascribed to absorption on particulate (also organic) matter: when plotted against potential temperature, the curve of Cu concentration has a concave shape, while the other constituents have a convex one. The Cu content of water for any intermediate temperature is less than would be expected from mixing of overlying and underlying water, then there must be a net

removal of Cu from the interior of the deep water column and a net addition of the other constituents [1].
Some elements that not have an essential biological role may follow the distribution type of their chemically similar more aboundant vertical neighbours in the Periodic Table, (uptake by analogy): this is the case of Ag respect to Cu.
Pb distribution in North Pacific and North Atlantic is unlike that of any other element, showing highest concentration in upper water that dramatically decrease with depth. Furthermore, the concentrations are unusually highest in the Atlantic. This situation depends on the measurable increase throughout possibly the entire ocean of Pb concentration, because of the atmospheric transport of anthropogenic lead (gasoline additives). The higher concentrations in the Atlantic are due to the high rate of burning of leaded gasolines in countries bordering the North Atlantic, compared to the Pacific, as well as to the greater volume of the Pacific and the average direction of the winds [9]. Stable lead is distributed with depth (adsorbed on particulate) similarly to anthropogenic tritium. Pb bulk is sufficient to impact sea chemistry, when tritium can also serve as a tracers of oceanic processes like water movement through the thermocline: Broeker [1] has provided an interesting interpretation of thermocline ventilation from anthropogenic tritium distribution.
Aluminum vertical profile shows a remarkable surface enrichment, then a mid-depth minimum and finally an increase with depth. This distribution is belived to be natural, without anthropogenic inputs, and depending on dissolution of atmospheric dust in surface water, with scavenging at mid-depth and further release at greater depth, but the processes involved are not understood. The lower concentration in Pacific respect to Atlantic water is ascribed to continuous scavenging [9]. This mid-depth minima profile, already noticed for Cu, is common also to Sn [8].
A number of minor and trace elements appear to be relatively conservative in seawater, with constant ratio to clorinity or salinity: Li, Rb, Cs, Tl, U, Au, and Mo [1, 8, 9].
The described profiles suggest that Cd and Zn are recycled many times in the ocean as the nutrient elements, even if this correlation may be affected by the signature of the inputs of these elements in the sea. Tritium and Pb differ greately, giving the apparence to be removed rapidly and irreversibily from the sea (i.e. very short residence time). The estimates of river inputs confirm that Cd, Cu and Zn reside in the sea longer than

the ocean mixing time. The residence time of the organic compounds of Ge in the ocean must be very long, because their concentration are so uniform, at least in the examined oceanic regions.

A further notation that underline the biological affinities of Cd, Zn, Ni, Cu, Se is their unexpected enrichment in atmospheric dust and aerosols: the release to the air could be either from terrestrial vegetation or from froth blow off of the sea surface [1].

5. The support of PIXE technique in the study of trace inorganic distribution patterns

High sensitivity analytical techniques and adequate procedures result of primary importance in order to perform reliable concentration measurements of trace inorganics in the marine environment.

The study of the described processes requires the examination of different media, such as air, water, sediment and living organisms. Among the analytical techniques that employ accelerators in the study of the environment [10], Particle Induced X-ray Emission (PIXE) has been proven an useful tool for the study of trace element distribution in ecosystems, very suitable to check trace inorganics diffusion in the sea. The major features of this technique are multi-element analysis, high sensitivity and excellent detection limits across a wide range of atomic numbers. Additional advantages is the speed of analysis of a wide variety of samples.

PIXE is an analytical method based upon X-ray spectrometry. A particle beam is used to eject inner-shell electrons from atoms in a specimen (target). When the resulting vacancies are filled by outer-shell electrons, characteristic X-ray whose energies identify the particular atom are emitted.

PIXE is performed by a small particle accelerator (the most widely used is the Van de Graaff machine at a 2-3 MeV voltage, more recently also small tandem accelerators), that provides a beam of protons (in few cases helium or heavier ions). The beam merging from the accelerator, stabilized and directed by electrostatic and magnetic steering elements, in a high vacuum line, enters the specimen chamber through suitable collimators. The X-ray detection system, housed in the chamber, is generally a Si(Li). This detector combines the advantge of high efficiency in the X-ray energy region of interest (typically 2-20 keV) with a good energy resolution (can fully resolve the Kα emissions of neighbouring transition elements)[11].

A typical PIXE spectrum consists of characteristic X-ray peaks superimposed on a background due to atomic bremsstrahlungs and nuclear reactions induced by the beam. The area of each peak is related directly to the concentration of the corrisponding element in the specimen. Obviously the total element concentration is measured, not distinguishing among different oxidation states. Various software codes has been developed to deconvolute spectra and to accurately calculate peak areas: from the basic physical quantities the absolute amounts of elements present in the specimen may be obtained, even if in some cases the experimental set-up is calibrated by means of standards (both internal and reference materials).

PIXE intrinsic detection limit is not very much below 1.p.p.m in a given matrix. It offers its maximum sensitivity (minimum detection limit) in the two atomic number (Z) regions ($20<Z<35$ and $75<Z<85$)[12]. In absolute terms it is an extremely sensitive method, but it must be stressed that adeguate sample preparation procedures are required to make full use of its capabilities. In fact, the concentrations of dissolved trace pollutant elements in sea water ranging from p.p.b. to fractions of p.p.t., a pre-concentration treatment is generally needed for these target preparation. Contamination must be obviously avoided during sample collection and handling.

Procedures of target preparation and suitable clean methodologies have been developed in Modena (Italy) to improve PIXE efficiency in analyses of non living environmental phases. This includes the study of the following media: sea and fresh water (where the suspended matter has been previously separated), sediment (the non lattice held fractions), pore water and atmospheric aerosol deposited in firn and ice cores [13].

Targets from natural water or solubilized (sediment) samples are prepared by deposition of carbamate elements on thin targets. This method involves minimal sample handling. The technique is based on the high stability of the carbamate of several elements and on the low solubility in water that is determined by pH: when it is kept at suitable values (typically 4), a large number of carbamate elements precipitate and can be collected by filtration on a thin backing (Nuclepore policarbonate filters, 0.4 µm pore size, 10 µm thickness, 1 mg/cm^2 areal density) using a co-precipitant agent. This procedure is suitable for natural water in which the trace elements are generally already in their more stable state, namely the one complexed by the carbamate chelant.

The methodology has been widely tested and calibrated with mono and multi element solutions (fresh and artificial sea water) and with reference materials for a group of elements (Ti, V, Cr, Mn, Fe, Co, Ni, Cu, Zn, Se, Hg, Pb, Bi, Mo, Y, Tl, Sb, Ag, Cd), in the p.p.t. - p.p.m. range, with pre-concentration of sample volumes from 500 to 15 ml. Satisfactory repetition results have been verified, with measurement errors generally lower than 10% and high recovery efficiency [14] for the most elements.

The detection limits [13] of the elements collected as carbamate on a Nuclepore filter are of some ng/cm^2 for Ti, V, Cr, Mn, Fe, Co, Ni, Cu, Zn, Se, Hg, Pb, Bi, Tl, of tenths for Sb, Mo, Y, and of about hundred for Ag and Cd.

The study of atmospheric aerosol from firn and ice samples involves also the determination of light constituents such as Na, Mg, Al, Si, P, S, Cl, K and Ca. These specimens are prepared by non boiling evaporation (lower than 60-70°C) ranging from 30 to 100 ml of the sample which has previously been melted at laboratory temperature. The sample is then spotted onto a thin polycarbonate backing and evaporated until dryness [15].

In addition to the already mentioned Na, Mg, Al, Si, P, S, Cl, K, Ca, also Ti, Cr, Mn, Fe, Co and Ni have been calibrated with this procedure, by means of external standards. Measurement errors are in the order of 15%, depending mainly on the slight variability of the proton flux.

The detection limits ranges from 15 ng/cm^2 for Na to 1 ng/cm^2 for Cr [15]. The recovery of heavier volatile elements is certainly impaired.

The PIXE set-up used for this research is at the INFN Laboratory at Legnaro (Padova, Italy). The incident particles are protons accelerated to 1.8 MeV by a Van de Graaff. The Si(Li) detector (resolution 150 eV at 5.9 keV, solid angle 0.0198 sr) is placed at 135° from the beam and in front of the target. Under these conditions the X-ray attenuation in the target is also limited. Various thickness mylar absorbers are placed in front of the detector. The PIXE measurement of a sea water sample typically requests about ten minutes. The set-up characteristics are detailed in [16].

The sensitivity of PIXE technique makes it suitable for the complete investigation of the environment, both in terms of pollution measurement monitoring as well as for geochemical and geophysical studies. At these extents, the technique is particularly useful, because it povides also a preliminary insight into the behaviour of the elements in the samples while showing the relative abundances of the elements present. Several environmental researches have been developed in the last twenty years by means of PIXE technique: for what concerns the authors of this lecture, the pollutant circulation at mainland interface in Venice Lagoon is under study [17]. An example of the effect of hydrodynamics on element distribution in the water column is shown in fig. 4, where the bottom current velocity trend along an estuarine mouth axis (in Venice Lagoon) is compared with the iron concentrations in water samples, collected at three different depths [18]. The Fe concentration in bottom samples increases following the current velocity behaviour, induced by tidal forcing, because remobilization of bottom material occurs. The effect is common to filtered and unfiltered samples, meaning that probably remobilization phenomena affect the whole water column.

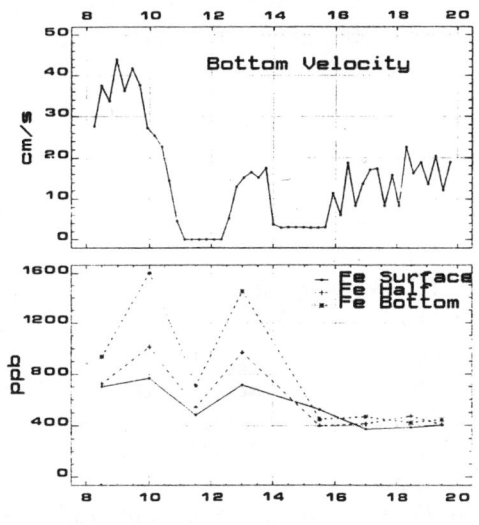

Figure 4 - The comparative behaviours of bottom current velocity along the axis of an estuarine mouth (in Venice Lagoon) and iron concentrations in water samples collected at three different depths.
Reprinted from [18] with kind permission of Elsevier Science - NL, Sara Burgerhartstraat 25, 1055 KV Amsterdam, The Netherlands.

The recent performance of microPIXE set-up allows to extend the PIXE capabilities to single particle investigation. This technique involves focusing the beam down to small areal dimension, typically a few micron, so that it can be scanned across the surface of a target and thus provide concentration data as a funcion of the position: in fact each irradiated point produces a PIXE spectrum. The ability to scan allows to obtain concentration maps of many elements in little portions of specimens: for example a single biological cell or an aerosol particle as well as the microscopic structure in minerals. The microPIXE set-up performed at 2 MeV Van de Graaff accelerator of INFN Legnaro Laboratory has a spatial resolution better than 1.5 µm with 100pA beam current [19]. These kinds of analysis obviously take longer time than a simple PIXE measurement.

The microPIXE is essentially analogous to a scanning electron microprobe, that uses a simple electron gun as injector; the increase in sensitivity achieved by proton microprobe may be of about 100, the detection limits of microPIXE beeing of the order of 10 p.p.m. of the examined specimen.

Development work on the proton microbeam is still in progress, but many applications both in research and in industrial field has been performed. The technique is largely used in biology and medicine (plants, bone and teeth, hair, skin, brain, various other organs and single cells are the investigated specimens) in geology (examination of small mineral or fluid inclusions) and environmental study (distribution of elements in single dust particles) as well as in materials analysis, metallurgy and solid state physics [20].

6. References

1 Broeker, W.S., Peng, T.H.(1982) *Tracers in the sea*, Eldigio Press, New York.
2 Mackenzie, F.T., Wollast, R.(1977) Models of Global processes in Goldberg, E.D. et al. (eds.) *The Sea* **6**, Wiley Interscience, New York , pp. 739 -785.
3 Albarède, F.(1995) *Introduction to Geochemical Modelling*, Cambridge University Press, New York.
4 Stumm, W., Morgan, J.J. (1981) *Aquatic Chemistry*, Wiley Interscience, New York.
5 Craig, H.(1969) Abyssal carbon and radiocarbon in the Pacific, *Jour. Geoph. Res.* 74, 5491 - 5506.
6 Defant, A.(1961) *Physical Oceanography* **1**, Pergamon Press, Oxford
7 Millero, F.J, Sohn, M.L.(1992) *Chemical Oceanography*, CRC Press, Boca Raton.
8 Craig, P.J.(1980) Metal Cycles and Biological Methylation, in O.Huitzinger (ed.), *The Handbook of Environmental Chemistry* **1A**, Springer Verlag Berlin Heidelberg , pp.169-227.

9   Pilson M.E.Q.(1989) Chemical Oceanography - Lecture Notes, Graduate School of Oceanography, University of Rhode Island .
10 Cecchi, R. , Ghermandi, G.(1986) The use of accelerators in environmental study, in S.Onori & E.Tabet (eds.) *Physics in Environmental and Biomedical Research*, World Scientific Publishing Co Pte Ltd., Singapore , pp.63-74.
11 Johansson, S.A.E., Campbell, J.L.(1988) *PIXE A Novel Technique for Elements Analysis*, John Wiley & Sons Ltd. , Chichester .
12 Johansson, S.A.E., Johansson, T.B(1976) Analytical application of PIXE, *Nucl. Instr. Meth.* **137**, 473-516
13 Ghermandi, G., Cecchi, R., Laj, P. (1996) Procedures of target preparation to improve PIXE efficiency in environmental research, *Nucl. Instr. Meth. B* **109- 110**, 63-70.
14 Cecchi, R., Ghermandi, G., Zonta, R.(1990) PIXE analysis to study metal diffusion and sedimentation phenomena in Venice Lagoon, *Nucl. Instr. Meth. B* **49**, 283-287.
15 Laj, P., Ghermandi, G., Cecchi, R.,Ceccato, D. (1996) Coupling PIXE and SEM/EDAX for characterizing atmospheric aerosols in ice-core, *Nucl. Instr. Meth. B* **109 - 110**, 252-257.
16 Aprilesi,G., Cecchi, R., Ghermandi, G., Magnoni, G., Santangelo, R. (1984) Calibration and errors in the detection of heavy metals in fresh and sea waters by PIXE in the p.p.b. - p.p.m. range, *Nucl. Instr. Meth.* (B3) 1-3 **231**, 172-176.
17 Cecchi, R., Ghermandi, G., Zonta, R.(1996) Geochemical processes in Venice Lagoon by PIXE technique: an overview, *Nucl. Instr. Meth. B* **109 - 110**, 407- 414.
18 Ghermandi, G., Cecchi, R., Costa, F., Zonta, R. (1991) Trace metal distribution in aquatic systems as studied by PIXE analysis of water and sediments, *Nucl. Instr. Meth. B* **56/57**, 677-682.
19 Boccaccio, P., Bollini, D., Ceccato, D., Egeni, G.P., Rossi, P., Rudello,V., Viviani, M.(1996) The LNL proton microprobe: original technical solutions and new developments, *Nucl. Instr. Meth. B* **109/110**, 94 -98.
20 Proc. VII Int. Conf. on PIXE and its Analytical Applications (1996), G.Moschini and V.Valcovic (eds.), *Nucl. Instr. Meth. B* **109/110**.

SEWAGE AND NUTRIENTS IN THE MARINE ENVIRONMENT: STIMULANTS FOR GOOD OR VECTORS FOR HARM?

Martin R. Preston
*Oceanography Laboratories, University of Liverpool,
Liverpool, L69 3BX, UK*

ABSTRACT

Sewage and associated nutrient elements are probably the most ubiquitous of marine contaminants. There are therefore comparatively few populated coastal areas where there is no evidence of sewage impact evidenced through elevated bacterial, viral or nutrient concentrations or, in extreme cases, by gross contamination by faecal solids or other sewage related debris. In addition, sewage, largely in the form of sludge, may be introduced to marine systems through dumping programmes (though in the EU this is to be phased out by 1998). This paper takes a wide ranging view of both the known effects of sewage contamination on marine systems and the consequences of using (and not using) the seas as a disposal option.

What are the effects of this contamination? Nutrient elements are essential for algal growth so increases in primary production might be expected to be a logical consequence of sewage discharges but, in excess, eutrophication may set in. This can have serious consequences for marine systems and industries dependent upon them such as fisheries and tourism. Sewage sludge dumping may cause organic solids to blanket the sea floor causing major changes in benthic communities; alternatively they may act as an additional food source to mid-water feeding fish. Sewage related, pathogenic bacteria may cause disease but without large numbers of degrading bacteria the persistence of organic matter is much greater. Coupled with these effects is the influence of sewage in introducing potentially harmful domestic or industrially used chemicals to marine systems and hence to marine food chains.

How may the effects of sewage pollution be assessed? The various alternatives for the scientific measurement of sewage pollution will be briefly reviewed. An example of a multi-parametric approach to sewage (and other contaminant) impact is described.

1. Introduction

The problems of human sewage disposal are as old as the human race itself but the option of direct land disposal has only really become a problem as centres of populations have evolved in which there is a low 'available land to population' ratio. This became a particular problem in Europe with outbreaks of plague, cholera and typhoid a relatively common feature of life in the Middle Ages. As populations grew so did increasing awareness of the aesthetic and health problems of sewage disposal and attention inevitably turned to the aquatic environment as a means of removing it from view.

The total population of the Earth is now close to 6 billion people and is continuing to rise at a rate at around 91 million people per year or 500 people every three minutes. A considerable proportion of these people live within coastal zones and the level and management of the impact of these pressures on coastal ecosystems is increasingly a matter for concern (1). Within Europe the rate of population growth is rather less than in many other parts of the world (2) but high coastal population densities are common particularly in summer months.

Inevitably, these populations produce large quantities of sewage (typically 200-300 L/person/day) which, after amounts of treatment which can range from none at all to advanced tertiary processing requires disposal may find its way indirectly or directly into coastal waters. Coupled with the liquid waste discharge has been, in some cases, the direct dumping of sewage sludge produced by treatment plants into the sea.

The consequences of sewage contamination of coastal zones can be considerable and range from aesthetic problems associated with faecal solids and other sanitary wastes, through diseases caused by pathogenic organisms to problems of eutrophication and disruption of marine ecosystems. In addition sewage discharges may act as a vector through which other contaminants (such as oil and agricultural and industrial wastes) may reach the seas.

Nevertheless, it is important to recognise that the effects of marine sewage discharges are not necessarily the worst possible options in waste management. There are cases where the use of marine disposal may, for economic, health or other factors may be, if not the best, the least worst of the available options.

2. The Nature of Sewage

Domestic sewage is 99.9% water containing a complex mixture of natural waste products (Figure 1). The organic components consist of nitrogenous compounds (protein and urea), carbohydrates (sugars, starches, cellulose), fats (cooking oils, soap, greases) together with increasingly large amounts of synthetic chemicals ranging from proprietary lavatory cleaners and disinfectants to medicines and/or their metabolites including synthetic hormones deriving from the use of the contraceptive pill. Inorganic components include trace metals and micronutrient elements. In addition there may be significant quantities of plastic or other non-biodegradable material such as condoms, feminine hygiene products and other debris from domestic sources.

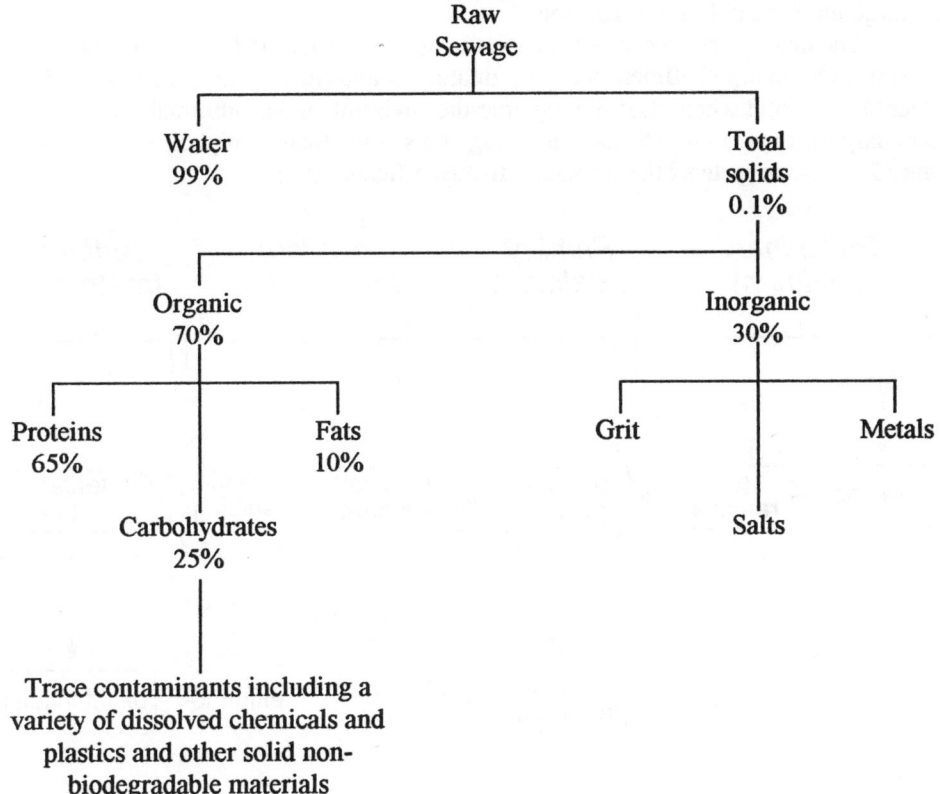

Figure 1 The composition of a typical raw sewage (Adapted from Lester, (3)).
Note that associated with all phases, but particularly organic solids, are very large number of pathogenic bacteria, viruses and other disease carrying organisms.

Where combined sewerage systems are used, there may be significant amounts of road grit, fuel and lubrication oils and a variety of miscellaneous debris which finds its way into the system.

2.1 SEWAGE TREATMENT AND DISPOSAL

The main problem associated with raw sewage apart from its aesthetic and pathogenic threat is the presence of large amounts of dissolved and particulate biodegradable organic matter which imposes a large Biological Oxygen Demand (BOD). Whilst BOD is technically a laboratory measurement of the amount of dissolved oxygen removed by bacterial oxidation of the organic matter in a sample during a 5 day incubation period, the effects of high BOD wastes on aquatic systems are considerable. In the case of raw sewage discharges to a river or estuarine environment the overall result may be the total removal of dissolved oxygen from the water and the death of very large numbers of living organisms.

The main objectives of sewage treatment are therefore to (i) separate solid material from the liquid effluent (primary treatment) and (ii) to reduce the BOD of the effluent to the point where its discharge into the environment has minimal impact (secondary treatment). A schematic drawing of a sewage treatment plant is shown in Figure 2. In some systems the secondary treated effluent may still

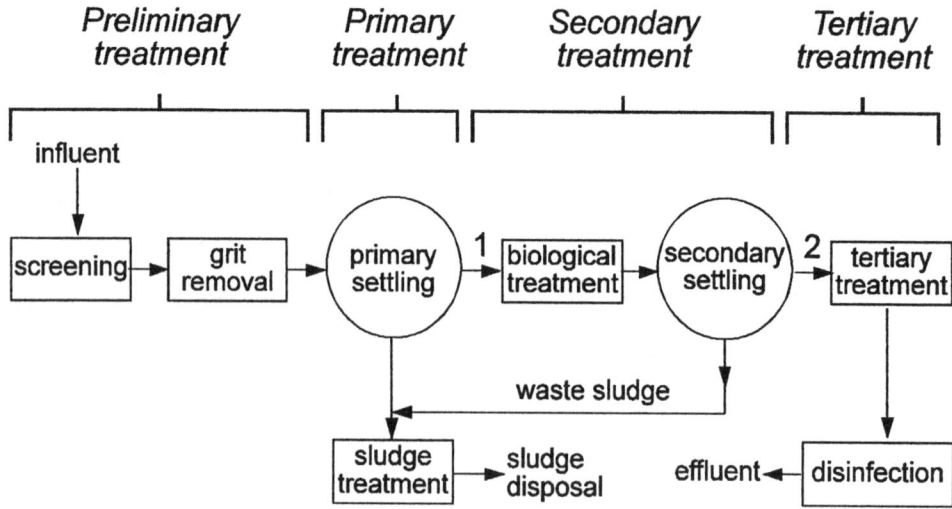

Figure 2 A schematic diagram of a sewage treatment plant. Assuming a raw sewage with a BOD load of 300 mg.L^{-1} and a suspended solids (SS) load of 400mg.L^{-1} the properties of the effluent at point (1) would be expected to be BOD ≈200 mg.L^{-1}, SS ≈150 mg.L^{-1} and at point (2) BOD ≈20 mg.L^{-1}, SS ≈30 mg.L^{-1}

contain components capable of producing an adverse impact (e.g. inorganic nitrogen and or phosphorus compounds). These may then have to be removed by a specialist (tertiary) treatment stage which may utilise purely chemical (e.g. coprecipitation) or biological (e.g. bacterial denitrification) to remove the problem.

The effects of differing degrees of treatment on the quality of receiving aquatic ecosystems have been summarised in an exercise which used the city of Athens as an example (Table 1). What is missing from this assessment exercise is a comparative study of the effects of the solid waste (sludge) disposal options where the terrestrial options are for (i) application to agricultural land (ii) landfill or (ii) incineration. Options (i) and (ii) may take place with or without digestion or sterilisation of the sludge. It is important to recognise the limitations of such procedures which include seasonal demand for sludge application on agricultural land,

TABLE 1 Treatment and disposal scheme options and environmental consequences for the sewage of Athens
(Sig - significant, Mod - moderate, Neg - negligible) (4)

Treatment and disposal scheme	Environmental impact with respect to:-						
	Dissolved oxygen depletion	Nutrients flux to W Gulf	Eutrophication and Transparency	Deposits	Dino-flagellate growth	Fish production	Comparative cost estimate
Primary treatment followed by discharge through a long outfall	Serious	Sig	Neg	Serious	Serious	Sig	100
Partial biological treatment followed by discharge through a long outfall	Sig	Sig	Neg	Sig	Sig	Sig	160
Biological treatment followed by discharge through a short outfall	Neg	Sig	Sig	Neg	Neg	Neg	180
Biological treatment and partial nitrogen removal followed by discharge through a short outfall	Neg	Mod	Neg	Neg	Neg	Mod	200

accumulation of toxic metals in treated soils and considerable public resistance to the construction of incinerators. The option of marine disposal of sewage sludge has been utilised by a number of countries around the world (Table 2) but is now, at least in Europe being phased out largely on both environmental and political grounds. The UK has undertaken to cease sea disposal of sludge by 1998 resulting the change of disposal patterns shown in Table 3. However it has to be recognised that options such as landfill are becoming increasingly difficult and expensive because of the lack of suitable sites, and that there is a greatly increasing awareness of the dangers of groundwater contamination by both persistent chemicals and pathogens.

TABLE 2 Sludge disposal in Europe (thousand tonnes of dry solids per annum; (4))

	Agriculture	Landfill	Incineration	Sea	Total
Austria	57	67	74	n/a	198
Belgium	8	15	6	0	29
Germany*	698	1286	196	0	2180
Denmark	57	39	35	0	131
France	234	446	170	0	850
Greece	0	15	0	0	15
Ireland	7	4	0	12	23
Italy	270	440	90	0	800
Luxembourg	12	3	0	0	15
Netherlands	127	55	6	11	199
Portugal	n/a	n/a	n/a	n/a	n/a
Spain	173	28	0	79	280
England and Wales	507	151	66	234	958
Sweden	108	72	n/a	n/a	180
Switzerland	113	80	57	n/a	198

*Figure applies to West Germany prior to reunification. n/a not available

TABLE 3 Actual and predicted levels for future UK sewage sludge production, re-use and disposal (thousand tonnes of dry solids; (4))

	1990-91	1999	2006
Agriculture	465	777	926
Dedicated disposal sites	25	30	30
Sea disposal	334	n/a	n/a
Landfill	88	35	46
Incineration	77	382	405
Other beneficial uses	68	137	161
Within curtilage	50	91	146
Uncertain	n/a	294	441
Total	1107	1746	2155

It therefore seems likely that over the next few decades there may have to be a fairly radical reassessment of the comparative merits of differing sludge disposal options.

2.2 SEWAGE AS A VECTOR FOR OTHER CONTAMINANTS

A variety of additional contaminants may be associated with the marine disposal of sewage. Of particular significance in this respect are a variety of trace metals and organochlorine compounds. Recently, there has also been increasing concern about the effects of oestrogenic chemicals or other disrupters of the endocrine system (5,6) and this is discussed in Section 3.2.3 below.

The significance of sludge dumping in the North Sea as a proportion of the total contaminant fluxes is indicated in Table 4 though it should be noted that natural inputs

TABLE 4 Estimates of inputs(t) to the North Sea via various pathways in 1990 (7)

	Pathway						
	Riverine[†]	Direct[†]	Atmosphere[†‡*]	Disposal at sea			
				Incineration	Industrial waste**	Sewage sludge	Dredged material[¶]
Cd	43	17	74	0.1	0.3	1.2	71
Hg	25	1.8	6.9	0.05	0.2	0.7	19
As	n.i.	n.i	220	0.05	n.i.	0.1	720
Cr	n.i.	n.i.	[180]	4.9	24	21	2800
Cu	1200	290	740	4.6	180	76	1300
Ni	n.i.	n.i.	400	5.0	64	11	1200
Pb	1000	160	1700	4.9	220	77	2700
Zn	6400	1300	5500	5.7	440	160	7900
CBs	2.2	0.2	n.i.	n.i.	n.i.	n.i.	0.6
HCH	0.9°	0.2°	[9.1]^∇	n.i.	n.i.	n.i.	n.i.
N	910000	120000	520000	n.i.	n.i.	6300	n.i.
P	48000	7100	n.i.	n.i.	n.i.	570	n.i.

* estimates of atmospheric inputs cover the North Sea to 59°N, including the Kattegat and Skagerrak but not the Channel. [†] maximum (upper) estimates. [‡] based on deposition measurements at coastal stations. ** chemical wastes, slurries, fly ash, minestones and colliery tailings. [¶] dredged material from harbours, estuaries and navigation channels. ° γ-HCH. ^∇ γ-HCH (8.1t) + α-HCH (1t)

of chemicals from the North Atlantic through both the Channel and from the North of the North Sea, are not included in these estimates and that, for a number of the elements listed these represent the dominant sources.

Whilst there are direct impacts on water quality from sewage sludge associated contaminants the main effects are to be observed in sediments at, or close to, operational (or historical) sludge dumping sites. Thus, benthic organisms in, for example, the Thames or Humber estuaries, or the German Bight have somewhat higher contaminant levels that those from remoter areas.

3. The Impact of Sewage Inputs on Marine Systems

3.1 OXYGEN DEPLETION

The primary impact of sewage (and other high BOD wastes is, as indicated above, the rapid, and possibly complete, removal of dissolved oxygen from the receiving waters. If water is completely saturated with oxygen it has the capacity to oxidise a BOD of between 8.0 and 8.5 mg.L^{-1} (8). Raw sewage, with a BOD of several hundred mg.L^{-1} exceeds this threshold value by a large amount and therefore has a very high capacity to cause damage. However, sewage is by no means the most dangerous of high BOD effluents. Agricultural wastes such as slurry or silage effluent, vegetable processing wastes and even apparently benign materials such as beer or milk all have BOD values that may be >100 times greater than raw sewage. Such discharges may be deliberate but there are many instances each year of accidents on farms, or involving tanker vehicles, that result in the discharge of such material to waterways.

The extent of the adverse effects depends markedly on the capacity of the receiving environment to (i) dilute the waste and (ii) replenish the dissolved oxygen supply. It therefore follows that the vulnerability of different aquatic systems to high BOD wastes decreases in the order freshwaters <estuaries < coastal seas <open ocean. Within semi-enclosed systems the levels of turbulence, tidal excursion and basin morphology all influence oxygen replenishment rates. The dissolved oxygen profile of an estuary influenced by high BOD inputs is shown in Figure 3.

Even large estuaries such as Chesapeake Bay and Long Island Sound are vulnerable to periods of hypoxia and anoxia resulting in a variety of adverse effects including fish kills and destruction of shellfish beds (9,10). In Europe significant parts of the Mediterranean and estuaries such as the Mersey, Thames, Humber, Scheldt, Seine and Elbe have all suffered from the effects of sewage pollution including oxygen depletion in the past (7,8). In some of these cases, improvements in effluent treatment have led to recent improvements in water quality (e.g. Thames & Seine(8)). Other marine areas such as the Kattegat, eastern Skagerrak, the central North Sea and various Norwegian fjords all suffer from periodic low oxygen levels though the extent to which these result from anthropogenic as opposed to natural processes remains unclear.

3.2 ECOLOGICAL CHANGES

3.2.1 Inshore Waters

The impact of oxygen depletion on aquatic ecosystems can be severe. Extreme cases will result in almost total removal of living organisms with only sewage fungi, bacteria and, perhaps tubificid worms present in significant quantities. Infact, the living biomass in sewage affected areas may be much higher than in the surrounding, uncontaminated areas. In saline systems, most terrestrial bacteria have relatively short lifetimes because of the salt content, the presence on natural antibiotics in sea water and exposure to UV light. Historical opinions that there is no significant long term threat from pathogenic bacteria in sea water have had to be modified following the discovery of a number of non-platable bacteria species which may remain active for extended periods.

The recent discovery of the impact of oestrogen mimicking compounds in aquatic systems is described in section *3.2.3* below.

3.2.2 Sludge Dumping Sites

Offshore sewage sludge dumping sites represent a special case of marine sewage disposal. For the most part the bulk of this activity has taken place in the North, Irish and Celtic Seas rather than other regional seas. Of the various nations which have used the practice in the past, only the UK continues the practice with a total of $\approx 5.4 \times 10^6$ wet tonnes dumped in the North Sea 1990 (out of a total of some 136×10^6 tonnes of material dumped in the North Sea by 8 coastal nations; (7) and a total of $\approx 9.3 \times 10^6$

tonnes dumped by the UK in total in that year, (11)). The use of the major site in the German Bight was discontinued in 1980.

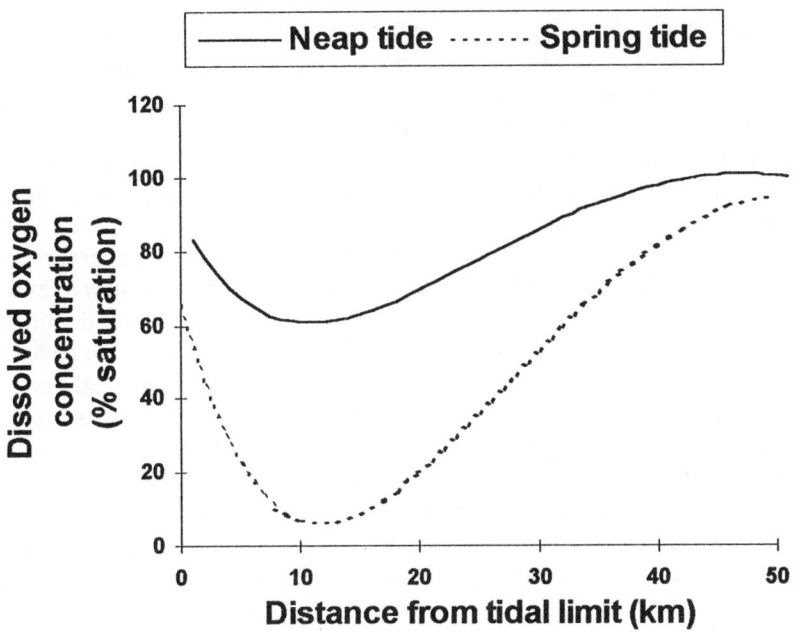

Figure 3 The dissolved oxygen profile of an estuary affected by high BOD wastes

The effects of sludge dumping depend to a considerable extent on the hydrodynamic characteristics of the dump site. At quiescent sites (e.g. the Garroch Head site in the Firth of Clyde) effects are comparatively large but limited in extent (somewhat analogous to a terrestrial landfill site). There is significant organic enrichment of sediments with a highly modified benthic fauna dominated by the polychaete *Capitella*. Contaminant levels in the sediments are also high though fish and shellfish in the area do not show enhanced metal levels (8). At the other extreme, highly dynamic sites (e.g. the Thames estuary site) are essentially dispersive in character with little or no impact on benthic ecology and it has been suggested that those effects that do exist in areas of minor sewage contamination may actually be beneficial (12,13). Other sites such as that in the New York Bight are intermediate in character and, in this case in particular, this has led to a variety of problems with coastal zone pollution by both sludge associated solids and pathogenic bacteria which has led to its closure (14).

3.2.3 Eutrophication
The artificial stimulation of primary production by the introduction of abnormal quantities of nutrient elements from both sewage and agricultural sources, is increasingly being recognised as a problem in certain areas which include Swedish coastal waters and areas of the Mediterranean. In these cases major algal blooms have led to serious damage to aquaculture units through oxygen depletion and the presence of toxins, and adverse impacts of the tourist industry through the presence of algal scums on beaches.

Of particular concern are the effects of sewage pollution on coral reefs. Corals rely on the presence of clear water for healthy growth. Addition of nutrients to such systems may stimulate phytoplankton growth and hence increase the turbidity of the water with detrimental effects on the corals. When these effects are added to thos imposed by other ecological stressees such as the impacts of tourism and destructive fishing techniques the prognosis for many coarl reefs in the world's seas is poor.

It is not only the presence of nutrients that can lead to adverse environmental changes. Anthropogenically induced modifications to nutrient ratios, particularly the N:P ratio (normally $\approx$ 16:1 in sea water), can lead to changes in the dominant phytoplankton species and species succession. For example a succession of unusual phytoplankton blooms (and effects) in the North Sea have been (7). These involve species such as *Noctiluca* spp (water discoloration), *Phaeocystis* spp. (foam production), *Chrysochromulina* spp., *Gymnodinium* spp.(fish or invertebrate mortality) and *Alexandrium* spp., *Dynophysis* spp.(toxicity to humans). However, links between these events and sewage/wastewater related nutrient inputs are hard to establish.

3.2.3 Problems associated with oestrogenic or other hormone-mimicking chemicals
Amongst the many chemicals associated with sewage are a number which have recently been identified as having oestrogenic or other endocrine disrupting properties (5, 15). Chemicals so far identified with such properties are shown in Table 5.

So far, effects have been seen only in freshwater systems such as Lake Apopoka, Florida where an alligator population had been exposed to a bad spill of dicofol (a DDT-like pesticide). Affected alligators showed a variety of reproductive problems including: reduced fertility, fewer eggs hatching and reduced survival rates of offspring. In addition, males showed reduced penis size and the juvenile gonads were dysfunctional or non-functional (16).

In the UK a study of the properties of the outfalls of a large number of sewage treatment works showed elevated vitellogen (female egg yolk protein) concentrations in caged male rainbow trout at all sites where tests were successful (17). These results have been confirmed in later tests conducted by the responsible government ministry (18). Environmental oestrogens have also been linked to various human ailments including reduced sperm counts and increased levels of breast cancer. However, as yet, the links remain tenuous and it remains to be seen whether they are indeed significant.

Similarly, the importance of these chemicals in marine systems has yet to be established.

TABLE 5 Some endocrine disrupting chemicals

Chemical	Source/use
Diethylstilboestrol } Ethynyloestradiol }	Synthetic oestrogens
DDT } DDE } Dieldrin } Chordecone } Endosulfan }	Organochlorine pesticides
PCBs	Transformer fluids
PCDDs and PCDFs	Dioxins and furans
Phthalates	Plasticisers
Nonylphenol	Antioxidant
Bisphenol-A	Dental amalgam and tin can lining

3.3 HUMAN HEALTH RISKS

In practice the greatest risks of marine sewage contamination are in regions where there are direct, raw sewage outfalls to beaches or close inshore. Such situations are not rare on the global scale and are a fairly common occurrence in respect of tourist hotels in developing countries and in any coastal village/town which has large seasonal influxes of tourists which overload the indigenous treatment facilities. People swimming /surfing /diving in sewage contaminated waters are all at risk to some extent, as are people who unknowingly each sewage contaminated fish and particularly shellfish. Each year there are many reports of outbreaks of diseases such as cholera or typhoid resulting from exposure to sewage contamination. There have been recent major outbreaks in West Africa and South America which form part of the seventh cholera pandemic which started in Indonesia in 1961 and has already passed through Asia, the Middle East, Southern Europe, India and Japan (19).

Other sewage related diseases include gastro-enteritis, salmonella, botulism, staphylococcal infections, infectious hepatitis and bacterial skin infections (20, 21).

4 The Assessment Of Sewage Pollution

A considerable variety of techniques have been developed for the assessment of sewage pollution. Most statutory regulatory measurements are based around the measurment of faecal coliforms (*E coli*) and faecal streptococci. Viruses such as coliphages can also be used as indicators of faecal contamination but these measurements are not commonly conducted. In the cases of both bacterial and viral assays results tend to be highly variable and the difficulty of obtaining statistically meaningful data cannot be underestimated.

Alternative assays may vary considerably in their degree of sophistication. At the simplest level, the counting of tomato seeds in sediments has been used as a measure of faecal contamination. Tomato seeds are not digested in the human body and are excreted intact. Their presence in sediments may therefore be taken as a measure of the presence of sewage solids.

The most widely used chemical marker of sewage contamination is probably coprostanol (5β-cholestan-3β-ol) (Fig 4a) which is a product of the metabolism of cholesterol in the digestive systems of higher mammals. It is not uniquely a human product (it is also produced by whales for example) but in most cases, where coprostanol is present, it is far more likely to have come from a sewage related source (22- 24). For example, in a multi-component investigation of both natural and contaminant components of the River Tamar Estuary, Readman *et al.* (23) (Figure 5) were clearly able to demonstrate the links between coprostanol and other contaminants in the sediments such as cholesterol and the UCM (uncharacterised material indicative of chronic oil contamination). There were also interesting relationships between contaminant sterols and various phaeophorbides reflecting perhaps active herbiverous grazing of chlorophylls in the most sewage affected areas which may possibly be stimulated by the quantities of organic matter in these sites.

Other markers of sewage contamination include 1-tocopheryl acetate (vitamin E acetate) which has been used in an analagous way to coprostanol (25), silicones (26, 27), linear alkyl benzene sulphonates (28, 29) and trialkyl amines (30).

Whilst the markers mentioned above enable the impact of sewage on sediments and suspended matter to be assessed, they do not permit any insights into the distribution of the water soluble components that make up the greater part of the BOD. To address this limitation Fitzsimons *et al.* (31) have used an assay based on the determination of 2-amino-propanone (Fig 5b), a water soluble component of urine, to the extent of sewage contamination on the dissolved phase. The lifetime of this ketone is sufficiently long ($t_{1/2}$= ca. 8-10 days) to parallel the normal decomposition rate of dissolved sewage related compounds in oxygenated water.

5 Sewage And Nutrients As A Stimulant For Good Or Vector For Harm?

The foregoing review concentrates very largely on the detrimental effects of sewage on marine systems and how these may be assessed. However, in summing up the use of the marine environment as a disposal option for sewage and related nutrients it is important to recognise that choosing *not* to use the seas in this way also has environmental consequences.

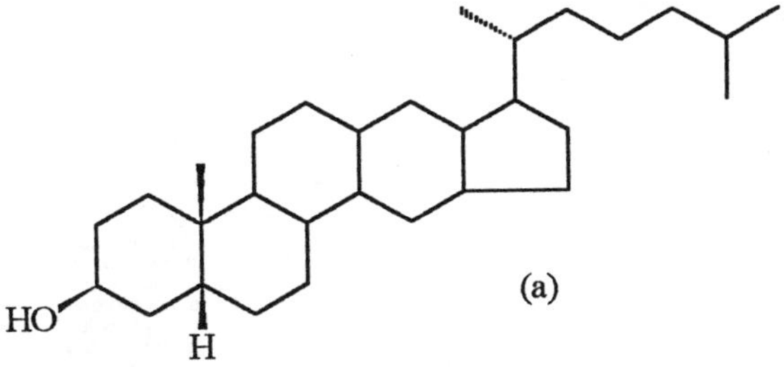

Figure 4 The chemical structures of (a) coprostanol and (b) 1-aminopropanone

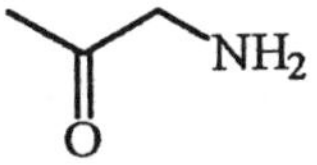

Other options for sewage disposal include landfill, uses as an agricultural fertiliser or soil conditioner and incineration and there are many environmental objections to each of these alternatives. For example, landfill is increasingly being recognised as a source of groundwater pollution. Contamination of groundwater, and other freshwaters may result in significant reductions of potable water supplies and lack of such waters probably represents the greatest environmental hazard that the human race will face in the next millenium. Application of sewage to land results in the transfer of both pathogens and persistent inorganic and organic components to soils and crops. This will happen with sewage only from from human sources, without any additional industrial contaminants which may also have to be considered. Repeated applications of sludge may therefore result in the buildup of concentrations of metals such as cadmium and zinc which not only represent a direct threat to human health but which may also act as plant toxins reducing soil fertility. The use of incinerators faces very considerably public opposition resulting from concerns over the waste products of such operations which, by there very nature, have to be sited in regions close to the major sources in heavily populated areas.

As is almost always the case, there is no ideal solution for the problems of sewage disposal. As the population of the world continues to increase rapidly, and given the general propensity for high population densities in coastal areas, there seems to be no doubt that on a global scale the pressures of sewage disposal to sea will continue to increase in a similar manner with all of the attendant consequences.

Further reviews of the effects of sewage pollution may be found in references 32-37.

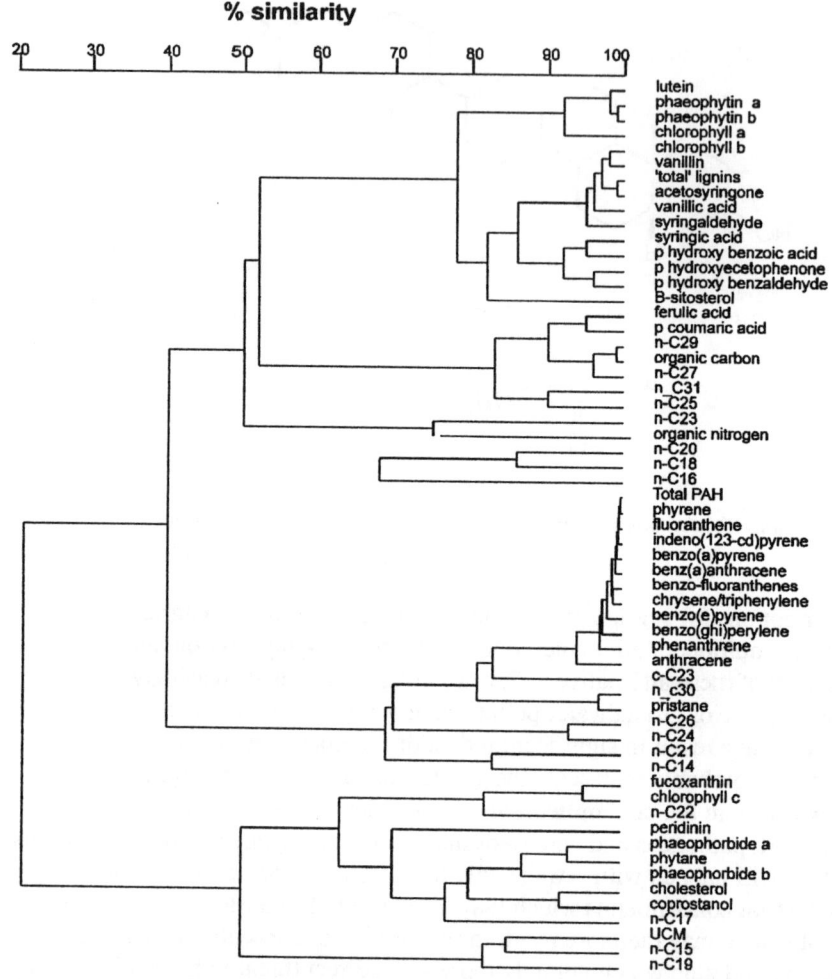

Figure 5 Cluster analysis dendrogram of natural and contaminant components of sediments from the River Tamar Estuary (23)

References

1. GESAMP (1996). IMO/FAO/UNESCO-IOC/WMO/WHO/IAEA/UN/UNEP Joint Group of Experts on the Scientific Aspects of Marine Environmental Protection, Reports and Studies No. 61 (in press).
2. OECD (1991). OECD Environment Indicators, 1991, p65.
3. Lester, J.N. (1990). Sewage and Sewage Sludge Treatment *In* Pollution - Causes, Effects & Control (2nd Ed.) R.M. Harrison (Ed) Royal Society of Chemistry, Cambridge pp33-62.
4. Newman, O. and Foster A. (1993) European Environmental Statistics Handbook. Gale Research International Ltd, Andover, UK, 436pp.

5. IEH, (1995) Environmental Oestrogens: Consequences to Human Health and Wildlife. Institute for Environmental Health, Leicester, 107pp.
6. Colborn, T., Myers, J.P. and Dumanoski, D. (1996). Natural History 105, 42-49.
7. NSQSR (1993). North Sea Quality Status Report 1993. Oslo and Paris Commissions, International Council for the Exploration of the Sea, London, 132pp.
8. Clarke, R.B. (1992). Marine Pollution (3rd Edition) Oxford Science Publications. Oxford, 172pp.
9. Seliger, H.H., Boggs, J.A. and Biggley, W.H. (1985) Science, 228, 70.
10. Correll, D.L. (1987). Nutrients in Chesapeake Bay *In* Contaminant Problems and management of Living Chesapeake Bay Resources, Majumdar,S.K., Hall, L.W. J. and Austin H.M. (Eds) Pennsylvania Academy of Science, Easton PA p298.
11. DOE (1995). Digest of Environmental Statistics No 17, HMSO, 213pp.
12. Gunnerson, C.G. (1981) In Marine Environmental Pollution Vol. 2 (R.A.Geyer, ed.) Pergamon Press, Oxford p 416.
13. Segar, D.A., Stanman, E. and Davis, P.G., (1985). Mar. Poll. Bull. 16, 186.
14. Mayer, G.F (ed.)(1982). Ecological Stress and the New York Bight: Science and Management, Estuarine Research Federation, Columbia, South Carolina.
15. Stone, R. (1994). Environmental Estrogens Stir Debate. Science, 265, 308-310.
16. Guillette, L.J., Gross, T.S., Masson, G.R., Matter, J.M., Percival, H.J. and Woodward, A.R. (1994). Developmental abnormalities of the gonad and abnormal sex hormone concentrations in juvenile alligators from contaminated and control lakes in Florida. Environ. Health Perspect. 102, 680-688.
17. Purdom, C.E., Hardiman, P.A., Bye, V.J., Eno, N.C., Tyler, C.R. and Sumpter, J.P. (1994). Estrogenic effects of effluents from sewage treatment works. Chem. Ecol., 8, 275-285.
18. MAFF (1994). Monitoring and surveillance on non-radioactive contaminants in the aquatic environment and activities regulating the disposals of wastes at sea, 1992 . Aquatic environment monitoring Report No. 40, Lowestoft, MAFF Directorate of Fisheries Research
19. Behrens, R. (1992) *In* Travellers' Health, Richard Dawood, Oxford University Press, Oxford, pp33-36.
20. GESAMP (1976). IMO/FAO/UNESCO-IOC/WMO/WHO/IAEA/UN/UNEP Joint Group of Scientific Experts for the Scientific Aspects of Marine Pollution. Reports and Studies, No 5.
21. Kennish, M.J. (1994). Practical Handbook of Marine Science (2nd Ed.) CRC Press Boca Raton, 566pp.
22. Readman, J.W., Preston, M.R. and Mantoura, R.F.C. (1986). Mar. Poll. Bull. 17, 298.
23. Readman, J.W. Mantoura, R.F.C., Preston, M.R., Llewllyn, C.A. and Reeves, A.D. (1987). Int. J. Environ. Anal. Chem. 27, 29-54.
24. Grimalt, J.O., Fernández, P., Bayona, J.M. and Albaigés, J (1990). Environ. Sci. Technol. 24, 357-363.
25. Eganhouse, R.P. and Kaplan, I.R. (1985). Environ. Sci. Technol., 19, 282.
26. Pellenbarg, R.E. (1979a). Mar. Poll. Bull. 10, 267-269.
27. Pellenbarg , R.E. (1979b). Environ. Sci. Technol. 13, 565-569.
28. Eganhouse, R.P., Blumfield, D.L. and Kaplan, I.R. (1983). Environ. Sci. Technol. 17, 523-530.
29. Brunner, P.H., Capri, S., Marcomini, A. and Giger, W. (1988) Wat. Res., 22, 955-961.
30. Valls, M., Bayona, J.M. and Albaigés, J. (1989). Nature, 337, 722-724.
31. Fitzsimons, M.F., Abdul Rashid, M.K., Riley, J.P. and Wolff, G.A. Mar. Poll. Bull 30, 306-312
32. Preston, M.R. (1987). Marine Pollution. In 'Chemical Oceanography' Vol 9,J.P. Riley and R. Chester (Eds). Academic Press, pp.55-196.
33. Preston, M.R. (1992). Marine Pollution. In Encyclopedia of Earth System Science, Academic Press, Vol. 3 pp 125-137.
34. Preston, M.R. (1996) Marine Pollution In Encyclopedia of Environmental Analysis and Remediation (in press)
35. Preston, M.R. and Chester, R. (1996) Marine Pollution - 3rd Edition of Pollution - Causes, Effects and Controls, R. Harrison (ed) (in press).
37. Bishop, P.L. (1983). Marine Pollution and its Control. McGraw-Hill, NY, 357pp.

ARSENIC IN THE MARINE ENVIRONMENT

Margaret E. FARAGO

Centre for Environmental Technology Imperial College of Science, Technology and Medicine Royal School of Mines Prince Consort Road London SW7 2BP, UK

1. Introduction: total arsenic concentrations in marine waters

There are many analytical methods for the determination of total arsenic in environmental samples, although to assign typical values for various media is more difficult. There are reports of the concentrations of total arsenic (As_T) in natural waters and sediments. Early data has been summarised by Onishi [1] who reported an average concentration of 2 mg/L in sea water. Examples of some later data for sea waters are given in Table 1, these include some literature values for oceans [1-5], coastal regions and estuaries [6-10]. Data for hydrothermal waters [11-13] are given in Table 2, and for sediments [14-20] in Table 3. Recent data [5] on the total arsenic concentrations of surface waters from a number of seas and oceans shows that these were relatively low (< 1 mg/L), only on the southwest Pacific and Antarctic oceans were the values above 1 mg/L. The authors suggest that less arsenic in these two regions is incorporated into biogenous matter.

1.2 HYDROTHERMAL WATERS AND SEDIMENTS

It can be seen that arsenic is enriched in hydrothermal waters and sediments (Tables 2 and 3). There has been some disagreement in the literature over the source of arsenic in marine metalliferous sediments. Boström and Valdes [14], Varnavas and Cronan [13] and Cronan [16] have suggested that the arsenic source, together with a number of other metals, is the hydrothermal exhalations. Neal et al. [15] have suggested that while on active ridges this hydrothermal derivation is possible, in sediments from mid ocean, freshly precipitated iron oxyhydroxides scavenge arsenic from sea water. Varnavas and Cronan [13] have demonstrated that in the Santorini hydrothermal field the majority of the arsenic is of hydrothermal origin. The enrichment in this area is in agreement with the hypothesis [21] that H_2S, released from rock fractures and oxidised to H_2SO_4 which in turn attacks the rocks resulting in hydrothermal solutions that are enriched with constituents derived from these rock-water interactions.

TABLE 1. Examples of total arsenic concentrations (As_T in mg/L) in sea water and submarine hydrothermal waters

Concentration	Location	Reference
2	Normal sea water	[1]
~2	Normal sea water	[2]
1.2 (mean)	Sargasso Sea (surface)	[3]
1.6 (20 nM)	North Atlantic Ocean	[4]
0.948	Northwest Pacific Ocean	[5]
0.64	China Sea	[5]
0.715	Indian Archipelago	[5]
0.849	North Indian Ocean	[5]
0.867	East Indian Ocean	[5]
1.167	Southwest Pacific Ocean	[5]
1.078	Antarctic Ocean	[5]
1.722	San Diego Trough	[6]
1.366-1.906	La Jolla, California	[6]
1.773-2.28	Tampa Bay, Florida	[7]
1.136-1.408 (15.17-18.8 nM)	Charlotte Harbor, Florida (estuarine)	[8]
1.504-3.045	California coast (surface)	[9]
2.55 (34 nM)	Davis Creek, California (surface)	[10]
7.42 (99 nM)	Salinas, California (estuarine)	[10]

It has been suggested that arsenic in hydrothermal waters in Yellowstone National Park, USA, is derived by this type of process [22] The subsequent behaviour of the released arsenic in hydrothermal waters will depend on the abundance of Fe-oxides, the pH and Eh and temperature [13].

TABLE 2. Examples of total arsenic concentrations (As_T in mg/L) in hydrothermal waters

Concentration	Location	Reference
0-20	Matupi Harbour, New Britain Hydrothermal waters	[11]
~2-~35	East Pacific Rise at 21° N Submarine hydrothermal solutions	[2]
~22-83	Guaymas Basin, Gulf of California Submarine hydrothermal solutions	[12]
7.5-16.5	Santorini Greece Hydrothermal waters	[13]

2. Anthropogenic Sources of Arsenic

2.1 MINING AND SMELTING

Input of arsenic derived from mine wastes is of particular importance in the case of river borne arsenic into estuaries, which is mostly in the dissolved phase. The proportion of this arsenic that reaches the open ocean is dependent on a number of factors, including the rapid changes in suspended load concentrations in low salinity regions [21]. South West Britain is extensively contaminated with arsenic from historic mining and smelting [24-26]. Total dissolved arsenic from two estuarine areas from this region; Rostronguet Creek and the Tamar Estuary, has been reported [23] to be in the ranges 5-100 mg/L and 0.5-16 mg/L respectively, showing enrichment from the mine waste.

On the East coast of Britain the Humber Estuary has been contaminated by the discharge of 1000 kg/day of arsenic from a copper smelter until August 1991 [20]. Mean As_T concentration in solution was 13.1 ± 0.7 mg/L (axial water samples), while that in suspended particulate matter was 51.9 ± 14.1 mg/g. Mean arsenic concentrations in SPM at an anchor station in the mouth of the estuary were 37.2 and 37.9 mg/g. The enrichments in the sediments could provide a major source of arsenic from the Humber to the North Sea [20].

TABLE 3. Ranges and means of total arsenic concentrations (in mg/g) in marine sediments and suspended particulate matter (SPM)

Concentration range	Mean	Location	Reference
11-26	18	North Atlantic Ocean	[14]
2-56	20	North Atlantic Ocean	[15]
8-86	16	Indian Ocean	[14]
23-455	121	Equatorial Pacific Ocean	[14]
3-32	10	South Pacific Ocean	[14]
48-361	174	Mid-Atlantic Ridge	[16]
13-400	63	East Pacific Rise	[17]
320-870	490	East Pacific Rise	[18]
268-579	425	Palaea Kameni (Santorini)	[13]
283-927	493	Nea Kameni (Santorini)	[13]
64-200	145	Channel, Santorini	[13]
15.7 (1-2 cm)		Black sea	[19]
14.6 (6-7 cm)			
11.6 (11-12 cm)			
7.4-76.4 (SPM)	51.9	Humber Estuary, UK	[20]

2.2. WOOD PRESERVATIVE

The wooden structures along shorelines, such as pilings and docks are preserved against attack by boring organisms and decay. In recent years, particularly in the USA traditional preservatives, such as creosote and pentachlorophenol have been replaced by pressure treatment with chromated copper arsenate (CCA). A commonly used formulation contains 47.5% CrO_3, 18.5% CuO and 34% As_2O_3. The elements Cu, As and Cr are supposed to be "fixed" into the wood and not subject to leaching, however, studies in the marine environment have demonstrated leaching [27-29]. All three metals have been found to be leached in salt water conditions, with copper being the most mobile, and with leaching rates varying with salinity [30]. The leached contaminants can be taken up by marine organisms, highest uptake is expected by the epibiotic community living directly on the wood.

The green algae *Ulva lactuca* and *Enteromorhpa intestinalis* collected from treated wood in Southampton, NY, showed that their tissues had become enriched with all three metals in comparison with samples from nearby rocks [30]. Arsenic concentrations had increased from 1 to 4.7 mg/g. Contaminants leached from the wood were also found to be adsorbed onto fine sediment particles [31]. Both biota growing on the wood and in the sediments adjacent to the wood have reduced community diversity, reduced growth and elevated incidence of abnormalities. Similarly, algae growing on CCA treated wood in Santa Rosa Sound FL, had increased concentrations of both Cu and As [32].

2.3. PESTICIDES

Simple inorganic arsenic compounds, sodium arsenate, arsenic trioxide have been used as pesticides for many years, but these have been mainly replaced by organoarsenicals. The latter are applied at lower concentrations (2-4 kg/ha) compared with the inorganic compounds (10-1000 kg/ha) [33]. Although there is the potential for contaminated runoff reaching rivers, estuaries and the ocean, major pollution results only in unusual circumstances. Mariner *et al.*, [34] report on a site on the Hylebos Waterway, Commencement Bay, on Puget Sound, where from 1940-1971 sodium arsenite (used as a pesticide/herbicide called Penite) had been produced. Disposal of the arsenite wastes was in unlined pits generating a groundwater arsenic plume, which is characterised by high pH and high silica, extends to the shore of the Hylebos Waterway. Only limited seepage was found into the adjacent waterway since cementation had occurred on the seepage faces. The authors concluded that the mixing of the high pH, high silica groundwater with the seawater from the waterway produced precipitation reactions that cemented the sediments along the shore face. These processes will have considerably reduced the cumulative discharge of arsenic into the waterway.

2.4. OTHER ANTHROPOGENIC SOURCES OF ARSENIC

The Tejo Estuary in Portugal is contaminated by industrially-derived arsenic giving a concentration of 6 mg/L (80 nM) [35]. Other anthropogenic sources of arsenic in the environment with potential for contamination of marine systems are burning of high

As coal and use of As-rich phosphatic fertilisers [36]. One further marine source is from underwater nuclear explosions. Recently, Loring et al., [37] have investigated contaminants in sediments from the southeast part of the Barents Sea known as the Pechora Sea. They suggest that the most likely source for arsenic enrichment is from the deposition and post-depositional modification of arsenic-rich radioactive particulates derived from underwater nuclear explosions on the southwest coast of Novaya Zemlya during the 1950s and 1960s and dispersed into the Pechora Sea.

3. Arsenic Species in Marine Waters

Early measurements on seawater were able to determine only the inorganic arsenic species As(III) and AS(V). The ratio of these species: $As(III)/As_i$ (where $As_i = [As(III) + As(V)]$) was reported as 0.2 in 1970 [1]. Although biological methylation of arsenic had been known since the nineteenth century, [38] methods for the determination of low concentrations of methylated arsenic species became available in the early 1970s [6, 7]. Braman and Foreback in 1973 [7] reported the occurrence of four arsenic species in natural waters: arsenite; arsenate; methylarsonic acid (MMAA), $CH_3As(O)(OH)_2$, or the anion; and dimethylarsinic acid (DMAA), $(CH_3)_2As(O)(OH)$, or the anion. These four species have been measured in a number of marine systems, some examples are shown in Table 4.

TABLE 4. Arsenic forms and concentrations (mg/L) in marine surface waters

	As(V)	MMAA	DMAA	As_T	As(III)	%MeA	Location
0.027	1.08	0.004	0.108	1.219	0.024	9.19	California
nm[a]	1.27[b]	0.006	0.140	1.416	nm[a]	10.31	Coast [9]
nm[a]	1.17[b]	0.004	0.132	1.306	nm[a]	10.41	
0.660	0.57	0.014	0.260	1.504	0.537	18.22	
0.870	0.16	0.200	0.236	1.466	0.845	29.708	
0.056	1.21	0.014	0.187	1.467	0.044	13.701	
0.009	1.34	0.006	1.690	3.045	0.007	55.680	
0.023	0.007	0.007	0.117	1.387	0.018	8.940	
0.019	1.75	0.017	0.12	1.906	0.011	7.188	La Jolla,
0.034	1.2	0.012	0.12	1.366	0.028	9.663	California [6]
0.017	1.49	0.005	0.21	1.722	0.011	12.486	San Diego
0.12	1.45	0.002	0.2	1.772	0.076	11.400	Tampa Bay,
0.62	1.29	0.08	0.29	2.28	0.325	16.228	Florida [7]
0.172	0.34	0.016	0.085	0.613	0.336	16.466	China Sea [5]
0.232	0.452	0.05	0.032	0.766	0.339	10.705	E. Indian
1.045	0.003	0,023	0.007	1.078	0.997	2.783	Antarctic

[a] Not measured. [b] Sum of As(III) + As(V).

Of the organic arsenic compounds in the environment, MMAA and DMAA have been most studied [39]. although evidence exists that more complex compounds exist in the marine environment [40]. Both MMAA and DMAA are water soluble and they can form significant proportions of the total dissolved arsenic.

3.1. FACTORS INFLUENCING ARSENIC SPECIES IN MARINE SURFACE. WATERS AND ESTUARIES

3.1.1. Depth

Methylated forms of arsenic are found mainly in the photic zone [7, 9, 41], and to a much smaller extent below the thermocline. Andreae [9], in a study of Californian coastal waters, suggested that the methylated species are produced in the photic zone by phytoplankton or by subnanoplankton, or both, This hypothesis was supported by the observations that the concentrations of the methylated species in the photic zone correlated with the indicators of primary production, e.g. chlorophyll production ($r = 0.70$) and ^{14}C uptake ($r = 0.61$). Arsenate concentrations are lower in surface waters than those at 100 m depth Andreae [9] suggested that arsenate is removed from the surface waters and transported downward in biogenic particles and released below the euphotic zone. The removal of inorganic arsenic from solution by phytoplankton has been demonstrated recently in the North Sea [42]. In Californian waters, methylated arsenic species and As(III) decreased rapidly below the photic zone [9]. In the San Pedro Basin profiles, arsenite increased sharply below the euphotic zone reaching a maximum near 200 m. This maximum occurred at the zone of mixing between oxic surface waters from the south and oxygen depleted, nutrient rich deeper waters, allowing bioreduction to As(III). Santosa et al. [5] measured ocean profiles down to 5 km depth and reported that the variation of inorganic arsenic was small, with a maximum concentration around 2000 m. After a decrease in DMAA from the surface, there is a small increase also near 2000 m. Thus, inorganic arsenic depletion occurred both in surface and in bottom waters. In bottom waters the conditions allow for the coprecipitation of arsenic with hydrated heavy metal oxides. Dissolved arsenic may be scavenged from the water column by surface active particulate matter SPM, particularly surface SPM whose surface activity has been augmented by iron oxide coatings [43].

3.1.2. Temperature and Salinity

The concentrations of methylated arsenic species in surface waters increase with increase in temperature. It has been suggested [5] that the low percentage of methylated species (>3%, Table 4) found in the Antarctic Ocean is caused by the low water temperatures, which limit the activity of the microorganisms, although these are present in the waters as indicated by chlorophyll a. In temperate estuaries, formation of the methylated species normally occurs at high salinity and when the water temperature exceeds 12°C [44, 45]. Forms of arsenic are reported to be directly related to salinity, particularly in estuarine systems [20].

3.1.3. Phosphate Concentration

Andreae [9], reported a high degree of correlation ($r = 0.91$) between the concentration of DMAA and the ratio of arsenate to inorganic phosphate. Since arsenate and phosphate are chemically similar, methylation or reduction of arsenate may be a mechanism whereby the biota remove competing arsenate in low phosphate conditions. Thus it has been suggested that the arsenate/phosphate ratio, rather than the total arsenic concentration, may be the critical parameter in controlling the methylation. Santosa et al. [5] found that the closer sampling is to the mainland, the higher the percentages of organic arsenic. Waters close to the mainland are nutrient-rich environments and if there is competition for the assimilation of arsenate and phosphate by microorganisms, it would be expected that in high phosphate areas less organic arsenic would be found. This point is illustrated by the arsenic biogeochemistry in Charlotte Harbor, Florida, which is a phosphate rich estuary, where the maximum percentage of methylated As was 10% of As_T [8], while in the Beaulieu Estuary southern England [46] it reaches 70%. In these latter estuarine waters, Howard [44, 45], however, found no correlation between As-methylation and phosphate concentrations.

3.1.4 Presence of Biota

The cryptophyte *Chroomonas sp.* has been associated with the production of MMAA [47]. Relative production by phytoplankton of methylated arsenic compounds is species dependent. DMAA is produced by most organisms [35]. *Skeletonema costatum* produced DMAA and arsenite [48], however, some diatoms, *Cyclotella sp.*, *Thalassiosira pseudonana* [49], and coccolithiphorids [50] can produce MMAA and no DMAA under certain conditions. Bacteria have been implicated in the release of organoarsenics metabolically from As-containing compounds in the water, or by release from algae that are known to contain organoarsenic compounds [9, 51]. Howard et al. [52] point out that arsenic within organisms, which may be of varying structures, must eventually be released back into the water, unchanged or in an altered form. The release process on the death of organisms will be mediated by bacteria. The rapid grazing of one type of organism by another may result in the sudden release of a number of different types of compounds into the water column Seawater bacterial cultures have been demonstrated to release As(III) at high As concentrations [53] and the bio-utilisation of arsenate results in the release of As(III).

3.1.5. Seasonal Variation of Arsenic Species in Estuarine Waters

Seasonal blooming of some organisms may give rise to seasonal variations in the ratios of arsenic species present in marine waters [20 52, 54]. In both Southampton Water (UK) [52] and the Patuxent River Estuary [54] As(III) maxima are seen in the spring, in the former when *Skeletonema costatum* is replaced by *Rhizosolenia delicatula*, the latter being a possible source of As(III). The maxima of DMAA and MMAA are seen later and increased to broad maxima over the summer, when the organisms *Mesodimium rubrum*, *Scrippsiella trochoidia* and associated microflagellates peaked in Southampton Water [52]. Similar behaviour was noted in the Patuxent River Estuary [54] for MMAA, with a broad summer maximum. Millward et al. [55] have evaluated the relative contributions to the seasonal variation of methylated arsenic species in the Tamar Estuary, UK, from three important sources: phytoplankton. macroalgae and sediment porewaters. Their data are summarised in Table 5.

During the spring inorganic arsenic is taken up and methylated by the phytoplankton blooms in the lower Tamar Estuary. When the phytoplankton are grazed by the diatom *Sleletonema costatum* DMAA is released. It has been assumed that most of the DMAA in the water column originates in this way. However, it was found that the contribution to the DMAA from macroalgae to the water column could be as high as 60%. Although the concentrations of DMAA in the tissues of the macroalgae are significant (3.4 - 33.8 mg/g, dry weight) the DMAA is released to the water column in microbial degradation process. This process is not very active at 5°C. The authors suggest that at temperatures around 13°C that once inorganic arsenic has been taken up by the macroalgae, it is rapidly methylated to DMAA and released to the water column. The presence of large colonies of macroalgae mat then influence the cycling of arsenic in productive estuaries, such as Southampton Water.

TABLE 5. Potential contribution (in mg/L) to observed dissolved concentration of the methylated arsenic species to the waters of the lower Tamar Estuary, over a tidal cycle, in summer and in winter [55].

Source	DMAA		MMAA	
	Summer	Winter	Summer	Winter
Macroalgae	0.08	0.01	0.01	<0.01
Phytoplankton	0.04	0	0	0
Sediment porewaters	<0.01	<0.01	<0.01	0.03

It was also concluded that marine sediment porewaters are not significant contributors of DMAA to the water column in summer, however, estuary transport of DMAA from fresh water sediments has to be considered as potentially significant source of DMAA. It was concluded that MMAA is not a product of phytoplankton degradation and it is rapidly transformed to arsenate, inputs from macroalgae could be significant with porewaters as a minor source.

4. Arsenic Compounds in Marine Biota

Some examples of organic arsenic compounds found in marine waters, tissue extractions or culture exudates are shown in Table 6 and 7. Organoarsenic compounds found in marine biota contain an arsenic atom in the place of nitrogen, for example arsenobetaine has been found in western rock lobster, *Panulirus cygnus* [56,57]; in shrimp [58]; in plaice, *Pleuronectes platessa* [59]; and in various fish and shell fish [60]. Arsenocholine is expected to be a precursor of arsenobetaine and has been reported in fish and shell fish [60]; and in shrimps (*Pandalus borealis*) [61], Shibata and Morita, however, were not able to detect arsenocholine in a variety of marine biota, including fish, crustaceans and molluscs [62] although they identified a strongly basic species in some of the samples as tetramethylarsonium and have suggested that the compound identified as arsenocholine was in fact the tetramethylarsonium ion. Arsenosugars have been identified in a number of marine organisms [63-66], in brown kelp (*Eklonia radiata*) [63]; and in green seaweed, *Codium fragile* [66].

TABLE 6. Methyl arsenic compounds found in marine waters, tissue extractions or culture exudates

Name	Formula
monomethylarsonic acid (MMAA)	$CH_3As(O)(OH)_2$,
dimethylarsinic acid (DMAA)	$(CH_3)_2As(O)(OH)$
dimethylarsine (DMA)	$(CH_3)_2AsH$
trimethylarsine (TMA)	$(CH_3)_3As$
trimethylarsine oxide (TMAO)	$(CH_3)_3AsO$
tetramethylarsonium ion.	$(CH_3)_4As^+$
arsenobetaine	$(CH_3)_3As^+CH_2CO_2^-$
arsenocholine	$(CH_3)_3As^+CH_2CH_2OH$

Detailed examples of organoarsenic compounds in marine biota are given by Cullen and Reimer [39], Irgolic [67] and Francesconi and Edmonds [66]. Concentrations of arsenic in tissues and organs of marine biota have been collected by Cullen and Reimer [39]. Shibata and Morita [68] have pointed out that some arsenosugars were reported in all the macroalgae analysed, while none contained detectable amounts of arsenobetaine, so far found in all marine animals. While arsenobetaine is the sole or dominant arsenic species found in squid, cuttlefish, octopus and molluscs, the presence of arsenosugars in addition to arsenobetaine in bivalves has been reported [68,69]. Arsenic compounds have been identified in culture compound, including arsine (AsH_3), MMAA, DMAA TMAO and TMA [10, 39] (Table 6).

TABLE 7. Examples of arsenic-containing ribofuranosides in the marine environment

R	R'	R''
$(CH_3)_2As(O)-$	$-OH$	$-OH$
$(CH_3)_2As(O)-$	$-OH$	$-OP(O)_2^-OCH_2CH(OH)CH_2OH$
$(CH_3)_2As(O)-$	$-OH$	$-SO_3^-$
$(CH_3)_2As(O)-$	$-NH_3^+$	$-SO_3^-$
$(CH_3)_2As(O)-$	$-OH$	$-SO_4^-$
$(CH_3)_3As^+-$	$-OH$	$-SO_4^-$

R, R' and R'' are substituents in the alkylic chain of marine ribufuranose

There is controversy in the literature as to whether arsenic in these compounds is in the oxidation state (III) or (V); Cullen and Reimer [39] suggest that all these compounds contain As(V).

5. Arsenic compounds and marine sediments

Cullen and Reimer have reviewed the presence and associations of arsenic with other elements in sediments [39]. Arsenic may undergo chemical transformations and mobilisation due to redox and methylation within sediments [70] In some estuarine systems the concentrations of inorganic arsenic depend on sediment mobilisation [45]. In estuarine conditions precipitates of iron hydrous oxides form at the saltwater interface, and this can remove a variety of elements including arsenic [43, 46], arsenic in the North Atlantic deepsea sediments has been shown to be associated with iron [15]. Adsorption and co-precipitation of As with Fe phases have been discussed [71,72]. In an investigation of arsenic in porewaters from three cores from the deep northeast Pacific and the California coast Andreae [73] found that As(V) concentrations were greater than As(III) even in reducing conditions. It was suggested that arsenate diffused from the water column to the sediments while arsenite diffused into the water column. Brannon and Patrick [74], however found that in a variety of saline and freshwater sediment samples, As(V) was reduced to As(III) under aerobic conditions. Andreae [73] was not able to detect methylated species in the sediment porewaters, but more recently methylarsenic species have been reported in estuarine sediments [39]. Wong et al. [51]demonstrated that arsenic-contaminated sediments incubated at 20°C produced MMAA, DMAA and sometimes TMAO. More recently Hanaoka et al. [75] found that marine microorganisms can degrade arsenobetaine to DMAA, TMAO or to inorganic arsenic, this degradation forming part of the cycling of arsenic in the marine system.

6. Cycling of arsenic in marine systems

A scheme for the cycling of arsenic in the photic zone is shown in Figure 1. Anderson and Bruland [8] suggest that arsenate is assimilated by marine biota, organic arsenic compounds are then formed by the processes of excretion and degradation.
Hanaoka et al. [75] postulate a cycle which begins with the methylation of inorganic arsenic on the way to arsenobetaine and finally ends with the complete degradation of arsenobetaine to inorganic arsenic by the pathway:

$$\text{arsenobetaine} \rightarrow \text{TMAO} \rightarrow \text{DMAA} \rightarrow \text{MMAA} \rightarrow \text{inorganic arsenic}$$

These authors found two strains of common marine bacteria from the *Vibrio-Aeromonas* group common that had arsenobetaine-decomposing activity. It was also suggested that arsenosugars can also be degraded to inorganic arsenic by microorganisms without conversion to arsenobetaine. Andreae [9] suggested that arsenate is removed from the surface waters and transported downward in biogenic particles and released below the euphotic zone. This release would be by such degradation processes. Cutter [19] has pointed out that As(III) as arsenite, produced by bioreduction exists in oxygenated seawater, as thermodynamically unstable species, and that slow rates of conversion allow such species to exist, thus a kinetic approach, rather than a thermodynamic approach should be used when considering biogeochemical cycling.

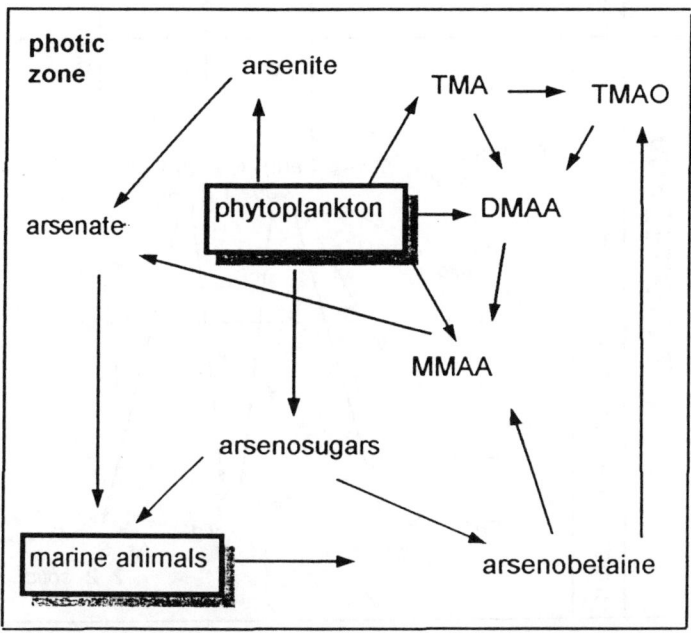

Figure 1. Model for the transformations of arsenic in the photic zone (adapted from Braman [41], Anderson and Bruland [8] and Hanaoka [75]).

A model for the cycling of arsenic in marine sediments is shown in Figure 2. Inorganic arsenic in the lower water column deposits in the sediment aerobic zone and is coprecipitated mainly as Fe(III) oxyhydroxy-arsenic complexes, manganese oxides may also be involved [37]. The iron and associated arsenic-phase becomes buried in a sedimentary redox gradient, leading to dissolution into the porewaters. Subsequent diffusion upwards into the water column results in coprecipitation or reoxidation. Diffusion downward most likely results in reaction of arsenic with sulphide leading to arsenic sulphides and arsenopyrite.

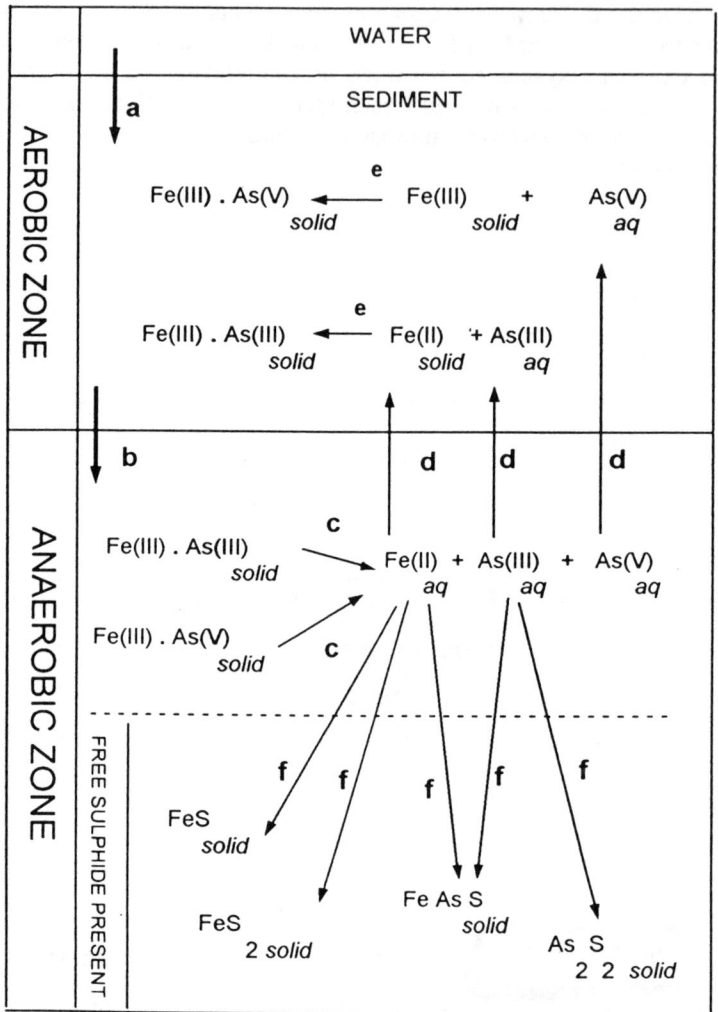

Figure 2. Model for the cycling of arsenic in marine and coastal sediments (after Cullen and Reimer [37]). Solid iron species are oxyhydroxides. Processes involved: **a**, sediment deposition; **b**, burial; **c**, diffusion; **e**, coprecipitation; **f**, diffusion and reaction with free sulphide.

7. Toxicity of arsenic to marine life

Most data is concerned with the adult stages of invertebrates [76], of these the sensitivity to arsenic is crustaceans > molluscs > annelids. Toxicity is also suggested to be in the order As(III) > Organoarsenic compounds > As(V) [76,77]. There appears to be no relationship between salinity and toxicity of As(V) to the estuarine invertebrates *Corophium volutator, Macoma balthica* and *Tubifex costatus* [78], whereas toxicity decreased with decrease in temperature. The reported 4 Day LC_{50} values for 5 species of fish exposed to As(III) are greater than the corresponding values for invertebrates, and range from 15 - 28 mg/L [76]. Data on toxicities of arsenic species to marine biota have been collected reviewed and by Mance [76].

8. Bioaccumulation of arsenic by aquatic species

Accumulation by aquatic species has been reviewed [79]. Förstner and Wittmann [80] suggested that arsenic and mercury, because of their organic chemistry were most likely to show biomagnification, where concentration increases progressively along the food chain. Such biomagnification was demonstrated in the marine food chain in waters near West Greenland [81] where concentrations of arsenic were (mg/g, dry weight): zooplankton, 6; mussels, 14.1-16.7; prawns, 62.9-80.2; fish, 43.4-188. Other studies had failed to show such biomagnification [82].
Seaweeds bioaccumulate trace elements and macrophytic brown algae have been used to monitor pollution in marine waters [83-85]. Trace element concentrations vary with environmental concentrations. It is reported that the alga *Sargassium filipendula* showed an affinity for As (along with Ag, Br and Sr) [85].

9. References

1. Onishi, H. (1970) Arsenic, in K.H. Wederpohl (ed.) *Handbook of Geochemistry, VolII/2*. Springer, Berlin, pp 3301-3305.
2. Von Damm, K.L., Edmond, J.M., Grant, B. and Measures, C.I. (1985) Chemistry of submarine hydrothermal solutions at 21° N, East Pacific Rise, *Geochim. Cosmochim. Acta* **41**, 2197-2220.
3. Waslenchuck, D.G. (1978) The budget and geochemistry of arsenic in a continental shelf environment, *Mar. Chem.* **7**, 39-52.
4. Middleburg, J.J., Hoede, D., Van der Sloot, H.A,, Van der Weijden C.H. and Wijkstra, J. (1988) Arsenic, antimony and vanadium in the North Atlantic Ocean, *Cosmochim Acta*, **52**, 2871-2878.
5. Santosa, S.J., Wada, S.W. and Tanaka S. (1994) Distribution and cycle of arsenic compounds in the Ocean, *App. Organomet. Chem.*, **8**, 273-283.
6 Andreae, M.O. (1977) Determination of arsenic species in natural waters, *Anal. Chem.* **49**, 820-823.
7. Braman, R.S. and Foreback, C.C. (1973) Methylated forms of arsenic in the environment, *Science*, **182**, 1247-1249.
8. Froelich, P.N., Kaul, L.W., Andreae M.O. and Roe, K.K. (1985) Arsenic, barium, germanium, tin, dimethylsulfide and nutrient biogeochemistry in Charlotte Harbour, Florida, a phosphorus-enriched estuary, *Estuarine, Coastal and Shelf Sci.*, **20**, 239-264.
9. Andreae, M.O. (1978) Distribution and speciation of arsenic in natural waters and some marine algae, *Deep-Sea Res.*, **25**, 391-402.
10. Anderson, L.C.D. and Bruland, K.W. (1991) Biogeochemistry of arsenic in natural waters: the importance of methylated species. *Environ. Sci. Technol.* **25**, 420-427.

11. Ferguson, J. and Lambert, I.B. (1979) Volcanic exhalations and metal enrichments at Matupi Harbour, New Britain. *Econ. Geol.* **67**, 25-37.
12. Von Damm, K.L., Edmond, J.M., Measures, C.I. and Grant, B. (1985) Chemistry of submarine hydrothermal solutions at Guaymas Basin, Gulf of California, *Geochim. Cosmochim. Acta* **49**, 2221-2237.
13. Varnavas, S.P. and Cronan. D.S. (1988) Arsenic, antimony and bismuth in sediments and waters from the Santorini hydrothermal field, Greece, *Chemical Geology*, **67**, 295-305.
14. Boström, K. and Valdes, S. (1969) Arsenic in Ocean floors, *Lithos*, **2**, 351-360.
15. Neal. C., Elderfield, H. and Chester, R. (1979) Arsenic in sediments of the North Atlantic Ocean and Eastern Mediterranean Sea, *Mar. Chem.*, **7**, 207-219.
16. Cronan, D.S. (1972) The Mid-Atlantic Ridge near 45°N, XVII, *Can. J. Earth Sci.*, **9**, 319-323.
17. Boström, K. and Peterson, M.N.A. (1969) Origin of aluminium poor ferromanganoan sediments in areas of high heat flow on the East Pacific Rise, *Mar. Geol.*, **7**, 427-447.
18. Rydell, H., Kraemer, T., Boström, K. and Joensuu, O. (1974) Post depositional injections of uranium rich solutions into Pacific Rise Sediments. *Mar. Geol.*, **17**, 151-164.
19. Cutter, G.A. (1992) Kinetic controls on metalloid speciation on seawater, *Mar. Chem.*, **40**, 65-80.
20. Kitts, H.J., Millward, G.E., Morris, A.W. and Ebdon, L. (1994) Arsenic biogeochemistry in the Humber Estuary, UK, *Estuarine, Coastal and Shelf Sci.*, **39**, 157-152.
21. Boström, K. and Widenfalk, L. (1984) The origin of the iron-rich muds in the Kameni Islands, Santorini, Greece, *Chem. Geol.*, **42**, 203-218.
22. Stauffer, R.E. and Thompson, S.P.(1984) Arsenic and antimony in geothermal waters of Yellowstone National Park, Wyoming, USA. Geochim. Cosmochim. Acta, 48, 2547-2561.
23. Millward, G.E. and Marsh J.G. (1986) Dissolved arsenic behaviour in estuaries receiving acid mine wastes, in: *Proceedings of .International Conference on Chemicals in the Environment, Lisbon, 1986*, J.N. Lester, R. Perry, R.M. Sterritt (eds), Selper, Ltd., London. pp 470-478.
24. Thornton, I. (1996) Sources and pathways of arsenic in the geochemical environment: health implications. In. *Environmental Geochemistry and Health* Geological Society Special Publication No.113, pp 153-161.
25. Thornton, I. 1994. Sources and pathways of arsenic in south-west England: health implications, in: W.R. Chappell, C.O. Abernathy, and C.R. Cothern, (eds.) *Arsenic Exposure and Health*, Science and Technology Letters, Northwood. pp 61-70.
26. Mitchell, P. and Barr, D. (1995). The nature and significance of public exposure to arsenic: a review of its relevance to South West England. *Env. Geochem. and Health*, **17**, 57-82.
27. Cherian, P.V., Sharma, M.N. and Cherian, C.J. (1979) A study on the leaching of copper-chrome-arsenic (CCA) from some common Indian timbers tested in Cochin Harbour waters. *J. Ind. Acad Wood Sci.*, **10**, 31-34.
28. Hegart, B.M. and Curran, P.W. (1986) Biodeterioration and microdistribution of copper-chrome-arsenic (CCA) in wood submerged in Irish coastal waters, *J. Inst Wood Sci.*, **10**, 245-253.
29. Sanders, J.G., Riedel, G.F. and Osman, R.W. (1994) Arsenic cycling and its impact on estuarine and coastal marine ecosystems, in: J. Nriagu (ed.) *Arsenic in the Environment*, Part I, *Cycling and Characterisation*, John Wiley, NY, pp 289-308.
30. Weis, J.S. and Weis, P. (1995) Transfer of contaminants from CCA treated lumber to aquatic biota *J. exp. Mar. Biol. Ecol.* **161**, 181-191.

31. Weis, J.S. and Weis, P. (1995) Effects of chromated copper arsenate (CCA) pressure treated wood in the aquatic environment, *Ambio*, **24**, 269-274.
32. Weis, J.S. and Lores, E. (1993) Uptake of metals from chromated copper arsenate (CCA) treated lumber by Epibiota, *Mar. Pollut. Bull.* **26**, 428-431.
33. Ashby, J.R. and Craig, P.J. (1990) Organometallic compound in the environment, in R.M. Harrison (ed.) *Pollution: Causes, Effects and Control*, Royal Society of Chemistry, London, pp 309-342.
34. Mariner, P.E., Holzmer, F.J., Jackson, R.E., Meinardus, H.W. and Wolf, F.G. (1996) Effects of high pH on arsenic mobility in a shallow sandy aquifer and on aquifer permeability along the adjacent shoreline, Commencement Bay superfund site, Tacoma, Washington. *Environ. Sci. Technol.* **30**, 1645-1651.
35. Andreae, M.O. (1983) Biotransformations of arsenic in the marine environment, in W.H. Lederer and R.J. Fernsterheim (eds.) *Arsenic: Industrial, biomedical, Environmental Perspectives*, Van Nostrand Reinhold, NY, pp 1-19.
36. Kebata-Pendias, A. (1994) Biogeochemistry of arsenic and selenium, in: A. Kebata-Pendias (ed.) *Arsenic and Selenium in the Environment: Ecological and Analytical Problems*, Polska Akademia Nauk, Warszawa, pp 9-16.
37. Loring, D.H., Naes, K., Dahle, S. Matishov, G.G. and Illin, G. (1995) Arsenic, trace metals and organic contaminants in sediments from the Pechora Sea, Russia, *Mar. Geol.*, **128**, 153-167.
38. Challenger F. (1945) Biological methylation, *Chem. Revs.*, **36** 315-361.
39. Cullen, W.R. and Reimer, K.J. (1989) Arsenic speciation in the environment, *Chem. Revs.*, **89**, 713-764 and references therein..
40. Howard, A.G. and Comber, S.D.W. (1989) The discovery of hidden arsenic species in coastal waters, *App. Organomet. Chem.*, **3**, 509-514.
41. Braman, R.S. (1983) Environmental reaction and analysis methods, in: B.Fowler (ed.) *Biological and Environmental Effects of Arsenic*, Elsevier, Amsterdam pp 141-154.
42. Millward, G.E., Kitts, H.J., Comber, S.D.W., Ebdon. I. and Howard, A.G. (1994) Methylated arsenic in the southern North Sea, *Coastal and shelf Sci.*
43. Pierce, M.I. and Moore, C.B. (1982) Adsorption of arsenite and arsenate on amorphous iron hydroxide, *Wat. Res.*, **16**, 1247-1253.
44. Howard, A.G., Arbab-Zavar, M.H, and Apte, S.C. (1982) Seasonal variability of biological arsenic methylation in the estuary of the River Beaulieu, *Mar. Chem.* **11**, 493-498.
45. Howard, A.G., Apte, S.C., Comber, S.D.W. and Morris, J.R. (1988) Biogeochemical control of the summer distribution and speciation of arsenic in the Tamar Estuary, *Estuarine, Coastal and Shelf Sci.*, **27**, 427-443.
46. Howard, A.G., Arbab-Zavar,M.H, and Apte, S.C. (1984) The behaviour of dissolved arsenic in the estuary of the River Beaulieu, *Estuarine, Coastal and Shelf Sci.*, **19**, 493-504.
47. Sanders, J.G. (1985) Arsenic geochemistry in Chesapeake Bay: dependence upon anthropogenic inputs and phytoplankton species composition. *Mar. Chem,*, **17**, 329-340.
48. Sanders, J.G. and Windom, H.L. (1980) The uptake and reduction of arsenic species by marine algae, *Estuarine and Coastal Mar. Sci.*, **10**, 555-567.
49. Sanders, J.G. (1983) Role of marine phytoplankton in determining the speciation and biogeochemical cycling of arsenic, *Can. J. Fish Aquatic Sci.*, **40**, 192-196.

50. Andreae, M.O. and Klumpp, I. (1979) Biosynthesis and release of organoarsenic compounds by marine algae, *Environ. Sci. Technol.*, **13**, 738-741.
51. Wong, P.T.S., Chau, Y.K., Luxon, I. and Bengert, G.A. (1977) Methylation of arsenic in the aquatic environment, *Trace Substances in Environmental Health*, **11**, 100-106.
52. Howard, A.G., Comber, S.D.W., Kifle, D., Antai, E.E. and Purdie, D.A. (1995) Arsenic speciation and seasonal changes in nutrient availability and microplankton abundance in Southampton Water, U.K., *Estuarine, Coastal and Shelf Sci.*, **40**, 435-450.
53. Johnson, D.I. (1972) Bacterial reduction of arsenate in seawater, *Nature*, **240**, 44.
54. Riedel. G. (1993) The annual cycle of arsenic in a temperate estuary. *Estuaries*, **16**, 533-540.
55. Millward, G.E., Ebdon, L. and Walton, A.P. (1993) Seasonality in estuarine sources of methylated arsenic, *App.Organomet. Chem.*, **7**, 499-511.
56. Edmonds, J.S., Francesconi, K.A., Cannon, J.R., Raston, C.L., Skelton, B.W. and White, A.H. (1977) Isolation, crystal structure and synthesis of arsenobetaine, the arsenic constituent of western rock lobster, *Panulirus lonpipes cygnus* George, *Tetrahedron Letters*, **18**, 1543-1546.
57. Cannon, J.R., Edmonds, J.S., Francesconi, K.A., Raston, C.L., Saunders, J.B., Skelton, B.W. and White, A.H. (1981) Isolation, crystal structure and synthesis of arsenobetaine, a constituent of western rock lobster, the dusky shark and some samples of human urine, *Australian J. Chem.*, **34**, 787-798.
58. Norin, H.and Christakopoulos, A. (1982) Evidence for the presence of arsenobetaine and another organoarsenical in shrimps, *Chemosphere*, **11**, 287-291.
59. Luten, J.B., Riekwel,-Booy, g. and Rauchbaar, A. (1982) Occurrence of arsenic in plaice *(Pleuronectes platessa)* nature of organic arsenic compound present and its excretion by man, *Environ. Health Perspect.*, **45**, 165-171.
60. Lawrence, J.F., Michalik, P., Tam, G. and Conacher, H.B.S. (1986) Identification of arsenobetaine and arsenocholine in Canadian fish and shellfish by HPLC with atomic absorption detection and confirmation by fast atom bombardment mass spectroscopy, *J. Agric. Food Chem.*, **34**, 314-319.
61. Norin, H., Ryhage, R., Christakopoulos, A. and Sandström, M. (1983) New evidence for the presence of arsenocholine in shrimps *(Pandalus borealis)* by use of pyrolysis GC-AAS/mass spectrometry, *Chemosphere*, **12**, 299-315.
62. Shibata, Y. and Morita M. (1989) Exchange of comments on identification and quantitation of arsenic species in a dogfish muscle reference material for trace elements. *Anal. Chem.* 61, 2116-2118.
63. Edmonds, J.S. and Francesconi, K.A. (1981) Arsenosugars from brown kelp *Ekloniaradiata*) as intermediates in cycling in a marine ecosystem, *Nature*, **289**, 602-604.
64. Jin, K., Hayashi, Y, Shibata, Y. and Morita M. (1988) Arsenic-containing ribofuranosides and dimethylarsenic acid in green seaweed, *Codium fragile*, *App. Organomet. Chem.*, **2**, 365-369.
65. Edmonds, J.S. and Francesconi, K.A. (1988) The origin of arsenobetaine in marine animals, *App. Organomet. Chem.*, **2**, 297-302.
66. Francesconi, K.A. and Edmonds, J.S. (1993) Arsenic in the sea, in: (eds.) A.D. Ansell, P.N. Gibson and M. Barnes, *Oceanography and Marine Biology*, an Annual Review, Vol 31, UCL Press, London, pp 111-123.
67. Irgolic, K.J. (1986) Arsenic in the environment, in: A.V. Xavier (ed.) *Frontiers in Bioinorganic Chemistry*, VCH, Wienheim, pp 300-408.

68. Shibata, Y. and Morita M. (1992) Characterization of organic arsenic compounds in bivalves, *App. Organomet. Chem.*, 6, 343-349.
69. Le, S.X.C., Cullen, W.R. and Reimer, K.J. (1994) Speciation of arsenic compounds in some marine organisms, *Environ. Sci. Technol.* 28, 1598-1604.
70. Seyler, P. and Martin, J.M. (1990) Distribution of arsenic and total dissolved arsenic in Major French estuaries: dependence on biogeochemical processes and anthropogenic inputs, *Mar.Chem.*, 29, 277-291.
71. Fuller, C.C., Davis, J.A. and Waychunas, G.A. (1993) Surface chemistry of ferrihydrite: Part 2. Kinetics of arsenate adsorption and coprecipitation, *Geochim. Cosmochim. Acta*, 57, 2272-2282.
72. Morse, J.W. and Arakaki, T. (1993) Adsorption and coprecipitation of divalent metals with mackinawite (FeS), *Geochim. Cosmochim. Acta*, 57, 2635-3640.
73. Andreae, M.O. (1979) Arsenic speciation in seawater and in interstitial waters. The influence of biological-chemical interactions on the speciation of a trace element, *Limnol. Oceanogr.*, 24, 421-433.
74. Brannon, J.M. and Patrick. W.P. (1987) Fixation, Transformation and mobilization of arsenic in sediments, *Env. Sci. Technol.*, 21, 450-459.
75. Hanaoka, K., Tagawa, S. and Kaise, T. (1992) The fate of organoarsenic compounds in marine ecosystems. *App. Organomet. Chem.*, 8, 139-146.
76. Mance, G. (1987) *Pollution Threat of Heavy Metals in Aqueous Environments*. Elsevier Applied Science, London.
77. Watling H.R. and Watling R.J. (1982) Comparative effects of metals on the filtering rates of the brown mussel (Perna perna), *Bull. Env. Contamin. Toxcol.*, 9, 651-57.
78. Bryant, V., McLusky, D.S., Campbell, R. and Newberry, D.M. (1985) Effect of temperature and salinity on the toxicity of arsenic to three estuarine invertebrate *(Corophium volutator, Macoma balthica* and *Tubifex costatus), Mar. Ecol. Prog. Ser.*, 24, 129-137.
79. Phillips, D.J.S. (1980) Quantitative Aquatic Biological Indicators, Applied Science, London.
80. Förstner, U. and Wittmann, G.T.W. (1979) *Metal Pollution in the Aquatic Environment*, Springer-Verlag, Heidelberg.
81. Bohn, A. (1975) Arsenic in marine organisms from West Greenland, *Mar. Pollut. Bull.*, 6, 87-89.
82. Guthrie, R.K., Davis, E.M., Cherry, D.S. and Murray, H.E. (1979) Biomagnification by organisms in a marine microcosm, *Bull. Environ. Contamin. Toxicol.*, 21, 53-54.
83. Shubert, L.L. (ed.) (1984) *Algae as Ecological Indicators*, Academic Press, London.
84. Stoeppler, M., Burow, M., Backhaus, F. Schramm. W. and Nürnberg, H.W. (1986) Arsenic in sea water and brown algae of the Baltic and North Sea, *Mar. Chem.* 18, 321-334.
85. Jayesekera, R. and Rossbach M. (1996) Use of seaweeeds for monitoring trace elements in coastalw aters, *Environ. Geochem. Health*, 18, 63-68.

THE INTRODUCTION OF SOME SELECTED PERSISTENT ORGANIC POLLUTANTS TO THE MARINE ENVIRONMENT AND ASPECTS OF THEIR SUBSEQUENT BEHAVIOUR

Martin R. Preston
*Oceanography Laboratories, University of Liverpool,
Liverpool, L69 3BX, UK*

ABSTRACT

Amongst the non-chlorinated, anthropogenic organic components in marine systems are the azaarenes, which are nitrogen substituted analogues of the better known polycyclic aromatic hydrocarbons (PAC). Like the PAC they are present in the combustion products of fossil fuel and, as is demonstrated in this paper, they are also found in the urban atmosphere and contaminated marine environments. Unlike the PAC, which frequently behave as a more or less homogeneous entity through their occlusion in particulate material (1,2), azaarenes are much more dynamic within the environment and undergo a variety of phase transitions between their point of release and their final deposition in sediments. These phase transitions show strong seasonal influences and it is believed that these can be accounted for through differences in production rates and the prevailing environmental conditions.

This paper examines the complex nature of the environmental behaviour of azaarenes and contrasts it with that of the PAC. Seasonal variations in both atmospheric and aquatic abundance show high winter and low summer concentrations reflecting variations in levels of combustion. Observed aquatic distributions indicate higher concentrations of low molecular weight species in colder conditions and this is also reflected in atmospheric distributions where much greater amounts of the more volatile compounds are detected in association with aerosol in winter months.

The factors influencing gas/solid phase transitions in the atmosphere and dissolved/particulate behaviour in aquatic systems are described as a function of temperature and pH respectively. This leads to a picture of significant seasonal variability and rapid post-depositional changes in the azaarene signal source strength to marine systems. In conclusion the usefulness of azaarenes as model compounds for the assessment of the transfer of semi-volatile compounds from land-based sources to marine systems is examined.

1. Introduction

During the course of this century there has been a steady increase in the number and quantity of organic chemicals produced and used by human populations (Fig. 1). These include not only synthetic products but also those derived from natural sources particularly fossil fuels. At present about 70000 synthetic organic chemicals are in regular use (3) although a rather smaller number than this (around 15000) are produced in sufficient quantities to represent potential environmental hazards. Of course, in addition to those chemicals which are deliberately synthesised there are very many more which are by- or waste products of both the manufacture and use of organic materials. Examples of chemicals in this category include the polycyclic aromatic hydrocarbons (PAC), polychlorinated dioxins and dibenzofurans (PCDD and PCDF) and the azaarenes which form the topic of much of this paper and which are formed by the combustion of organic matter or during the synthesis of chlorinated compounds.

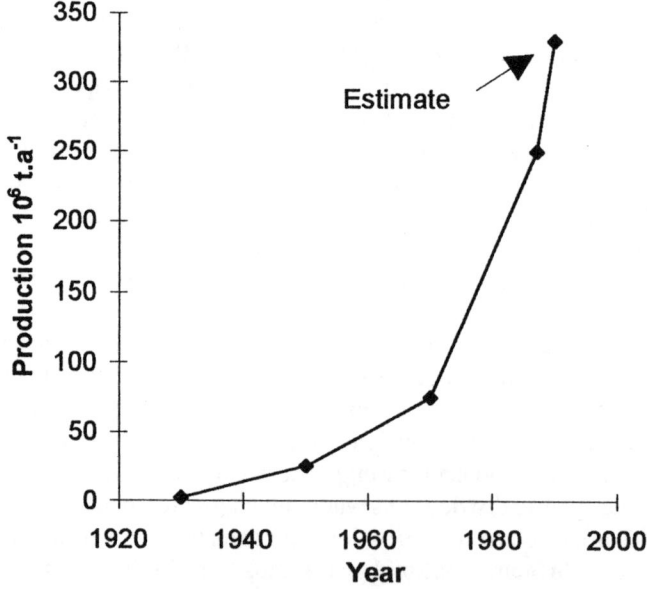

Figure 1 The change in production of synthetic organic chemicals since 1930 (3)

The influence of the relationship between material production and its influence on the environmental record is clearly shown in Figure 2 where the history of the rise and fall of DDT production is directly followed by the sedimentary deposition rates. Whilst detailed analyses of cores for other environmental contaminants are relatively rare, there are few reasons to suggest that the behaviour of DDT is in any way unique amongst relatively persistent organic environmental contaminants.

The mechanisms of transfer of organic contaminants from their point or region of origin to the aquatic environment and the sedimentary record is a complex one. Unlike most inorganic contaminants, most organic species have significant vapour pressures at ambient temperatures so the atmosphere plays a very significant role in their

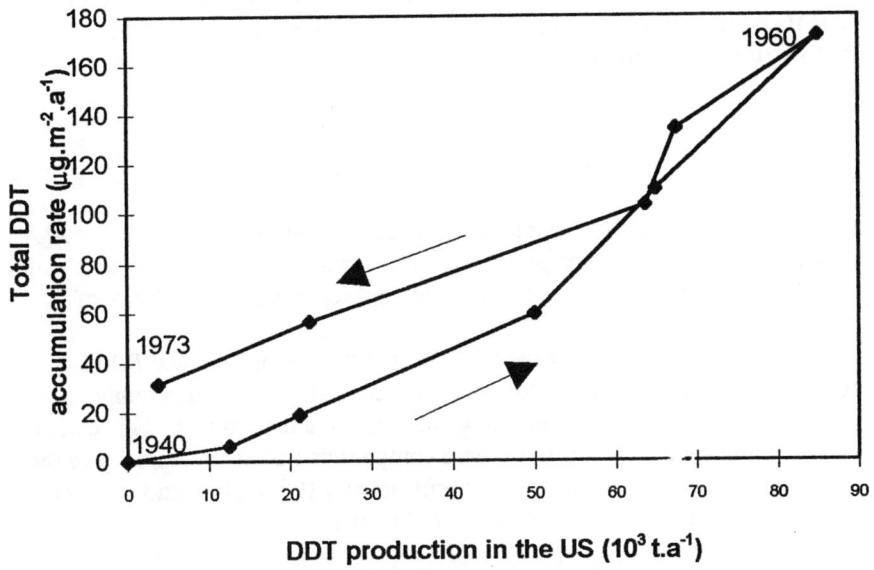

Figure 2 The relationship between DDT production rate and the accumulation rate of total DDT (= DDT+DDE+DDD) in a sediment core from Lake Ontario (data from Eisenreich *et al.*, (4))

transport. Evaporation may be followed by photolytic, adsorptive and/or desorptive reactions so that when the organic species is transferred to the aquatic environment (through wet or dry deposition or gas phase exchange) it may be significantly different from the parent material. Organic chemicals, may also have significant water solubilities which increase with increasing molecular polarity. Leaching of chemicals from land based sources by rain or groundwater may therefore also be of considerable significance and this may be assessed to some extent by the octanol/water partition coefficients of the species (log K_{ow} vary over the approximate range of 4.5-7.4 (5)). Similarly, water soluble species present in atmospheric aerosols will also be subjected to solubilsation during wet depositonal processes or after settling on water surfaces after dry deposition.

This paper addresses some aspects of the behaviour of polyaromatic species with particular attention to the azaarenes whose properties make them important examples of a number of these transformational processes.

2 Factors influencing partitioning behaviour

The factors influencing the partitioning behaviour of organic species between water/particle, air/particle, gas/particle, gas/water are complex and for a detailed treatment of the different aspects of this, the books by Stumm (5), Morel and Hering (7) or Schwarzenbach et al. (3) should be consulted. For the purposes of this paper only one aspect of the theoretical aspects of this behaviour will be briefly examined, namely the particle/air partitioning.

The vapour phase/solid phase partitioning of a number of organic species including PAC and azaarenes (see below) are well described by Pankow and Bidleman's partitioning coefficient K_p :-

$$K_p = m_p \frac{F/[TSP]}{AT} + b_p \qquad (1)$$

where F (ng.m^{-3}) (normalised to the total particle concentration TSP (μg.m^{-3})) and A (ng.m^{-3}) are the particulate and gas-phase concentrations respectively, T is the absolute temperature and m_p and b_p are the slope and intercept respectively of the plot of against $1/T$.

The values of m_p and b_p for the PAC are relatively constant (see Table 1) indicating that amongst many of the commonest species there is limited temperature dependent partitioning over the normal range of ambient temperatures. This can also be deduced from the vapour pressures of these compounds which are typically of the order of 10^{-9}-10^{-10} atm. at STP increasing to only around 10^{-8} at 50°C and generally following a relationship which approximates to the form:-

$$\ln P^o = -\frac{B}{T} + A \qquad (2)$$

where $B = \Delta H_{vap}/R$ (ΔH_{vap} = heat of vaporisation) and A is a constant. The linear nature of this relationship means that once B and A have been determined the vapour pressure at any other temperature may be calculated with a reasonable degree of accuracy. In practice, ΔH_{vap} is not independent of temperature and a more widely applicable relationship known as the Antoine equation (3) may be used using a third parameter C which provides the necessary temperature dependent correction. Values of A, B and C are available for many compounds in the *CRC Handbook of Chemistry and Physics*.

$$\ln P^o = -\frac{B}{T+C} + A \qquad (3)$$

TABLE 1 m_p and b_p parameters for PAC (8)

Compound(s)	m_p	b_p
PAC		
Phenanthrene/anthracene	4095	-18.38
Me-phenanthrene/Me-anthracene	3359	-15.44
Fluoranthene	4402	-18.44
Pyrene	4167	-17.50
Benzo(a)pyrene/benzo(b)fluorene)	4538	-18.44
Chrysene/benz(a)anthracene/triphenylene	5806	-21.83
Benzo(b)fluoranthene/benzo(k)fluoranthene	5677	-20.19
Benzo(a)pyrene/benzo(b)pyrene	4867	-17.00

At normal atmospheric temperatures, saturation concentrations are between about 1ng.m^{-3} and 100 µg.m^{-3} (5). These low vapour pressures still have a significant influence because PAC (in common with many other environmental contaminants) also have low water solubilities which results in high aqueous phase fugacities.

For more volatile species such as the azaarenes there is considerably more temperature dependent speciation variability, and when this is coupled with their higher molecular polarity (higher aqueous solubility) their environmental behaviour becomes considerably more complex.

3 Polycyclic Aromatic Hydrocarbons and Azaarenes

The PAC as environmental contaminants have long been of interest because of their toxicity and carcinogenic behaviour. They are present naturally in crude oils and are also commonly present in the combustion products of fossil fuels. Combustion produced PAC are characterised by a relatively high proportion of unsubstituted parent compounds (Readman et al.(2) and references therein). Some of the more important PAC are shown in Figure 3.

In terms of their aquatic environmental behaviour, the PAC behave with a considerable degree of coherence. This results from their mechanism of incorporation into particulate material which is frequently one of occlusion in such a manner that the compounds are not available for aqueous/particle partitioning as might be predicted from thermodynamic theory (1). Their behaviour in the atmosphere is also relatively consistent with only comparatively minor variations between species of differing volatility.

Azaarenes are nitrogen substituted analogues of PAC which are present in the combustion products of fossil fuels and therefore also in air and waters subject to anthropogenic influences. Structures of some of the commoner compounds are shown in Figure 4. The nitrogen substituent in the azaarene molecules not only changes their volatility but also makes them relatively basic in character and this feature is exploited in the selective isolation of these compounds during laboratory sample preparation. In addition to the parent molecules shown in the Figure, a considerable variety of methyl substituted species are also found in environmental samples.

The environmental behaviour of azaarenes has received little previous attention despite their fairly high toxicity. 2- and 3- ring azaarenes have a broad *in vivo* reactivity with threshold effects observed in membrane structure/function and respiratory processes (9). A number of species including quinoline, 4- and 8- methylquinoline, acridine, and benzoquinolines are active as tumour initiators on mouse skin (10-12). Human exposure arises through the inhalation of polluted air (including cigarette smoke) and ingestion of smoke contaminated food (10).

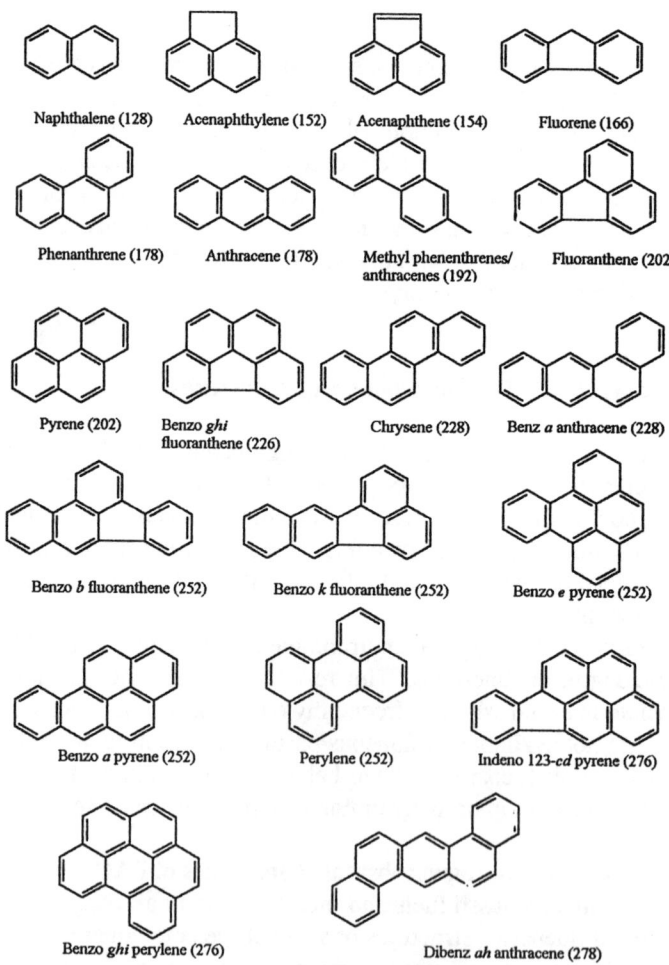

Figure 3 Some PAC (and their characteristic molecular ions m/z)

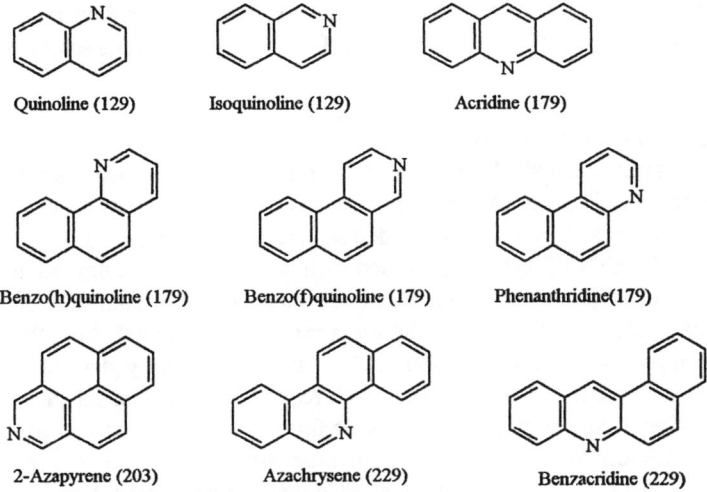

Figure 4 Some azaarenes (and their characteristic molecular ions m/z)

The azaarenes are therefore potentially important environmental contaminants which may also influence human health. The differences in physico-chemical properties between azaarenes and PAC means that their environmental behaviour is quite different and this paper addresses a number of these behavioural characteristics.

4 Analytical methodology

4.1 Sampling

All suspended sediment and most sediment samples were collected during a number of exercises conducted in 1993 and 1994 using either a custom built sampling unit (13) or a hand held grab respectively, using the National Rivers Authority boat *Sea Jet* as a sampling platform. A few sediment samples were collected from inter-tidal mud flats in places where these were accessible. Suspended sediment samples were collected in 2.5 litre glass Winchester bottles sealed with PTFE lined caps and pre-ashed aluminium foil. These samples were filtered as soon as possible onto pre-ashed (400°C, 8h) Whatman GF/F filters using a N_2 pressured (positive pressure) stainless steel filtration unit (Sartorius).

Sediments recovered with the grab were emptied onto a tray and transferred into glass jars with a solvent rinsed metal spatula. The jars were covered with pre-ashed aluminium foil before the cap was screwed on. Both sediment and suspended sediment filters were then freeze dried before being taken on to the extraction stage. In the case of the sediments the <125μm fraction was analysed.

Atmospheric samples were collected using a Sierra Andersson Hi-Vol sampling unit mounted on the roof of a 22m high building within the campus of the University of Liverpool. Samples were collected at a flow rate of 0.8 $m^3.min^{-1}$ onto pre-ashed glass fibre filters. Sampling periods were generally around 24h corresponding to a sampling volume of ca. 2300m^3. A Sierra Andersson PUF sampler unit was used for the collection of aerosol/vapour phase materials using a flow rate of 0.28 m^3·min^{-1} with a sample volume of ~ca. 2400m^3.

4.2 Glassware, filters and PUF plugs
All glassware was initially soaked overnight in Decon® detergent. It was then rinsed with 2M hydrochloric acid, distilled water and then soaked in 2M sodium hydroxide overnight. Finally, the glassware used for water/suspended sediment sampling (2.5 litre Winchester bottles) was rinsed with Milli-Q® water, sealed with pre-ashed aluminium foil and used within 48 hours. Glass jars used for sediments were treated as above but were oven dried (150°C, 2 hours) and then baked in a muffle furnace (350°C, 18 hours). After this, the jars were sealed with pre-ashed aluminium foil until required. Laboratory glassware was washed as above but, in the final stage, was extracted with redistilled extraction solvent.

Glass fibre filters were ashed in a muffle furnace (450°C overnight) then wrapped in pre-ashed aluminium foil until required. PUF plugs were exhaustively extracted with dichloromethane in a large soxhlet apparatus, dried then stored in pre-ashed foil. All filters and PUF plugs were loaded into and removed from the holders in a laminar flow cupboard.

4.3 Solvents
HPLC grade solvents (Rathburn Chemicals Limited, Walkerburn, Scotland and Fisons Scientific Equipment, Loughborough) were used throughout the study and were redistilled through an all glass (1m column) fractional distillation unit before use. Each solvent batch was tested before use by concentrating a 300mL aliquot to 50mL (initially by rotary evaporation to 1mL followed by blowdown to dryness using a gentle stream of purified nitrogen) and analysing the concentrate for impurities by GC-MS.

4.4 Standards
Standard azaarenes were obtained from the following sources: 6-methylquinoline (Sigma Chemical Co. Ltd, Fancy Road, Poole Dorset, UK) 8-methylquinoline, acridine, phenanthridine, 2,6,-di-*p*-tolylpyridine (Aldrich Chemical Co. Ltd, Gillingham, Dorset, UK); 2,6-dimethylquinoline, 7,8-benzoquinoline, 9-methylacridine (Janssen Chimica, Newton, Hyde, Cheshire, UK); 2-azapyrene, 2-azachrysene (NCI Chemical Carcinogen Repository, MRI Michigan, USA); d$_9$-acridine (MSD Isotopes, Cambrian Gases, K&K-Greef Ltd., Suffolk House, George St, Croydon CR9 3QL, UK).

4.6 Gas chromatography (GC) and gas chromatography-mass spectrometry (GC-MS)
All GC work was carried out on a Hewlett Packard HP5890A Series II machine with on-column injection, flame ionisation and nitrogen phosphorus detectors and a HP3396A integrator. GC-MS was conducted on a similar system (HP5890A Series I GC) coupled to a VG250 double focusing, magnetic sector mass spectrometer using selective ion monitoring mode. Data processing was performed using a networked pair of Vax 2000 Workstations. The main chromatography columns used were 5% phenyl/methylsilicone; either HP-5 (25m, 0.32mm i.d., 0.52mm phase thickness - Hewlett Packard) or DB5-MS (25m, 0.25mm i.d., 0.25mm phase thickness - J&W Scientific). These were programmed from an initial 60°C

(1 min. hold) at 10°C.min^{-1} to 300°C with a final 10min hold giving a total analysis time of 35min.

The compounds quantified (and the representative ions utilised were):- methylquinolines/methylisoquinolines (6 isomers; m/z 143), dimethylquinolines (11 isomers; m/z 157), trimethylquinoline isomers (10 isomers; m/z 171), benzoquinolines (2 isomers; m/z 179), acridine/phenanthridine (m/z 179), methylated 3-ring azaarenes (7 isomers; m/z 193), azafluoranthene (2 isomers, m/z 203), azapyrene (2 isomers m/z 203), benzoazafluorene (2 isomers; m/z 217) benzacridine/azachrysene (4 isomers; m/z 229). Total azaarenes were quantified as the sum of the above components.

4.7 Sample extraction and cleanup

The main analytical procedures utilised for azaarene analyses were based on those described by Brumley *et al.*, (14) as modified by Osborne (15).

Dried samples (sediments or suspended matter including filter) were extracted for 18h with dichloromethane in a Soxhlet apparatus. An extraction cycle of 12 cycles per hour was shown to be adequate to extract >92% of the azaarenes present (Osborne, 1995). The organic extract (ca. 250mL) was then concentrated to ca. 1mL by rotary evaporation and was quantitatively transferred to a pre-cleaned glass vial using 2x1mL aliquots of dichloromethane for washing the flask. If the clean-up procedure was not to be carried out immediately then the vials were tightly capped and stored at -24°C.

The concentrated extracts were further processed by first adding an internal standard (2,6,-di-*p*-tolylpyridine or d$_9$-acridine) and then a 1mL aliquot of 4M HCl. The mixture was then shaken for 10 minutes on a IKA Vibrax-VXR Shaker and the two phases allowed to separate. The acidified layer was removed and placed in a clean vial whilst the organic layer was extracted with acid two further times. All acidic extracts were then combined and cooled in an ice bath for 15 minutes. Sufficient 6M NaOH was then added to raise the pH to 14 and this basic aqueous phase was back extracted into dichloromethane (3x) using the same method as for the initial acid extraction. The organic layers were combined, dried by passage through a short column of anhydrous sodium sulphate and evaporated to approximately 1mL under a gentle stream of nitrogen while maintaining a low temperature. The samples were then either analysed immediately or stored at -24°C until analysis could be performed. The coefficient of variation of replicate sample analyses was within the range 2.6 to 9.0% with the exception of the most volatile methylquinolines where the coefficient of variation was 24%. For 'total azaarenes' (see above) the coefficient of variation was 7.1%. Corrections for recovery were made on the basis of recovery of standard compounds relative to the internal standard (60-85%).

5 Results

5.1 Atmospheric distributions and partitioning

The concentration of particulate azaarenes shows a very clear seasonal signal with high concentrations in winter months and much lower concentrations in summer (Fig.

5). There is also a clear relationship with temperature (Fig. 6) but this can potentially be explained in at least two distinct ways. The most obvious is that during warmer weather the production of azaarenes, which we believe at present to be largely due to combustion of oil and coal in domestic applications, is much less. An alternative explanation is that during warmer weather there is a change in the partitioning behaviour such that a much greater proportion of the total is in the vapour phase and hence uncollectable by a simple high-volume filtration system. In fact we are able to discount this second explanation as determining the total azaarene abundance through a series of samples in which both aerosol and vapour phase azaarenes were collected. This data (Fig. 7a) again shows the differences in concentration between winter and summer months whilst also indicating that there is indeed a change in partitioning behaviour as a function of temperature (Fig.7b).

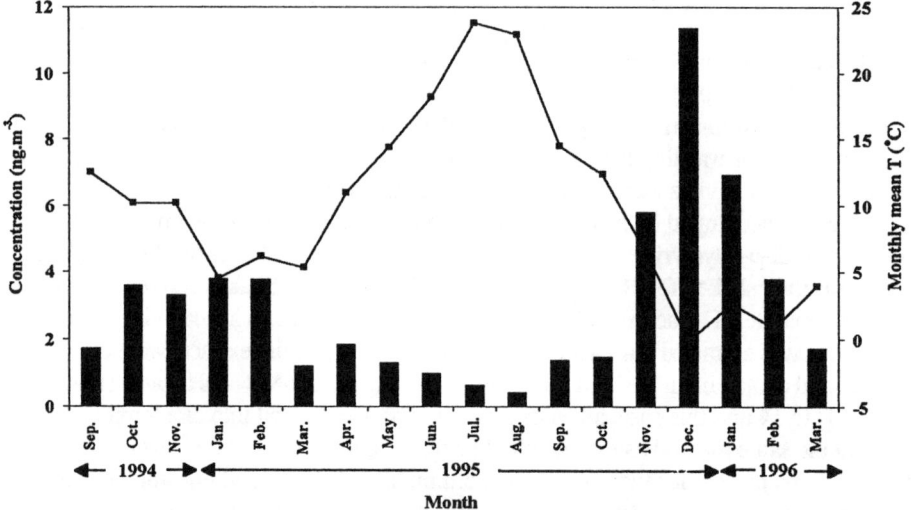

Figure 5 Monthly variations in aerosol azaarene concentrations and temperature

The partitioning behaviour is well explained by the Pankow and Bidleman relationship (eqn 1) as exemplified by the plot shown in Figure 8. Values of m_p and b_p for the azaarenes are given in Table 2. From this it can be seen that unlike the PAC (Table 1), there are considerable differences in the vapour/particle phase distributions within the normal ambient temperature range. It therefore follows that dry depositional fluxes of azaarenes to aquatic systems will be characterised by low concentrations of particles with a preponderance of higher molecular weight species in the warmer months, and higher concentrations of particles with a greater proportion of lower molecular weight species in the winter.

SOME SELECTED PERSISTENT ORGANIC POLLUTANTS

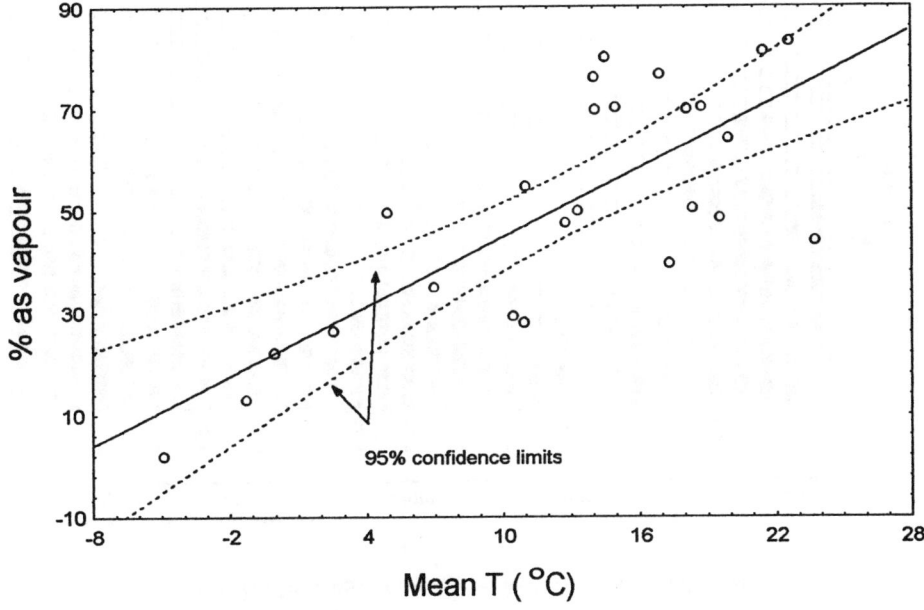

Figure 6 The % of Σazaarenes as vapour as a function of temperature

Knowledge of the m_p and b_p parameters, coupled with measurement of the ambient air temperature during sampling, means that it is now possible to predict the aerosol/vapour partitioning of azaarenes with a reasonable degree of reliability. This allows single measurements of particulate material to be extended to estimates of total atmospheric burdens.

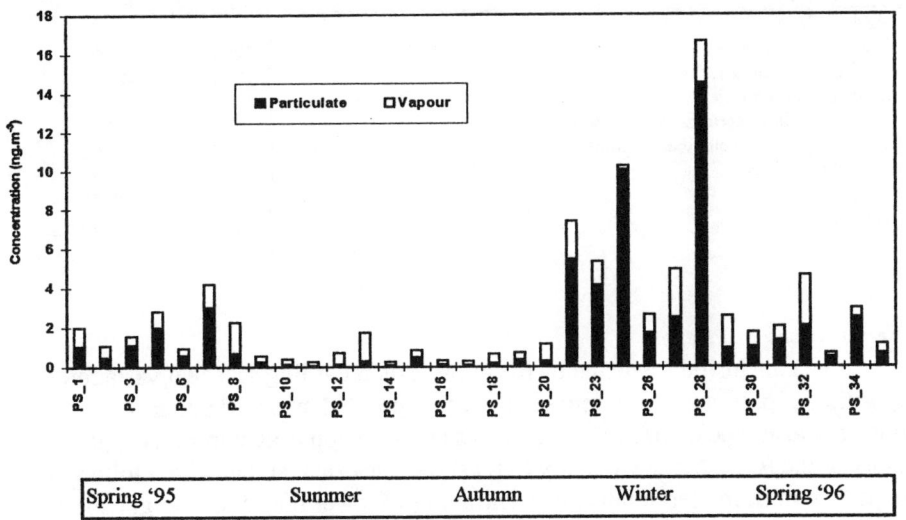

Figure 7(a) Vapour/particle phase partitioning with time

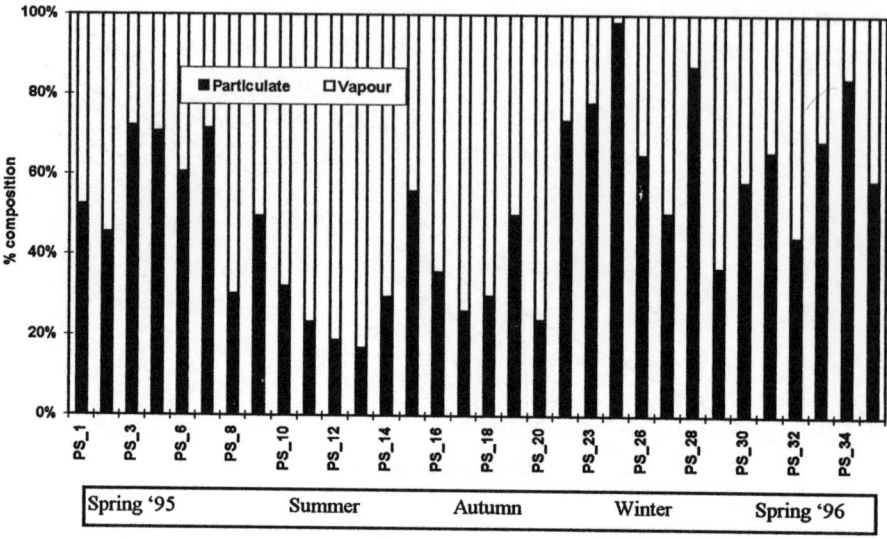

Figure 7(b) Normalised vapour/particle phase partitioning with time

TABLE 2 m_p and b_p parameters for azaarenes

Compound(s)	m_p	b_p
AZAARENES		
Methyl quinolines	2294	-9.89
Dimethyl quinolines	2800	-11.36
Trimethyl quinolines	3847	-14.60
Benzoquinolines/acridine/phenanthridine	3512	-13.88
Methyl acridines/methylbenzoquinilines	3914	-15.05
Phenylquinolines	3407	-13.66
Azapyrenes	5613	-20.59
Azachrysenes	5694	-20.35

5.2 Sedimentary azaarenes

The conclusion predicted by the arguments regarding seasonal variation in the composition of particulate azaarenes above are supported by observations of sedimentary azaarene distributions. A series of observations at sites in a canalised tributary of the River Mersey Estuary (UK) show a seasonal signal which follows the same pattern of decreased relative concentrations of lower molecular weight species in warmer months (Fig. 9) followed by an increase in the following winter. Within the main part of the estuary the very high tidal range (>10m) leads to large scale sediment movements which obscure this phenomenon. In addition, high tidal currents scour the main channels leaving rather course sand and gravel which is difficult to sample reliably.

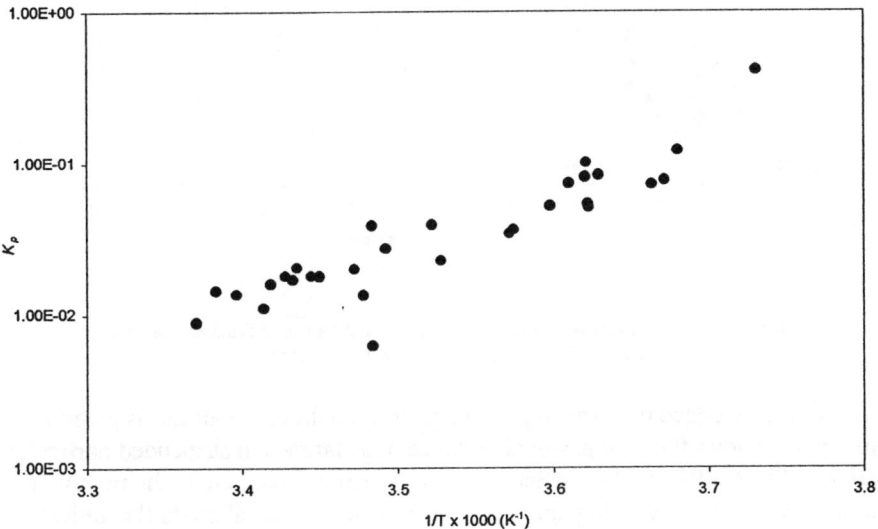

Figure 8 The Pankow and Bidleman (eqn 1) plot for compounds of methyl acridines and methylbenzoquinolines (m/z 193).

5.3 Aqueous solubility characteristics

The fact that azaarenes are considerably more polar molecules than the corresponding PAC has consequences for their aqueous solubility. In fact the pH dependency of azaarene solubility is exploited in the analytical methodology used to isolate them from environmental samples. The pH of natural waters generally ranges from ~4 in acidic fresh or rain waters to ~8.0 in sea water so it might be anticipated that azaarene behaviour would be strongly influenced by the movement from atmosphere to water or from fresh to sea water. Given that lower molecular weight species are frequently rather more soluble than higher molecular weight ones, differences in solubility could also influence the relative abundance of different azaarenes in the various components of aquatic systems (dissolved, suspended particulate, sedimentary, pore water dissolved etc.).

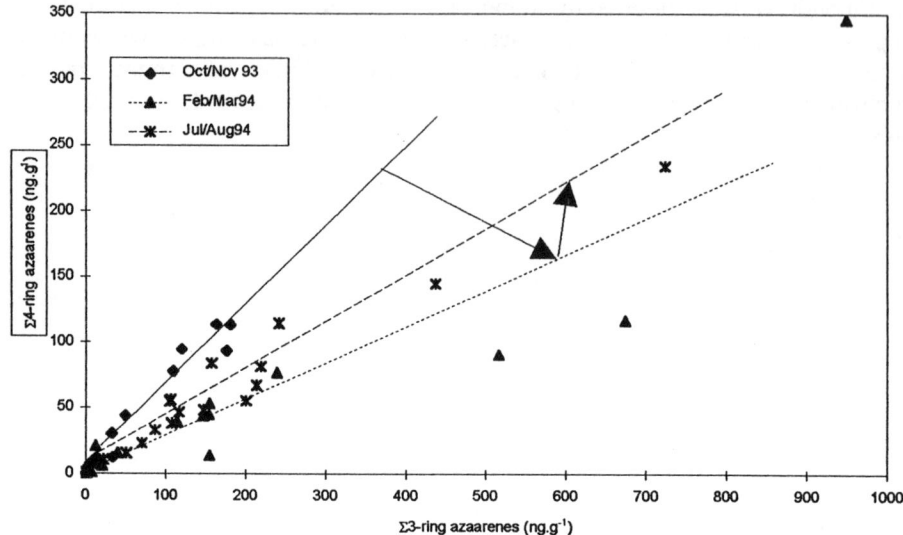

Figure 9 Seasonal variations in the concentrations of Σ4 ring v Σring azaarenes in sediments from the River Mersey Estuary system.

Some evidence than this type of behaviour could be important is given in Figure which shows the changes in abundance of azaarenes in suspended particulate material within the River Mersey Estuary. Of particular note here is the relatively rapid disappearance of two ring species. This selective removal could theoretically be due to one or the other (or both) of two mechanisms namely selective volatilisation or solubilisation.

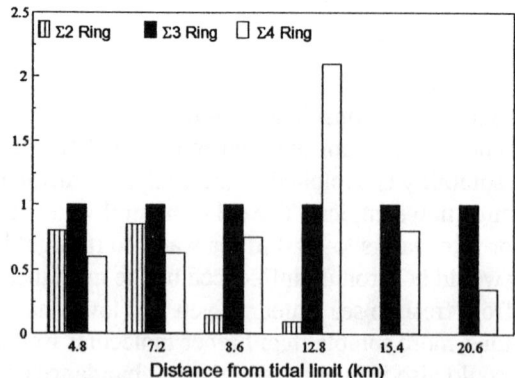

Figure 10 Changes in the relative amounts of different molecular size azaarenes in suspended matter from the River Mersey Estuary

In order to test the relevance of the latter process, a series of experiments were performed on aerosol samples. In these experiments, high-volume filter samples were divided into portions and exposed to potassium hydrogen phosphate buffer solutions at different pHs. The dissolution rate was then assessed by measuring the residual particulate azaarenes after various time intervals. The results (Fig. 1) indicate that rapid dissolution of all species occurs and, as expected, both the rate and extent of dissolution are enhanced at lower values of pH. Lower molecular weight species tend to dissolve rather more rapidly than higher molecular weight ones at a given pH.

The dissolution experiment does not entirely eliminate the differential volatilisation hypothesis which we have not yet tested, but dissolution does seem to us to be the dominant process.

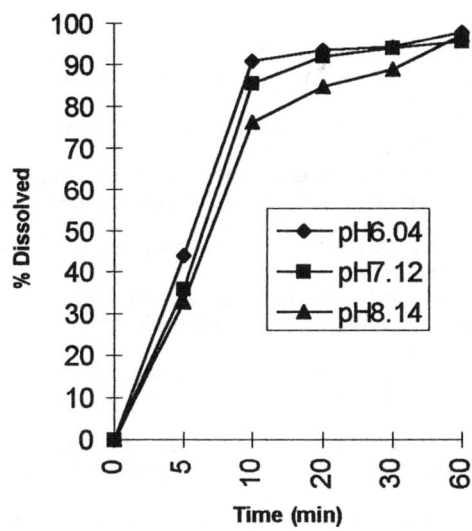

Figure 11 The dissolution rate of aerosol azaarenes at different pH

6. Conclusions

The studies discussed in this paper illuminate a variety of the aspects of the behaviour of environmental azaarenes. Their relative volatility and aqueous solubility results not only in rapid partitioning between environmental compartments but also fractionation between different species depending on their physico-chemical properties. Of particular interest is the post-depositional behaviour of azaarenes in the atmpsoheric aerosol where rapid removal of the more soluble species represents a significant difference in behaviour from, for example, PAC. The consequences of this solubilisation are unclear, but the azaarenes are unlikely to be the only nitrogen-containing species that behave in this manner and it may well be that contaminant organic nitrogen species are capable of making a aignificant contribution to the dissolved organic nitrogen content of rain identified by Cornell et al. (16)

There is a further aspect of the behaviour of azaarenes that is potential concern. Azaarenes are predominantly present in the atmosphere as fine particles (<10μmm) means that they are within the range capable of being inhaled. Particles introduced into the lungs will rapidly release the azaarene contaminant burden potentially resulting in localised high concentrations of these carcinogenic compounds. The importance of this process in the onset of lung cancers and other respiratory diseases remains to be determined.

References

1. Readman, J.W., Mantoura, R.F.C. and Rhead, M.M. (1984). Fresenius Z. Anal. Chem, **319**, 126.
2. Readman, J.W. Mantoura, R.F.C., Preston, M.R., Llewllyn, C.A. and Reeves, A.D. (1987). Int. J. Environ. Anal. Chem. **27**, 29-54.
3. Schwarzenbach, R.P., Gschwend, P.M. and Imboden, D.M. (1993) Environmental Organic Chemistry. John Wiley and Sons, Inc. NY, 681pp.
4. Eisenreich, S.J., Willford, W.A. and Strachan, W.M.J., (1989) "The role of atmospheric deposition in organic contaminant cycling in the Great Lakes" In Intermedia Pollutant Transport: Modelling and Field Measurements. Plenum NY. D.Allen (Ed.).
5. Afgan, B.K. and Chau, A.S.Y. (Eds) (1989). Analysis of Trace Organics in the Aquatic Environment, CRC Press, Boca Raton, FL, pp 220.
6. Stumm, W. (1992) Chemistry of the Solid-Water Interface, John Wiley and Sons Inc. NY, 428pp.
7. Morel, F.M.M. and Hering, J.G., (1993). Principles and Applications of Aquatic Chemistry, John Wiley and Sons Inc. NY,588pp
8. Pankow, J. F. and Bidleman, T. F. (1991), Effects of temperature, TSP and per cent non-exchangeable material in determining the gas-particle partitioning of organic compounds. Atmospheric Environment, **25**, 2241-2249.
9. Catallo, W. J., Hoover, D. G., and Vargas, D. (1994), Effects of an Environmental Toxicant on Ultrastructure in Paramecium Caudatum and the Protective Effect of Dissolved Calcium. Aquatic Toxicology, **29**, 291-303.
10. IARC (1983), IARC Monographs on the Evaluation of the Carcinogenic Risk of Chemicals to Humans: Polynuclear Aromatic Compounds, Part 1, Chemical, Environmental and Experimental Data, vol. 32, pp. 123-134, World Health Organization, France.
11. Vo-Dinh, T. (1989), Significance of Chemical, Analysis of Polycyclic Aromatic Compounds and Related Biological Systems. In "Chemical Analysis of Polycyclic Aromatic Compounds" (edited by Vo-Dinh, T.) Chapter 1, pp. 1-30, John Wiley & Sons, New York.
12. Matzner, R., and Bales, R. C. (1994), Transport of Acridine in Saturated Porous Media. Chemosphere, **29**, 1755-1773.
13. Preston, M.R. and Al-Omran, L.A. (1986). Dissolved and particulate phthalate esters in the River Mersey Estuary. Marine Pollution Bulletin **17**, 548-553.
14. Brumley, W.C., Brownrigg, C.M. and Brilis, G.M. (1991). J. Chromatogr.., 558, 223-233.
15. Osborne, P.J. (1995) Azaarenes in some merside sediments. Ph.D. Thesis, University of Liverpool, 257pp
16. Cornell,S., Rendell, A. & Jickells, T. Nature, **376**, 243-246 (1995).

THE INTERFACE AIR-WATER: OIL SPILLS AND TROPOSPHERIC CHEMISTRY. TECHNIQUES OF REMEDIATION

Ezio Bolzacchini, Simone Meinardi, Marco Orlandi and Bruno Rindone

*Department of Environmental Sciences,
University of Milano, Via Luigi Emanueli, 15,
I-20126 Milano, Italy.*

1.1 Hydrocarbons and sea

An important, but not easily recognizable, diffuse and chronic input of hydrocarbons into the sea is from refineries, land run-off and atmospheric deposition. The disposal of oil with waste water from refineries is generally well controlled. Other sources, related to urban activities, are not well understood. Polluted rivers and urban waste waters contain hydrocarbons in an unknown quantity and quality. Incomplete combustion of fuel oils on land is a major source of hydrocarbons deposited at sea. This might be a significant component of marine oil pollution. For comparison, 20% of the hydrocarbons (e.g.. the harmful PAH's) in the North Sea are deposited from the air.

Although oil is degraded in the marine environment, the various sources of oil pollution result in enhanced concentrations of hydrocarbons in sea water and the sediments. The high solar radiation promotes biodegradation, but the low nutrient concentrations in the open sea are a strong limiting factor.

1.2 Consequences of oil pollution

The consequences of large oil spills always cause considerable public concern, but the continuous loading of oil into the marine system can have an environmental impact as well. The problems of oil pollution are strongly related to the fate of oil in the marine environment. To a certain extent the environmental impact of oil pollution depends on seasonal or regional variation in the presence of sensitive species or life-stages.

1.1.1. Oil compounds that are dispersed or dissolved in sea water may give rise to toxic effects on marine biota, viz. fish-larvae, crustaceans (e.g. zoo plankton which is a main food source for fish) and filter feeding invertebrates (a.o. corals, sponges, sea-anemones and bivalves). In general, toxic effects have to be expected when the concentration of hydrocarbons in the water exceeds 50-100 µg/l. The impact depends on dilution, degradation, volatilization and the time that the floating slicks persists.

1.2.2 Oil compounds that are accumulating in the sediments may inhibit the juvenile settlement and condition of bottom dwelling animals, especially the longer living conservative species representing a main food source for the marine food web. In sediments containing more than 10-150 mg/kg of hydrocarbons these species are replaced by small, short living opportunists that have limited food value. Also in this case the impact depends on biodegradation, leaching and resuspension.

1.2.3 Floating oil slicks are initially mainly a risk for seabirds, that might become smothered with oil. For birds that come in contact with slicks of more than 1 mm thickness, the oil smoothing is generally lethal due to the fact that the thermal isolation by the feathers is reduced, or by ingestion of toxic oil compounds by cleaning the feathers. The risk of smothering can be reduced by removal of floating oil (mechanically or chemically).

1.2.4 Floating oil that is stranded in intertidal coastal areas (e.g. beaches, rocky shores, salt-marshes, tidal flats) causes substantial damage to the sedentary biota (algae, plants and animals) when more than 40-100 mg of hydrocarbons per m^2 is covering the biological community. The impact is related to suffocation and toxicity. Generally, the recovery time of intertidal communities is relatively slow, and restoration can take years depending on the rate of oil degradation and the reproduction or recolonization capacity of the local species. The impact can be reduced by enhancement of biodegradation (bioremediation) and recolonization.

1.2.5 Oil pollution of beaches also has consequences for tourism and recreation. Stranded oil slicks or tar balls (the final end product of marine oil spills and thus proportional to the oil pollution rate) are not appreciated and polluted beaches are avoided.

1.3 Potential solution to petroleum hydrocarbons in the sea

1.3.1 Prevention of petroleum input is the first priority. However, due to the tremendous traffic in transportation of oil it seems highly probable that some spills or other accidents will occur in the future, as has occurred in the past. Prevention includes enforcement of existing regulations and disposal policies both at the national and international levels in order to prevent input from normal operations involved in the movement and use of petroleum and its products.

1.3.2 Mechanical containment and recovery of spills from incidents at sea or in the harbors is a primary response method in shallow waters with temperatures above 6 °C. The technology currently available for mechanical oil spill cleanup has many limitations and only a very small percentage of the oil is actually recovered. The basic technology had not changed in the past years. The technology is not applicable to shallow in-shore areas and estuaries where many spills tend to occur. It does not appear that there will be much improvement in these methods in the near future for in shore areas. More costly larger systems may be soon available for large off-shore spills.

1.3.3 In-situ burning of spilled oil is used especially for highly flammable constituents. In certain situations, where no significant problems can be anticipated due to flammable or otherwise dangerous situations the techniques can be used but is problematic. The problems are primarily resulting from the toxic compounds that are produced during the frequently incomplete combustion of the burned oil. Some of these products causes serious air-pollution problems.

1.3.4 Absorbance and polymerization is used when control of the spread of spills and not direct clean-up of the spilled oil is required. Many of the available products require use in large volume inputs or result in sizable volumes of dense materials to be picked up for disposal by some other methods. Absorbants such as straw used on beaches are considered appropriate, although somewhat primitive. Likewise other disposal methods, usually incineration is necessary to get rid of the oil. No significant improvements in this technology is expected over the next few years.

1.3.5 Dispersants are used in spill response especially in off-shore incidents where it is highly desirable that the spill be prevented from reaching the shoreline. It is believed that limited environmental problems result from use of dispersants in off-shore, highly dilute conditions. However, broad applications have not occurred due to concerns about the toxicity of dispersed oil in shallow waters with a limited capacity for dilution. Additionally, improved biodegradation has not been proved. In some cases where dispersants have been used, hydrocarbons actually reach the sediments without significant biodegradation. The sedimentation of oil is even enhanced by dispersion, resulting in a long lasting accumulation of oil in the marine environment. The use of dispersants in shallow waters with a limited capacity for dilution should be avoided. Figure 1 shows the effect of the use of monorhamnolipid acid, dirhamnolipid acid and dirhamnolipid methyl ester on the dispersion of hexadecane.

1.3.6 Bioremediation is the use of microorganisms to eliminate or detoxify environmental contaminants by degrading hydrocarbons using their innate or natural abilities to metabolize these compounds. The catalytic cycle of cytochrome P-450, one of the most important enzymes involved in this task is shown in figure 2. Here, a protoporphyrin-IX-complexed iron(III) is transformed into a formal iron(V) by reaction with dioxygen and two reduction steps. Fe(V) will insert one oxygen atom into an organic compound and this will lead to more oxidised, more metabolizable products which are then transformed into biomass.

Monorhamnolipide (R = H or CH_3)

Dirhamnolipide (R = H or CH_3; R' = Rhamnose)

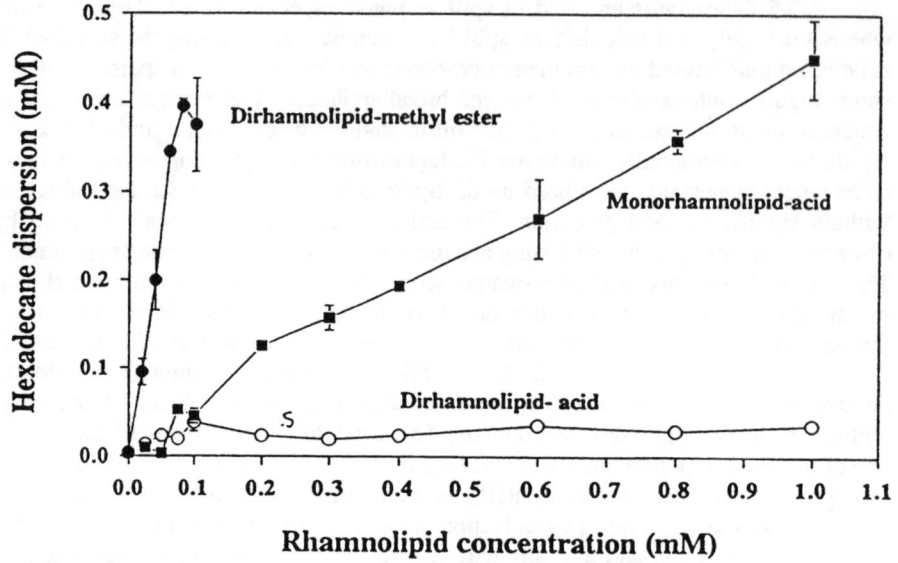

Figure 1. The effect of the use of monorhamnolipid acid, dirhamnolipid acid and dirhamnolipid methyl ester on the dispersion of hexadecane.

Bioremediation appears to offer the most promise for improved cost-effective treatment of oil spills, at least in in-shore or contained areas.

2.1 Oxygen species in the atmosphere

A huge amount of material evaporates and enters into the chemical oxidative cycles of the troposphere, genarting photochemicals smog, noxious compounds and long range effects.

Ozone is photolyzed in the atmosphere and both singlet and triplet oxygen atoms are formed.

$$O_3 + h\nu \Rightarrow O(^1D) + O_2 \quad (h\nu < 310 \text{ nm})$$

$$O_3 + h\nu \Rightarrow O(^3P) + O_2 \quad (h\nu < 1180 \text{ nm})$$

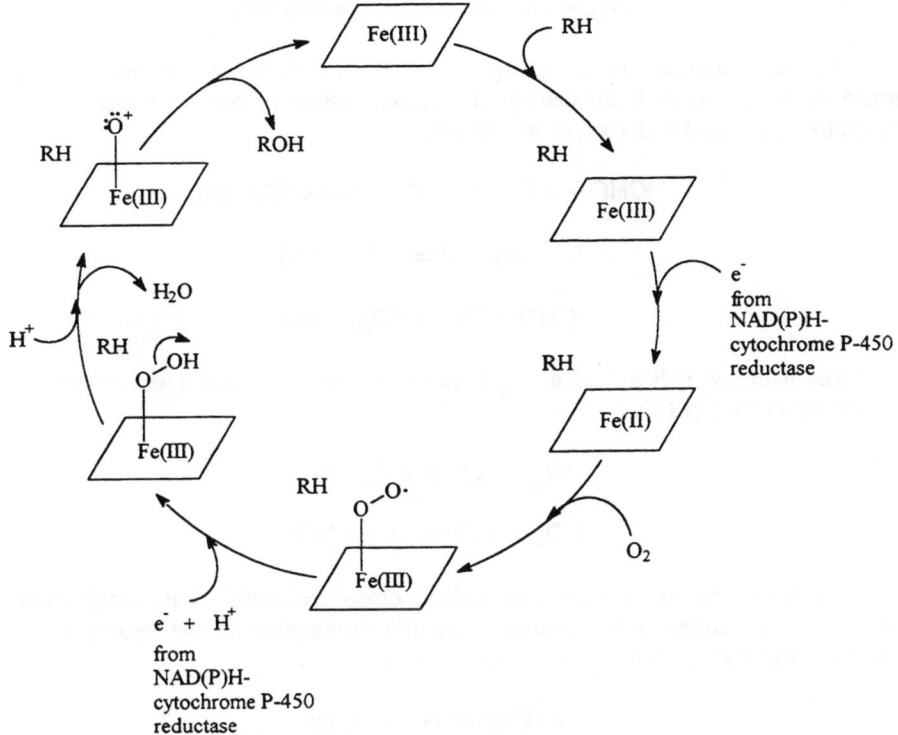

Figure 2. The catalytic cycle of cytochrome P-450

These oxygen atoms will react with a molecule of oxygen in the presence of a third body and will form back ozone. The photostationary concentration of ozone depends from these reactions.

$$O(^1D) + M \Rightarrow O(^3P) + M \quad (M=N_2, O_2)$$

$$O(^3P) + O_2 + M \Rightarrow O_3 + M$$

Oxygen atoms will also react with water and will form the most important oxidizing species in the atmosphere: the hydroxyl radical. Other sources of this species are the photolysis of nitrous acid and the photolysis of hydrogen peroxide.

$$O(^1D) + H_2O \Rightarrow 2\ HO\cdot$$

$$HONO + h\nu \Rightarrow HO\cdot + NO \quad (h\nu < 400\ nm)$$

$$H_2O_2 + h\nu \Rightarrow 2\ HO\cdot\ (h\nu < 360\ nm)$$

Another important oxidising species is the hydroperoxyl radical. This is formed by photolysis of formaldehyde and recombination of both hydrogen atoms and the formyl radical with oxygen molecules.

$$HCHO + h\nu \Rightarrow H\cdot + CHO\ \ (h\nu < 370\ nm)$$

$$H\cdot + O_2 + M \Rightarrow HO_2\cdot + M$$

$$\cdot CHO + O_2 \Rightarrow HO_2\cdot + CO$$

The hydroxyl radical and the hydroperoxyl radical will also interconvert via reaction with ozone or NO.

$$HO_2\cdot + O_3 \Rightarrow 2\ O_2 + OH\cdot$$

$$HO_2\cdot + NO \Rightarrow OH\cdot + NO_2$$

All these reaction are steps of radical chain reactions. Chain-termination reactions in conditions of low pollution are the dismutation of the hydroperoxyl radical and the reaction of $HO_2\cdot$ with $HO\cdot$.

$$2\ HO_2\cdot \Rightarrow H_2O_2 + O_2$$

$$HO_2\cdot + OH\cdot \Rightarrow H_2O + O_2$$

In the presence of high concentrations of NO_x ($NO + NO_2$) the chain termination reactions produce nitric acid.

$$OH\cdot + NO_2 \Rightarrow HNO_3$$

$$HO_2\cdot + NO \Rightarrow HNO_3$$

Table 1 shows the concentration of oxidising species in the troposphere.

2.2 The fate of the hydroxyl radical HO·.

The removal of HO· in the free troposphere is 80% due to CO and $HO_2\cdot$ is formed.

$$CO + HO\cdot \Rightarrow CO_2 + H\cdot$$

$$H\cdot + O_2 + M \Rightarrow HO_2\cdot + M$$

THE INTERFACE AIR-WATER

Table 1

Molecule	Concentration (molecules cm^{-3})
OH·	$5 \times 10^5 - 5 \times 10^6$
HO$_2$·	$1 \times 10^7 - 5 \times 10^8$
O	$1 \times 10^4 - 1 \times 10^5$
CH$_3$O$_2$·	$3 \times 10^6 - 5 \times 10^8$
NO$_3$ (nighttime)	$2 \times 10^9 - 1 \times 10^{10}$
O$_3$	$5 \times 10^{11} - 5 \times 10^{12}$
H$_2$O$_2$	$5 \times 10^9 - 5 \times 10^{10}$

In the polluted troposphere HO$_2$· is removed by NO and ozone is formed.

$$HO_2· + NO \Rightarrow HO· + NO_2$$

$$NO_2 + h\nu \Rightarrow NO + O(^3P) \quad (h\nu < 410 \text{ nm})$$

$$O(^3P) + O_2 + M \Rightarrow O_3 + M$$

The net result of the removal of HO· is the formation of ozone.

$$CO + 2 O_2 \Rightarrow CO_2 + O_3$$

The removal of HO· in the free troposphere is 20% due to CH$_4$. The methyl radical is first formed. This reacts with oxygen and forms the methylperoxyl radical which is reduced to the methoxyl radical hich gives formaldehyde.

$$CH_4 + HO· \Rightarrow CH_3· + H_2O$$

$$CH_3· + O_2 + M \Rightarrow CH_3O_2·$$

$$CH_3O_2· + NO \Rightarrow CH_3O· + NO_2$$

$$CH_3O· + O_2 \Rightarrow HCHO + HO_2·$$

The net result of the removal of HO· in this case is:

$$CH_4 + 4 O_2 \Rightarrow HCHO + H_2O + 2 O_3$$

Hence, concerning the ozone budget, in the presence of NO$_x$ ozone is formed.

$$CH_4 + 4 O_2 \Rightarrow HCHO + H_2O + 2 O_3$$

$$CO + 2 O_2 \Rightarrow CO_2 + O_3$$

In the absence of NO$_x$ ozone is consumed.

$$HO· + O_3 \Rightarrow HO_2· + O_2$$

$$CO + O_3 \Rightarrow CO_2 + O_2$$

Moreover,

$$2\ HO_2\cdot + HO_2 \Rightarrow H_2O_2 + O_2$$

$$CH_3O_2\cdot + HO_2\cdot \Rightarrow CH_3OOH + O_2$$

Ozone is invariant in these last reactions

2.3 The nitrogen oxide cycle

Figure 3 shows the nitrogen oxide cycle.

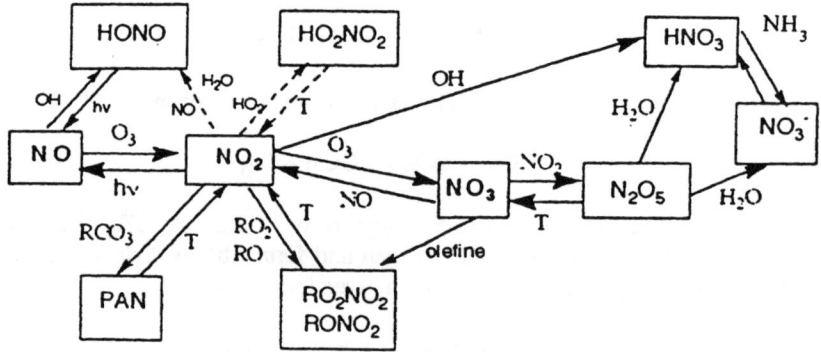

Figure 3: The nitrogen oxide cycle.

NO is emitted in all combustion phenomena via the reaction:

$$N_2 + O_2 \rightleftharpoons 2\ NO$$

NO is then transformed into NO_2 via reaction with ozone.

$$NO + O_3 \Rightarrow NO_2 + O_2$$

NO_2 is easily photolyzed, and has $t_{1/2}$ 85 sec. at 40° di latitude.

$$NO_2 + h\nu \Rightarrow NO + O \quad (295 < \lambda < 430)$$

Hence, the photostationary concentration of ozone is defined as:

$$[O_3] = \frac{J\ [NO_2]}{k\ [NO]}$$

where J_{NO_2} is the photolysis rate; k is the rate of the reaction of NO with ozone.

Hydrocarbons perturb this cycle introducing other reactions which convert NO into NO_2 without consuming ozone. This increases tropospheric ozone and generates photochemical smog.

2.4 Nighttime chemistry

The generation of the nitrate radical via reaction of NO_2 with ozone:

$$NO_2 + O_3 \Rightarrow NO_3 + O_2$$

has k (298 °K) = 3,2 x 10^{-17} cm^3 molecole^{-1} s^{-1}. These species give rise to an equilibrium with N_2O_5.

$$NO_2 + NO_3 + M \rightleftharpoons N_2O_5 + M$$

$$N_2O_5 + h\nu \Rightarrow NO_2 + NO_3 \quad (\lambda < 330 \text{ nm})$$

The photostationary equilibrium is:

$$K_{eq} = \frac{[N_2O_5]}{[NO_2][NO_3]} = 1.19 \; 10^{-27} \exp(11180/T)$$

NO_3 is rapidly photolyzed during the day according to the reactions:

$$NO_3 + h\nu \Rightarrow NO_2 + O \quad (\lambda < 630 \text{ nm})$$

$$NO_3 + h\nu \Rightarrow NO + O_2 \quad (\lambda < 630 \text{ nm})$$

2.5 The fate of organic compounds in the troposphere

The oxidizing power of the troposphere transforms organic compounds by reaction with OH· (during the day), NO_3·(during the night), O_3 (24 hours) or by photolysis (during the day). Table 2 shows reaction rates and lifetimes of some representative volatile organic compounds.

2.6 Hydrocarbons in the troposphere

A high percent of hydrocarbons from oil spills evaporate and are transformed by the oxidising power of the troposphere. Alkanes are degraded by reaction with OH·to give alkyl radicals which react with oxygen and form peroxyl radicals: These will be reduced by NO to alkoxy radicals.

Table 2
Reaction rates and half lives of some representative volatile organic compounds.

	night		day			
	kNO_3	Lifetime (h)	kOH	Lifetime (h)	kO_3	Lifetime (h)
Ethane	8×10^{-18}		3×10^{-13}			
Ethene	2×10^{-16}		8×10^{-12}			
n-Butane	6×10^{-17}		2×10^{-12}		$< 10^{-23}$	
1-Butene	3.6×10^{-13}	3	5.6×10^{-11}	200	2×10^{-16}	1.5
2,3-Dimethylbutane	4×10^{-16}		6×10^{-12}			
2,3-Dimethylbut-2-ene	4.5×10^{-11}	0.026	1×10^{-10}	2.1	1×10^{-15}	0.24
α-Pinene	6×10^{-12}	0.18	5×10^{-11}	4.3	8×10^{-17}	3.4
β-Pinene	1.1×10^{-12}	0.5	8×10^{-11}	2.9	3×10^{-17}	9.8
Isoprene	8×10^{-13}	1.4	1×10^{-10}	2.03	1.4×10^{-17}	20
Dimethysulphide	1×10^{-12}	1.1	1.5×10^{-11}	37	$< 8 \times 10^{-19}$	> 350
Acetaldehyde	2.8×10^{-15}		1.5×10^{-11}			
SO_2	$< 10^{-20}$	1.5×10^{-4}	10^{-12}	240	$< 10^{-23}$	
Toluene	4×10^{-12}	0.3	6×10^{-12}	38	$< 1 \times 10^{-20}$	$> 2 \times 10^4$
Phenol	1.3×10^{-12}		3×10^{-11}	9	$< 5 \times 10^{-19}$	> 480
NO_2		equilibrium	1×10^{-11}	1	3.2×10^{-17}	8

$$R\text{-}H + OH\cdot \Rightarrow R\cdot + H_2O;$$

$$R\cdot + O_2 \Rightarrow RO_2\cdot;$$

$$RO_2\cdot + NO \Rightarrow RO\cdot + NO_2.$$

RO· will be degraded to reaction products (carbonyls, alcohols etc.).

The peroxyl radical will also react with NO_2 forming alkyl nitrates.

$$RO_2\cdot + NO_2 \Rightarrow RO_2NO_2$$

Aromatics are attacked by OH· to give a benzyl radical (10% of the total reaction with toluene) and a hydroxycyclohexadienyl radical (90% of the total reaction with toluene).

Aromatic aldehydes, alcohols and nitrates are formed via the benzyl radical.

The hydroxycyclohexadienyl radical will form a phenol or ring cleavage products such as methylmuconaldehyde from toluene.

These will be further cleaved to small dicarbonyl compounds (glyoxal, methylglyoxal).

In the case of ortho-xylene also biacetyl and maleic anhydride are formed.

THE INTERFACE AIR-WATER

[Reaction scheme showing OH· radical reactions with toluene-derived intermediates producing various CHO-containing compounds and maleic anhydride]

Halflife for toluene in the troposphere is 38 h. Halflife for alkanes is several weeks. Table 3 shows the reaction products and yields in the laboratory reaction of OH and toluene or p-xylene.

Also tropospheric ozone formation via the reaction of photogenerated oxygen atoms with dioxygen is important:

$$NO_2 + h\nu \Rightarrow NO + O$$

$$O + O_2 \Rightarrow O_3$$

Here, NO_2 derives from the reduction of hydroperoxyl radicals with NO. Hence, tropospheric ozone formation depends from the concentration of both volatile organic compounds and NO (photochemical smog).

Table 3
Reaction products and yields in the laboratory reaction of OH and toluene or p-xylene

Products	Yield from Toluene	Yield from p-Xylene
Carbon monoxide	2.7 ± 1.4	8.0 ± 1.4
Carbon dioxide	3.5 ± 1.8	4.5 ± 0.4
Formaldehyde	2.3 ± 0.6	1.2 ± 0.2
Formic acid	12.9 ± 3.2	5.1 ± 1.0
Methanol	1.4 ± 0.4	+
Methylhydroperoxide	1.2 ± 0.3	+
Ketene	0.5 0.1	0.3 ± 0.1
Acetilene	0.7 ± 0.2	-
Acetic acid	5.2 ± 1.3	3.8 ± 0.3
Glyoxal	3.7 ± 0.9	4.0 ± 1.7
Methylglioxal	4.4 ± 1.1	2.2 ± 0.5
Maleic anhydride	4.2 ± 1.1	-
Methylmaleic anhydride	-	+

Table 3: continued

4-Oxo-pent-2-enal	2	+
Es-3-en-2,5-dione		8.3 ± 1.9
Benzaldehyde	7.1 ± 1.8	-
p-Tolualdehyde	-	6.4 ± 1.5
2-Hydroxytolualdeide	-	+
o-Cresol	+	-
m-Cresol	+	-
p-Cresol	+	-
Phenol	+	-
2,5-Dimethylphenol	-	8.0
Benzyl alcohol	+	-
4-Methylbenzylalcohol	-	+
Benzoic acid	+	+
4-Methylbenzoic acid	-	+

OH· will add to alkenes:

The resulting alkoxy radical will be transformed into carbonyl compounds.

Isoprene will also be attacked. The final reaction products are believed to be methyl vinyl ketone, metacrolein and 3-methylfuran.

Atmospheric transport of hydrocarbons from the site of an oil spill may generate these effects in regions very far apart from the sea. Also atmospheric transport of long living species (e. g: peroxyacetyl nitrate) may be important.

Here the formation and the decomposition of a peroxyacylnitrate is shown. An acyl radical derived from hydrogen abstraction from an aldehyde reacts with oxygen and gives a acylperoxy radical which reacts with NO_2 to give a peroxyacyl nitrate.

$$R-CHO \xrightarrow{OH\cdot} R-\overset{O}{C}\cdot \xrightarrow{O_2} R-C(=O)-O-O\cdot$$

$$\downarrow NO_2$$

$$R-C(=O)-O-O-NO_2$$

This species is enough stable to be transported elsewhere. There, photolysis will generate radicals and start a new cycle.

$$R-C(=O)-O-O-NO_2 \longrightarrow R-C(=O)-O-O\cdot + NO_2$$

$$R-C(=O)-O-O\cdot + NO \longrightarrow R-C(=O)-O\cdot + NO_2$$

$$R-C(=O)-O\cdot \longrightarrow R\cdot + CO_2$$

2.7 Oxidations with ozone

Ozone is toxic to organisms and oxidizes important biogenic alkenes such as ethylene and isoprene. An ozonide will be primarily formed and will cleave to carbonyl compounds.

2.8 Reactions of NO_3

Also NO_3 reacts with hydrocarbons via hydrogen atom abstraction or, in the case of aromatics and alkenes, via addition to the aromatic nucleus:

$$NO_3\cdot + RH \Rightarrow HNO_3 + R\cdot$$

Carbonyl compounds and nitrates are formed therefrom.

2.9 The chemistry of dimethylsulphide

Dimethylsulphide (DMS) is a gas emitted during decomposition of biota in marine environments. The decomposition of methionine will form dimethylsulphonium propionate which will decompose to DMS and acrylic acid.

CH$_3$S-CH$_2$-CH$_2$CH(NH$_2$)-COOH
Methionine

↓

CH$_3$S-CH$_2$-CH$_2$COOH

↓

(CH$_3$)$_2$S$^+$-CH$_2$-CH$_2$COOH
Dimethylsulphonium propionate

↓

(CH$_3$)$_2$S + CH$_2$=CH$_2$COOH
DMS

It will react with OH· by hydrogen abstraction and will finally form methanesulphonic acid.

$$CH_3\text{-}S\text{-}CH_3 + HO\cdot \Rightarrow H_3C\text{-}S\text{-}CH_2\cdot$$

$$H_3C\text{-}S\text{-}CH_2\cdot + O_2 \Rightarrow H_3C\text{-}S\text{-}CH_2\text{-}O\text{-}O\cdot$$

$$H_3C\text{-}S\text{-}CH_2\text{-}O\text{-}O\cdot \Rightarrow H_3C\text{-}SO_3H + CH_2O$$

Further reaction products will be dimethylsulphoxide, dimethylsulphone, SO$_2$, H$_2$SO$_4$, as shown in figure 4. Hence, DMS has an acidifying effect on the troposphere.

THE INTERFACE AIR-WATER

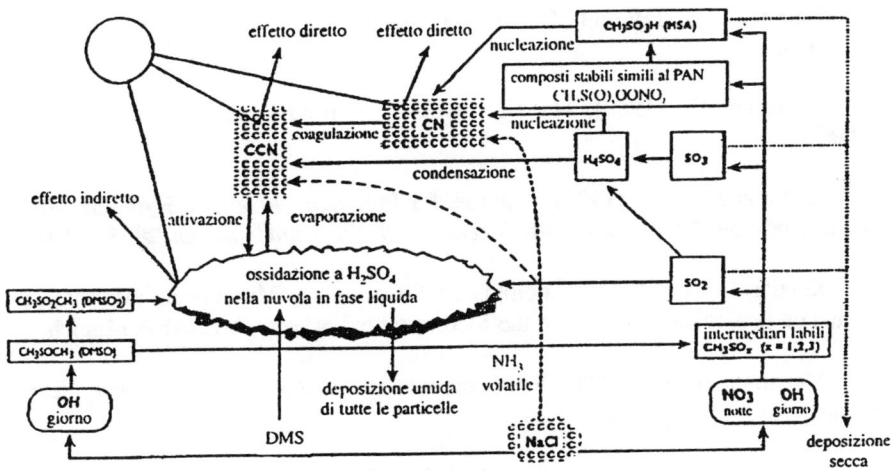

Figure 4. The chemistry of dimethylsulphide

References

Atkinson, R. (1994), Gas Phase Tropospheric Chemistry of Organic Compounds, American Chemical Society.

Boin, B, Cook, R. b:, (1983), Major Biogeochemical Cycles and Their Interactions, John Wiley.

Chianelli R.R. et al. (1991) Bioremediation technology development and application to the Alaskan Spill. In: Proc. Internat. Oil Spill Conference, API, San Diego, pp. 549-558.

Cohen, G., Greenwald, R. A., (1983), Oxy Radicals and Their Scavenger Systems, Elsevier.

Craig, P. J.. (1980), The Natural Environment and the Biogeochemical Cycles, Springer Verlag.

Floodgate G.D. (1984) The fate of petroleum in marine ecosystems. In: *Petroleum Microbiology* (R.M. Atlas ed.), pp. 355-397, Mc Millan Publ. Company, New York.

Graedel, T. E., (1978), Chemical Compounds in the Atmosphere, Academic Press.

Hansen, L. D., Eatough, D. J. (1991), Organic Chemistry of the Atmosphere, CRC Press.

Hutzinger, O., (1988), The Handbook of Environmental Chemistry, Springer Verlag.

Iosifidou H. et al. (1982) Analyses for Polycyclic Aromatic Hydrocarbons in Mussels from the Thermaicos Gulf, Greece. *Bull. Environm.Contam.* **28**, 535-541.

Kotzias D. (1980) Die Ökologische Situation im Mittelmeer. Information - Meeting on ecological situation in the Mediterranean Waters, MESAEP, Munich.

Manahan, S. E., (1993), Fundamentals of Environmental Chemistry, Lewis Publishers.

Oudot J. (1984) Rates of microbial degradation of petroleum components as determined by the computerized Gas Chromatography and computerized Mass Spectroscopy. *Marine Emvironmental Research* **13**, pp. 277-302.

Preston M.R. (1989) Marine Pollution. In: Chemical Oceanography, Vol. 9, Ch. 50, pp. 53-196. Academic Press, London.

Seinfeld, J. H., (1986), Atmospheric Chemistry and Physics or Air Pollution, John Wiley and Sons.

Sirvins, A., M. Angles, (1987)- Biodegradation of Petroleum Hydrocarbons, NATO ASI Series, Volume 69, Reidel, Dordrecht.

Sveum P. A. Ladousse (1989) Biodegradation of Oil in the Artic: Enhancement by Oil-Soluble Fertilizer Application. Proceeding of the 1989 Oil Spill Conference, San Antonio, Texas pp. 439-446.

Vallini G. (1992) Considerations About in Situ Bioremediation of Oil Polluted Marine Enviroments. Internal Report of the Biodegradation and Biotransformation Group, CNR - Soil Microbiology Center, Pisa, October/92.

Warneck, P., (1988), Chemistry of the Natural Atmosphere, Academic Press.

Wayne, R. P., (1991), The Chemistry of the Atmospheres, Oxford Science Publications.

INTERACTIONS OF MARINE BIOGEOCHEMICAL CYCLES AND THE PHOTODEGRADATION OF DISSOLVED ORGANIC CARBON AND DISSOLVED ORGANIC NITROGEN

R. G. Zepp
Rosenstiel School of Marine and Atmospheric Sciences
University of Miami, Miami, Florida, USA

Abstract

Recent human-induced changes in atmospheric composition are having major effects on UV radiation reaching the Earth's surface. Experimental and modeling results discussed in this paper indicate that these changes in UV radiation and accompanying changes in photochemical reactions in the sea are an important element of global change that impacts marine biogeochemical cycles. Exposure to solar radiation enhances the oxidation of marine dissolved organic carbon (DOC) resulting in fading in the UV and blue spectral region, loss of fluorescence, and formation of trace carbon gases including carbon dioxide, carbon monoxide, and carbonyl sulfide. Photochemical production of such trace gases occurs most rapidly in coastal regions of the sea. In addition to the direct production of trace gases, photodegradation of biologically-refractory organic matter produces organic substrates that can be readily mineralized or taken up by microbiota. Photodegradation of organic matter also converts biologically-resistant dissolved organic nitrogen (DON) to ammonium and other forms of labile nitrogen. This process can increase fertility and productivity of certain N-limited ecosystems. Action spectra indicate that organic matter photodegradation is primarily induced by the UV part of the solar spectral irradiance in seawater. Methods that are being used to study these reactions in the laboratory and field are discussed as are approaches to modeling the effects of these photoreactions on marine biogeochemical cycles.

1. Introduction

As evidence of global changes in climate and solar ultraviolet radiation has emerged over the past few decades, so has interest in the interactions of these changes with the Earth's biogeochemical cycles. The ocean, with its multiple reservoirs of energy and matter, is a major focus of research in this area. Marine biogeochemical cycles, the complex interaction of biological, chemical, and physical processes that control the exchange and recycling of matter of key chemical constituents and energy, are essential to the maintenance of life in the sea. In addition, marine biogeochemical cycles influence the composition of the atmosphere, including the concentrations of many trace gases that affect the Earth's radiation budget or that participate in chemical reactions in the troposphere and stratosphere.

Marine biogeochemical cycles are linked with global changes that are occurring on the continents as well. Human perturbations of biogeochemical cycles on land go hand-in-hand with expanding populations. Coastal regions are at the land-sea interface and thus they bear the brunt of the impacts of these perturbations. For example, as human activities put increased amounts of nitrogen in the atmosphere and into terrestrial ecosystems, nitrogen inputs into coastal areas are swelling. Moreover, coastal regions are increasingly exposed to an assortment of pollutants that enter the sea through rivers and atmospheric deposition. The fate and distribution of marine pollutants are intimately intertwined with biogeochemical cycles and thus to the global changes noted earlier.

In addition to human activities, other factors can affect marine biogeochemical cycles. These include natural variations in the Earth's orbit around the sun, sunspot activity and internal oscillations on Earth such as volcanic eruptions and biological evolution. Earth system models and new observational tools are being developed to help understand and predict how these natural and human perturbations interact.

Much of our current understanding of the Earth system has derived from measurements of the chemical and physical properties of the ocean, coupled with understanding of the observed variability in terms of fundamental biological, chemical and physical principles and the verification and extrapolation of results using various models. Development of new measurement techniques has permitted observations of the sea that have provided new, fresh perspectives. For example, as discussed in this paper, sensitive new analytical techniques that permit the analysis of trace organic constituents and reactive oxidants in the sea, have provided insights about the photochemical effects of changing solar ultraviolet radiation on marine biogeochemical cycles. Moreover, measurements of the spectral properties of dissolved organic carbon (DOC) in the sea are providing the basis for remote sensing of the DOC and a better understanding of underwater visibility.

During the 1980s scientists began to seriously discuss various approaches to the study of the effects of UV light on chemical and biological processes in the ocean [1-3]. Comparisons of experimental results revealed that DOC plays a central, multifaceted role in marine photochemistry and photobiology. On the one hand, DOC is the most important UV-absorbing substance in the sea and thus it regulates the penetration of biologically-harmful UV radiation. On the other hand, DOC is a photoreactive material that undergoes spectral changes on exposure to solar UV radiation, at the same time producing atmospherically-important trace gases and biologically-available carbon- and nitrogen-containing compounds, and initiating free radical and photosensitized reactions that affect oceanic composition.

Since the 1980s, more detailed and comprehensive field experiments on the photochemical effects of solar UV radiation have been conducted in marine biomes that span a wide range of biological activity, including Antarctica where high exposures to UV radiation occur now on a regular basis. New remote sensing techniques have been developed to measure marine DOC distributions across large regions of the sea and satellite observations are providing valuable new data on UV-absorbing constituents of the atmosphere that can be used to evaluate global changes in solar UV radiation. Ground-based observations of UV radiation are expanding as well These studies have laid the groundwork for the reliable extrapolation of laboratory and process-oriented local field experiments to larger regional and global scales.

This paper presents a sampling of the burgeoning recent research on photochemical reactions of DOC and DON. First, the paper briefly describes selected current research on the changes that are taking place in solar UV radiation reaching the Earth's surface. Then, recent studies of photochemical reactions of DOC and DON are considered, including spectral changes in DOC, photoproduction of atmospheric trace gases in the sea, and the conversion of persistent DOC and DON to readily-assimilable photoproducts (labile DOC and DON and ammonium). The paper is not intended to provide a comprehensive review of marine organic photochemistry but rather to provide the reader with a taste of the exciting studies that are taking place in this area and the types of measurement approaches and concepts that are being applied. To better appreciate the biogeochemical implications of this paper, readers are invited to peruse several recent overviews of global biogeochemical cycles such as books written by Butcher et al. [5] and Schlesinger [6]. UV effects on biogeochemical cycles are more specifically discussed in several recent papers [7-9].

2. Changes in Total Ozone and Solar UV Radiation

Atmospheric composition is changing and, going hand in hand with these changes, are changes in the amount and distribution of gaseous and particulate species that absorb solar ultraviolet radiation. For the purposes of discussion here, UV-B refers to radiation in the 280 to 315 nm spectral region and UV-A to that in the 315 to 400 nm region. As used here, the term UV-R refers to all UV radiation reaching the Earth's surface, the sum of UV-A and UV-B radiation. Visible radiation extends from 400 to 700 nm. Atmospheric ozone absorbs radiation most strongly at wavelengths <315 nm and it is the most important determinant of solar UV-B radiation that reaches the sea surface. Total ozone is the sum of ozone in the stratosphere and troposphere, but, since the latter is less than 10% of the former, changes in stratospheric ozone are by far the most important with respect to effects on UV radiation reaching the Earth's surface. In addition to ozone, clouds and aerosols also have important effects on the transmission of UV radiation. The latter have received less attention thus far, in part because their effects on solar radiation are less selective, *i.e.,* they also affect UV-A as well as visible radiation. Nonetheless, because clouds and aerosols are likely to be strongly affected by future changes in climate, future research will likely focus more closely on their effects on solar UV radiation reaching the sea surface.

Changes in total ozone since the development of the "hole" over the Antarctic have recently been compiled by a detailed scientific assessment of ozone depletion that was sponsored by the World Meteorological Organization [10]. The measurements have involved usage of ground-based networks of Dobson instruments and, more recently, satellite-based instruments, especially the Total Ozone Mapping Spectrometer (TOMS) and the Solar Ultraviolet spectrometers (SBUV). Total ozone has been measured since the 1920s using Dobson instruments and by the 1960s reasonably good global coverage was achieved by Dobson monitoring stations located around the world. Truly global coverage via satellite measurements commenced during November, 1978 via the TOMS and SBUV instruments mounted on the Nimbus 7 satellite. The Nimbus 7 TOMS finally failed on May 6, 1993. During its period of service it produced a valuable global daily record of the remarkable changes that have occurred in total column ozone over Antarctica and elsewhere over a 14-year period. Another TOMS instrument was launched on the Russian satellite, Meteor 3, during August, 1991, but, due to the satellite's drifting orbit, geographic coverage of this TOMS is much more limited. The SBUV instrument on Nimbus 7 has also provided a useful long-term record over the same period and another SBUV instrument, called the SBUV/2, was launched aboard the NOAA-11 satellite during January 1989.

Total ozone trends (% per decade) at various latitudes based on measurements by the Dobson and SBUV instruments are illustrated in Figure 1. The data indicate that ozone depletion has occurred at all latitudes. The most pronounced depletion has occurred over the Southern Hemisphere, but significant depletion has taken place over the mid-and high-latitudes of the Northern Hemisphere as well.

Increasing numbers of UV measurements at various locations around the globe have confirmed that ozone depletion, as expected based on theory, is accompanied by increases in UV-B radiation reaching the Earth's surface [9]. The estimated trends in biologically-effective UV doses, expressed as DNA damage, is shown in Figure 2. In this case, the fractional increases in damaging UV-B radiation are amplified compared to the changes in total ozone. The degree of amplification, or "radiation amplification factor (RAF)", is defined by eq. (1):

$$(E)_2/(E)_1 = [(O_3)_1/(O_3)_2]^{RAF} \qquad (1)$$

where $(E)_2$ and $(E)_1$ are the "weighted" irradiances that correspond, respectively, to total ozone amounts $(O_3)_1$ and $(O_3)_2$. The weighted irradiance is simply the integrated cross-product of the spectral irradiance and some weighting function that quantifies the biological or chemical effect of the radiation at various wavelengths [7, 11]. Plots of the weighting function versus wavelength are referred to as "action spectra." Action spectra vary for a variety of reasons. In the case of biota, the action is affected by factors such as pigmentation that helps shield the DNA from UV-B damage, as well as repair mechanisms that involve longer-wavelength UV-A radiation that is little affected by ozone depletion. Marine photoreactions involving DOC also exhibit great variability in RAF as demonstrated in Figure 3. These RAF values indicate that total ozone depletion since 1979 may have significantly enhanced photochemical DOC

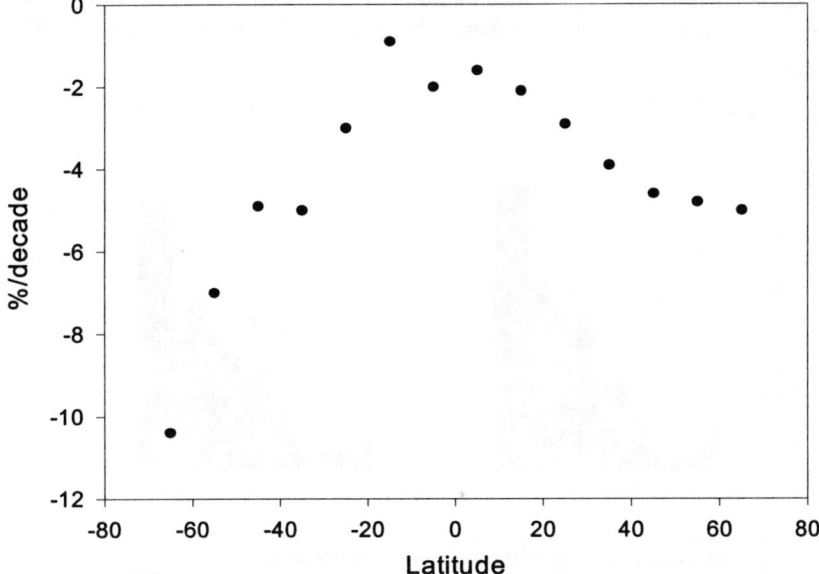

Figure 1. Total ozone trends in percent per decade by latitude for the period from 1979 to 1994. From Harris *et al.* [10].

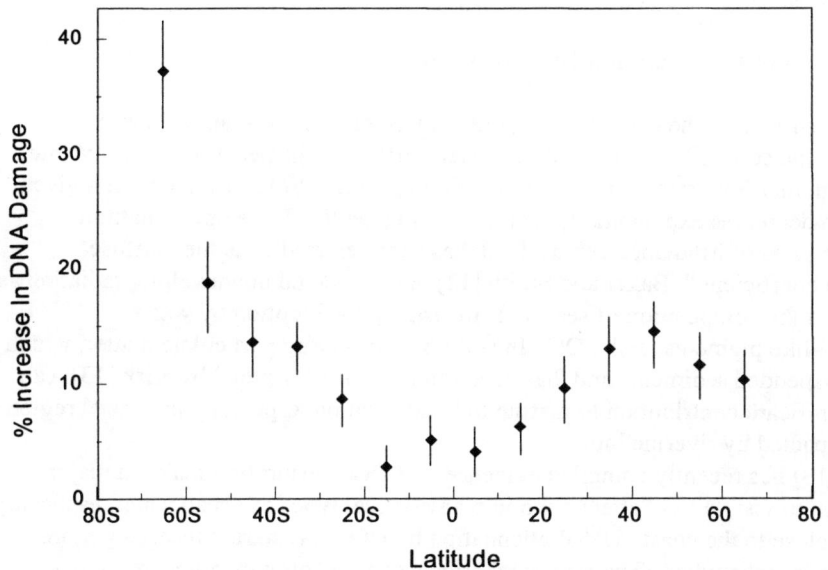

Figure 2. Calculated increases in annual UV exposures by latitude for the period 1979 to 1994. The action spectrum for DNA damage was used in estimating UV exposure. From Madronich *et al.* [11].

turnover in some parts of the ocean. The overall impact is likely to be greatest at mid- to-high latitudes in the Southern Hemisphere where both UV-B increases and ocean area are largest.

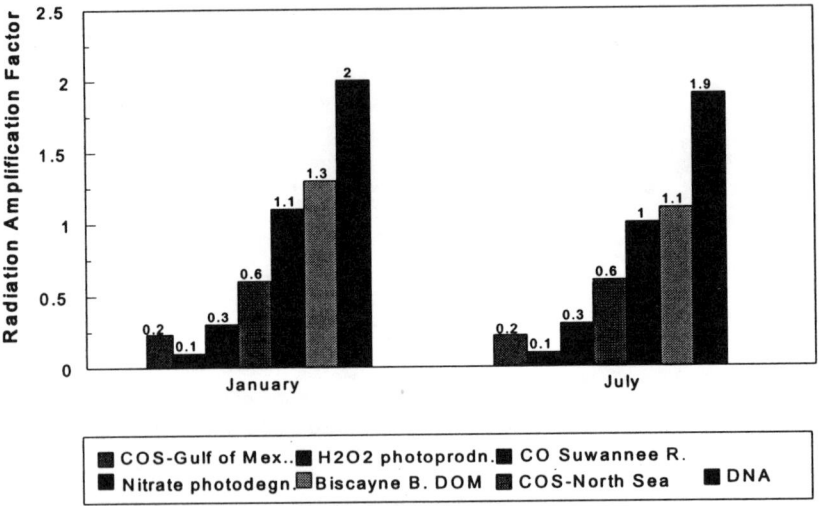

Figure 3. Radiation amplification factors computed for the effects of ozone depletion on selected DOC photoreactions and various biological responses [7, 11].

3. Absorption of UV Radiation By Marine DOC

The discussion in the preceding section focused on effects of atmospheric composition on solar UV radiation at the ocean surface. The penetration of UV into the sea is another important determinant of UV exposure. Solar irradiance at a given wavelength decreases exponentially with increasing depth. The slope of natural logarithmic plots of irradiance versus depth has been referred to as the "diffuse attenuation coefficient." Baker and Smith [12] have modeled downwelling radiation as a function of the composition of seawater, focusing on absorption by water, chlorophyll-like pigments, and DOC. In addition, non-living particulate matter, which includes suspended sediments and has been referred to as "tripton" by Kirk [13], can make a significant contribution to marine light attenuation, especially in coastal regions that are impacted by riverine inputs.

Kirk [13] has recently compiled evidence that DOC absorption makes a major contribution to visible light attenuation in many optical types of ocean water, especially in regions close to the coast. UV-R attenuation by DOC has started receiving more attention in recent studies. The important effects of DOC on light attenuation in a coastal region of the sea have been recently demonstrated by Vodacek *et al.* [14] (Figure 4). In this study DOC absorption was found to be the predominant contributor to attenuation in the UV-B and UV-A spectral region. DOC in both coastal and freshwaters contains a significant fraction of humic substances [15]. These substances, which are isolated by sorption onto hydrophobic resins, include a large part of the sunlight-absorbing components of DOC. Studies of dissolved organic matter isolated from freshwater ecosystems, however, indicate that hydrophilic acids also absorb solar radiation and are photoreactive [16]. Given the fact that hydrophilic acids make up more than half of marine DOC [15], it should not be assumed that humic substances are the only, or indeed even the predominant, photoreactive component. Thus, in the remainder of the paper I will refer to the colored or chromophoric dissolved organic matter in seawater as CDOM, a term coined by Blough and co-workers [17]. CDOM, which has often been referred to as "gelbstoffe" ("yellow substance") or "gilvin" by marine scientists, is a chemically complex and poorly-characterized mixture of anionic organic oligoelectrolytes known to contain phenolic moieties and exhibit surface active properties [15].

Light absorption by CDOM typically decreases in an approximately exponential fashion with increasing wavelength [17-21]. The absorptivity, $k_h(\lambda)$, can be represented by an equation of the following form [19]:

$$k_h(\lambda) = k_h(\lambda_o) e^{S(\lambda_o - \lambda)} \, 1 \, (\text{mg DOC})^{-1} \, \text{m}^{-1} \qquad (2)$$

where S is the negative slope of a natural log plot of $k_h(\lambda)$ versus wavelength λ(nm), $k_h(\lambda)$ and $k_h(\lambda_o)$ are the absorption coefficients at wavelength λ and reference wavelength λ_o. S is a parameter that characterizes how rapidly the absorption decreases with increasing wavelength. The absorptivity, $k_h(\lambda)$ is defined by the relation:

$$k_h(\lambda) = 2.303 \, A(\lambda) / \, (l) \, [DOC] \qquad (3)$$

where $A(\lambda)$ is the absorbance, l is the pathlength in meters, and $[DOC]$ is the DOC concentration expressed as mg C/L.

The coefficients, $k_h(\lambda)$ and S, vary both spatially and temporally [17] although the variation is not large for CDOM in freshwaters [19] and coastal waters [20]. Generally, values of $k_h(\lambda)$ in the UV-B region have been observed to be lower for very clear oligotrophic seawaters than for coastal seawaters strongly influenced by terrestrial input. S appears to increase with decreasing $k_h(\lambda)$ ranging from as low as 0.012-0.013 nm^{-1} for some highly absorbing coastal waters to over 0.02 nm^{-1} for weakly absorbing oligotrophic waters. Carder et al. [20] and Stewart and Wetzel [26] have related spectral properties of humic substances to their molecular weights. Moreover, as average molecular weight of the humic substance increases, absorptivity in the visible region also increases and the change in absorptivity with wavelength decreases. The magnitude of $k_h(\lambda_o)$, the pre-exponential factor, increases with increased degree of polymerization of the humic substance. Thus, humic acids have higher values of $k_h(\lambda_o)$, but lower values of S, than the lower-molecular-weight fulvic acids.

Comparisons by Kirk [13] of the absorptivities of pure water to the attenuation

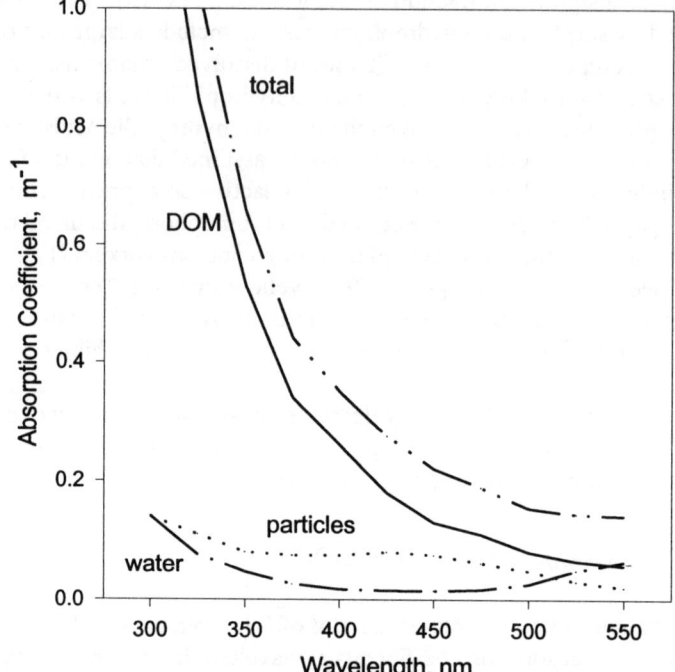

Figure 4. Comparison of the effects of DOC, particulate matter, and water absorption on the attenuation coefficients of coastal seawater (Mid-Atlantic Bight near the United States). From Vodacek and co-workers [14].

coefficients of Smith and Baker [22] for the clearest ocean water show that the pure water absorbs an order of magnitude more weakly in the UV-B region. These comparisons could indicate that, even in highly oligotrophic ocean water, absorption by DOC is mainly responsible for UV-B light attenuation. Other constituents of air-saturated low-productivity seawater, such as molecular oxygen and phytoplankton, absorb UV-B far too weakly to account for the difference. If this analysis is validated by future research, it indicates that phytoplankton, the main source of DOC in the open ocean, have a major influence on their exposure to UV-B radiation. Another important factor in that regard may be DOC photoreactions which decrease UV light attenuation by DOC. Such reactions are briefly discussed in the remainder of this section.

The susceptibility of CDOM spectral properties to changes on exposure to sunlight have been known for close to a century. Most of these studies have focused on freshwater lakes and ponds. Scientists interested in the aesthetics of drinking water were the first to demonstrate that the color of DOM was bleached by the action of sunlight [23, 24]. In 1970, Gjessing and Gjerdahl [24] estimated that about 20% of the reduction in color in a Norwegian lake is due to bleaching by natural UV radiation. In like fashion a few years later, Strome and Miller [25] provided evidence that the optical properties of DOC changed in a high-latitude pond over the course of a summer. The changes were consistent with continual photochemical alteration of organic matter by sunlight. Neither of these studies, however, included adequate controls that conclusively demonstrated that the spectral changes in the water bodies were attributable to photoreaction. Other possible effects such as precipitation of humic substances or dilution by groundwater could have significantly altered the DOC optical characteristics.

However, other studies under controlled conditions in the laboratory and with adequate controls in the field have provided ample evidence that the UV absorptivity of CDOM is reduced on exposure to short-wavelength UV light from artificial light sources [24] as well as from solar radiation [25-37]. To indicate the broad scope of this phenomenon, studies have examined photoreactions of DOC in freshwater, coastal

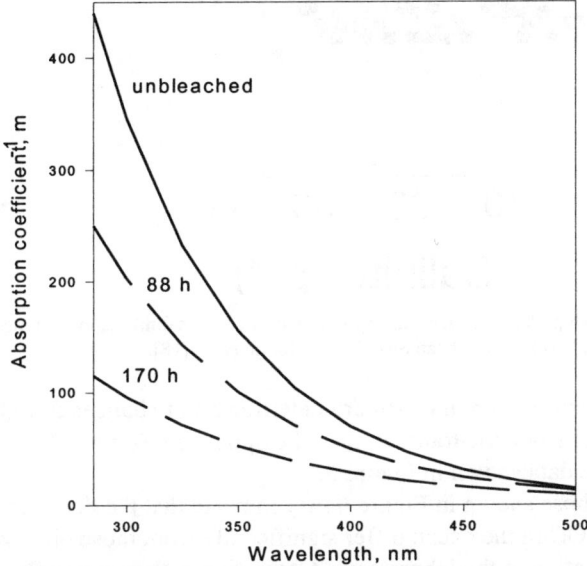

Figure 5. Changes in the absorption spectrum of irradiated water sample from Suwannee River, Georgia, USA on exposure to simulated solar radiation (total irradiance 650 W m^{-2}). From Miller and Zepp [59].

seawater, and open ocean water are cited. Because the fate of riverine DOC that enters the sea is believed to be strongly impacted by photochemical reactions, I demonstrate the photochemical spectral changes using DOC isolated from the Suwannee River, a river in the southeastern United States that drains into the Gulf of Mexico. The spectral changes observed in the Suwannee DOC on irradiation involve fractional decreases in the UV-R region that exceed those observed in the visible region (Figure 5). The net result is a decrease in the slope S (eq. 2) for the water. This result indicates that DOC photoreactions should tend to enhance the penetration depth of solar UV radiation into the sea. However, the results obtained by Blough *et al.* during field studies [17, 18] have shown that the slope S for the absorption spectrum of DOC increases with increasing distance from the mouth of rivers with high CDOM concentrations, such as the Orinoco River (Figure 6).

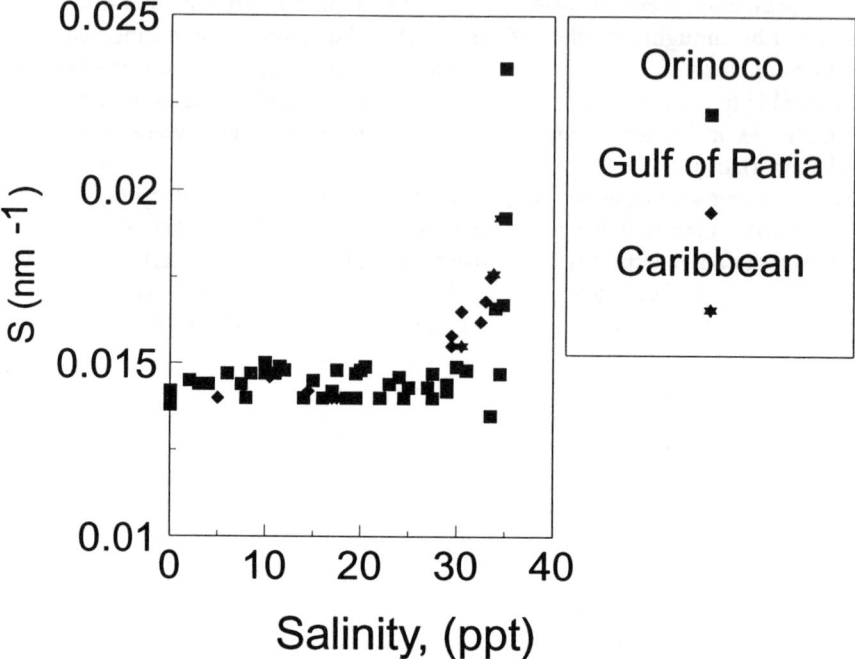

Figure 6. Changes in the slope S of the DOC absorption spectra versus salinity along a transect from the Orinoco River, Venezuela, into the Caribbean Sea. From Blough et al. [18].

If photochemical reactions were the primary determinant of changes in CDOM spectral characteristics along the transect, then the opposite result would have been expected based on the data shown in Figure 5.

Results such as those shown in Figure 6 may indicate that the depth-integrated spectral changes in DOC in the ocean differ significantly from those observed with full exposure to solar radiation in the laboratory. Alternatively, they may indicate that other, algal-derived sources of DOC are gradually becoming dominant along the transect, perhaps as the riverine DOC is removed by photochemical and biological processes. The latter explanation is supported by studies that have compared the isotopic composition of riverine DOC and open ocean DOC. The ^{13}C isotopic composition of DOC in the open ocean, -20‰ to -21‰, is about the same is that found in marine plankton, but is significantly (5‰ to 7‰) lower than that of riverine DOC [38, 39]. These isotopic studies likely indicate that phytoplankton are the primary source of the DOC in the open ocean. Additional research is required to better understand the interacting effects of DOC photoreactions and marine biological processes on the spectral properties of marine DOC.

In addition to absorption changes, some studies have indicated that the fluorescence of DOC also changes on exposure to solar radiation [26, 32, 40-46]. These data indicate that the photodegradation of the fluorescent component of DOC, *i.e.*, the DOC fluorophores, on exposure to full-spectrum solar radiation occurs more rapidly than the loss of UV absorption. Field studies of fluorescence distributions in the sea indicate significantly lower fluorescence efficiencies in near-surface than deeper waters, possibly due to fluorophore photodegradation in the photic zone. Models have been developed that use rates of fluorophore photodegradation, diffuse attenuation coefficients, simulated incident solar spectral irradiance, and wind speeds to simulate observations of fluorescence distributions in marine waters [32].

Studies in lakes by Stewart and Wetzel [26] have indicated decreases in the fluorescence to absorption ratios for DOC in the upper layers of small lakes compared to deeper water, but this may have been attributable to other factors such as precipitation of the CDOM. Højerslev similarly observed that fluorescence to "yellow substance" (i.e. CDOM) ratios decreased with decreasing depth in poorly-mixed regions of the coastal Baltic Sea [40]. He computed the concentration of the "yellow substance" from the attenuation coefficients of the ocean water in the UV-R and blue (400 to 450 nm) spectral region. The depth dependence was attributed to increased production of the fluorophores on degradation of the yellow substance at greater depths, although it is more likely that near-surface photodegradation of the fluorophores was mainly responsible. However, the ratio was well-correlated at various depths in well-mixed waters with > 15 ‰ salinities off the coast of Denmark. Based on these studies, Højerslev suggested that fluorescence:yellow substance ratio should be a useful index for classification of coastal water types. Blough and Green [17] have pointed out that fluorescence-based remote sensing is a promising technique for the regional mapping of DOC in coastal regions [17].

4. Trace Gas Photoproduction from DOC

The spectral changes that occur on irradiation of DOC are accompanied by the formation of several volatile inorganic photoproducts. The formation of the trace gases, carbon monoxide (CO), carbonyl sulfide (COS), and carbon dioxide has been demonstrated through recent research. In addition, marine DOC plays an important role in the photochemical oxidation of the volatile organosulfur compound, dimethylsulfide (DMS). Because DMS photooxidation and COS photoproduction both involve photosensitized reactions that have important effects on sulfur cycling in the sea, both are discussed together.

4.1. CO PHOTOPRODUCTION

DOC photoreactions are believed to be the main source of CO in seawater; CO loss has been ascribed primarily to exchange to the atmosphere and microbial metabolism. As a result of these processes, CO emissions from the sea follow a diurnal pattern with maximum near surface ocean concentrations greatly exceeding saturation during daylight. Although the sea is thought to be a net source of CO, great uncertainty exists regarding the strength of this source. Recent estimates range from 13

Tg/yr up to 1200 Tg/yr [43-48]. The latter estimate by Zuo and Jones [47] is based on numerous measurements of photoproduction rates of CO, coupled with observations that uptake by chemoautotrophic bacteria are quite slow under conditions that exist in most regions of the ocean. These studies indicate that most of the CO that is produced should escape to the atmosphere. On the other hand, Bates et al.[48], using extensive data on CO concentrations in near-surface seawater and the air above in the Pacific Ocean, have estimated that the annual amount of CO emitted from the sea to the atmosphere is about two orders of magnitude lower than the estimates of Zuo and Jones [47]. The flux estimates of Bates and co-workers are derived from air-sea exchange equations that use measured CO concentrations and exchange coefficients that are related to wind speed.

The large difference between CO flux estimates derived from these two approaches is not understood at this point. One possibility is that there must be a major oceanic CO sink that has not been previously identified. Alternatively, the CO concentrations determined on the cruises may not have been measured at depths and locations appropriate for computing realistic fluxes. For example, as discussed below, other studies have indicated that rates of DOC photoreactions in natural waters, including near-surface CO photoproduction, are approximately proportional to the UV absorptivity of the water and thus to CDOM concentrations [29-32, 37, 44, 47] (e.g., Figure 7). Coastal and upwelling waters have higher UV absorptivities and thus likely have higher CO photoproduction rates and greater CO concentrations during daylight near the water surface [44, 47] than oligotrophic ocean water. However, because these waters also attenuate UV radiation to a much greater extent than open ocean waters, CO photoproduction is confined to depths very close to the surface. Depth-integrated CO production per unit area is estimated to be about the same in both oceanic water types (see below discussion). It is possible that the ship's bow intake for the CO water sampling may have been positioned well below the depth of maximum CO photoproduction, especially in the most productive waters where near-surface CO photoproduction is highest but UV light penetration is lowest (see subsequent discussion later in this paper). If so, the computed fluxes would be greatly in error. Additional studies should be conducted to check out these possible artifacts..

Various studies have been conducted using monochromatic radiation under controlled laboratory conditions to determine the wavelengths responsible for photoproduction of CO and other DOC photoproducts in natural waters. Such data are required in models that extrapolate measured CO fluxes to unmeasured situations and estimate the changes in CO photoproduction in response to changes in atmospheric composition such as ozone depletion. The data obtained in these studies has been presented as both quantum yields and action spectra. Action spectra for formation can be expressed as plots of "response functions" $[X_{CO}(\lambda)]$, for each wavelength versus wavelength. The quantum yield, $\Phi_{CO}(\lambda)$, which is the fraction of absorbed radiation that results in formation (unitless) of the labile N species, is defined by the following equation:

$$\Phi_{CO}(\lambda) = \frac{[\text{Rate}(\lambda)]}{I_\lambda F_\lambda} \tag{4}$$

where Rate(λ) is the observed formation rate of CO (in molecules cm^{-3} s^{-1}), F_λ is the fraction of light absorbed at wavelength λ, and I_λ is the number of photons that enter the photoreaction cell per unit volume and unit time (units of photons cm^{-3} s^{-1}). The latter will be determined by chemical actinometry or using a calibrated spectroradiometer. It can be shown that the quantum yield is related to the response function by the following simple relationship:

$$X_{CO}(\lambda) = (a_\lambda) [\Phi_{CO}(\lambda)] \tag{5}$$

where a_λ is the mean absorption coefficient of the water sample (e.g. in units of m^{-1}) during the irradiation period. When defined using these units, the response function has units of m^{-1}. The photoproduction rate at wavelength λ is the cross product of the irradiance and the response function.

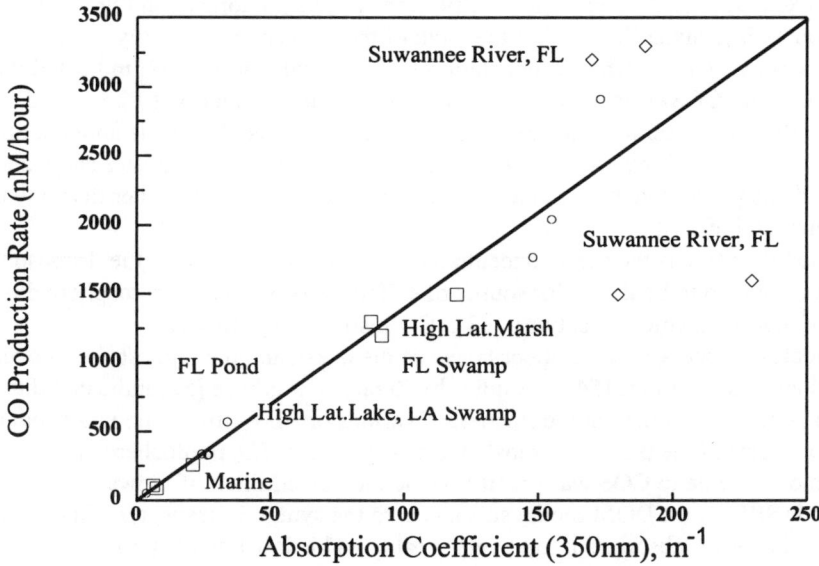

Figure 7. Plot of CO photoproduction rates versus absorption coefficients (350 nm) for natural water samples and humic substance solutions using same light source and intensity noted in Figure 5 [44, 59].

Using this experimental approach, action spectra and quantum yields for CO photoproduction were determined for the DOC in a variety of water samples that were obtained from several wetland and near coastal sites in North America. For wavelengths > 300 nm, the greatest action for CO production was in the UV-B region, but the spectra tailed out well into the UV-A region [44]. The quantum efficiencies for CO photoproduction and, consequently, photoproduction rates normalized to DOC absorptivities at 350 nm (Figure 7), were quite similar for DOC in freshwaters and in coastal regions.

4.2. DOC PHOTOSENSITIZATION OF ORGANOSULFUR REACTIONS

Sulfur-containing species strongly affect the radiative balance of the atmosphere [49]. Marine emissions of dimethylsulfide (DMS) and subsequent transformation of DMS into sulfate particles may influence the atmospheric radiative balance, especially in remote regions of the ocean that are not heavily impacted by aerosols derived from human activities. DMS production in the ocean is derived mainly from the decomposition of dimethylsulfonium propionate (DMSP), which is produced most efficiently by selected marine phytoplankton such as coccolithophorids [50]. DMS emissions to the atmosphere compete with its microbial and photochemical decomposition in the upper ocean. Recent studies by Kieber et al. [51] indicate that DMS photoreactions accounted for 7% to 40% of the total turnover of DMS in the surface mixed layer of the equatorial Pacific Ocean. The photoreaction involved conversion to dimethylsulfoxide, but the yield of the conversion was only 14%, much less than was expected. Other recent studies by Zepp and Andreae[52] indicated that carbonyl sulfide (COS) was a minor product of DMS photooxidation (0.2 % yield). DMS absorbs little or no light at wavelengths >295 nm, so its photolysis under sunlight in pure water is very slow. Apparently, natural substances in the seawater catalyzed or "sensitized" its photooxidation. Such "indirect photoreactions" are further discussed later in this section.

Carbonyl sulfide is the most concentrated sulfur-containing gas in the troposphere and it is considered to be the major source of sulfate aerosols in the stratosphere during periods of quiescent volcanic activity [53]. COS is formed primarily by photooxidation processes in the upper layers of the ocean and the highest fluxes occur in coastal and shelf regions [54]. Studies by Zepp and Andreae [52] indicated that COS can be formed by the photooxidation of various organosulfur compounds that are known to be part of the protein of most marine organisms. The photochemical oxidation of cysteine to COS was greatly accelerated on addition of Suwannee River fulvic acid (SRFA), a CDOM model substance, to the synthetic sea water (Figure 8). Cysteine, like DMS, absorbs sunlight very weakly. This low rate of sunlight absorption helps to explain why its direct photolysis to form COS is so slow. Likewise, the Suwannee River humic substance itself produced no COS. Thus, the data in Figure 8 indicate that the presence of the riverine humic substance promoted, i.e. "sensitized," the photoproduction of COS from cysteine. Other studies showed that addition of cysteine to coastal water from the North Sea that was visibly colored (about the same optical properties as those of 3.4 mg/L SRFA)

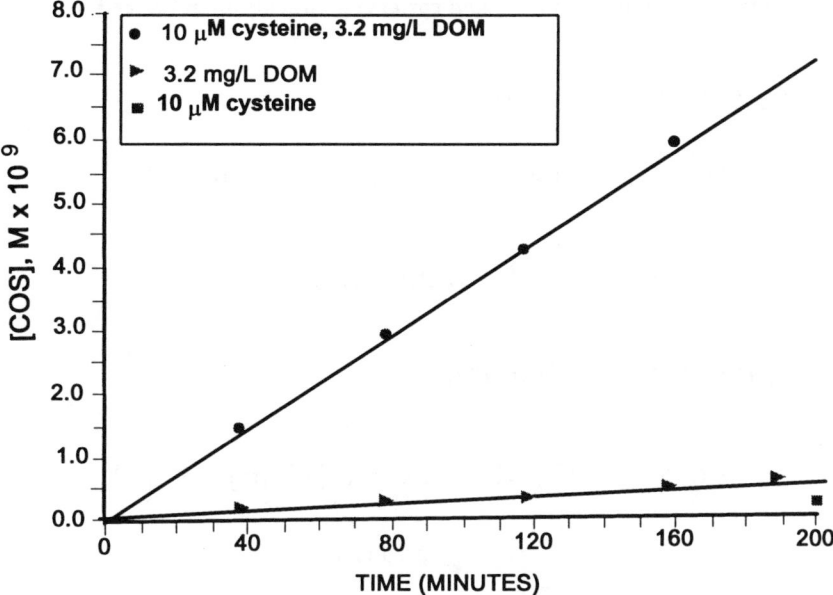

Figure 8. Effect of added Suwannee River humic substance on the photooxidation of cysteine to COS. The results indicate that humic substances can photosensitize the oxidation of organosulfur compounds to COS. Taken from Zepp and Andreae [52].

resulted in an increase in the photochemical formation rate of COS. Addition of the same concentration of cysteine to a North Sea sample that was obtained farther away from the coast and that absorbed sunlight much more weakly in the ultraviolet region resulted in COS production, but at a rate that was at least an order of magnitude lower. Clearly then, the presence of CDOM in seawater can greatly enhance the photooxidation rates of organosulfur compounds to COS.

Marine photoreactions generally fall into two classes that are described in Figure 9. Direct photoreactions involve light absorption by the reactive substance followed by some chemical reaction that is mediated by its excited state (s). The other indirect photoreaction pathway is initiated through light absorption by a substance other than the reactant (Figure 9). Both the DMS photooxidation and the organosulfur photooxidations that results in COS formation are predominantly "indirect" photoreactions. Indirect photoreactions are mediated by reactive transients that are produced on absorption of light by natural substances, with DOC playing a key role as the source of the transients [55-57]. These transients include free radicals produced by electron transfer, hydrogen atom transfer, or fragmentation of the DOC, and excited states of the DOC. Also, reactive oxygen species such as hydroxyl radicals (.OH), organoperoxyl radicals (RO_2.), hydroperoxyl/superoxide radicals, singlet molecular oxygen ($^1O_2$), dibromine radical anions, and carbonate radicals are produced in

Direct Photoreactions

$$\text{rate} = I_a \phi_r = [P] \int k_a(\lambda) \phi_r \, d\lambda = k_p [P]$$

$$P + \text{light} \longrightarrow P^* \longrightarrow \text{products}$$

Indirect Photoreactions

$$\text{rate} = \int E(\lambda) \, X(\lambda) \, d\lambda = k_T [T]_{ss} [P] = k_i [P]$$

$$S + \text{light} \longrightarrow T \begin{array}{c} \overset{P}{\nearrow} \text{products} \\ \searrow \text{other decay pathways} \end{array}$$

Figure 9. Pathways and kinetic equations for photoreactions in the sea.

irradiated systems solutions of CDOM in seawater. In the case of the reactions involving reactions of organosulfur compounds to produce COS, it was proposed that electron transfer from a bivalent S atom to carbonate radicals may initiate reaction.

The COS photoreaction appears to involve primarily UV-B radiation [52], but the DMS reaction in the Pacific Ocean mainly involved longer-wavelength visible radiation [51]. The latter action spectrum is the first report of a substantial involvement of visible light in an indirect photoreaction in seawater. The action spectrum possibly could indicate involvement of chlorophyll a like pigments at that location, although the involvement of red (600-700 nm) radiation was not reported.

4.3. DIC FORMATION

Marine scientists have long known that intense short-wavelength UV radiation can mineralize DOC. Indeed, this technique has provided a useful tool for the quantification of marine DOC. However, mineralizations obtained under these extreme conditions cannot be reliably extrapolated to natural conditions. Only very recently have several reports indicated that DOC in freshwaters and seawater can be directly mineralized on exposure to sunlight [37, 58-61]. Miles and Brezonik [58] were first to report this reaction in a natural freshwater system. They presented evidence that this process involved photoreactions of DOC-iron complexes. Recent studies by Faust and Zepp [62] and Voelker and Sulzberger [63] have provided a more detailed understanding of photoreactions involving iron and hydrogen peroxide in the natural photooxidation of DOC. Appreciable iron concentrations are sometimes found in high-DOC, acidic rivers that flow into the ocean. It seems likely that future research will demonstrate an important role for iron in enhancing photochemical mineralization of DOC in freshwaters. The role of iron in the photochemical mineralization of DOC in the ocean is unknown. Studies by Miller and Zepp [37, 59] indicated that manipluations of the pH of coastal seawater had little effect on photooxidation rates of DOC to DIC, although photoreactions involving iron complexes usually exhibit pH dependence. Other pathways for decarboxylation of organic acids that don't involve iron also are available, however [64].

Details on the kinetics of DOC photooxidation to DIC and related photoreactions are beginning to appear. The rates of dissolved inorganic carbon (DIC) photoproduction in water obtained from the Mississippi River plume, Sapelo Island (a Georgia, U.S. coastal wetland),and the Okefenokee Swamp, Georgia, U.S. were about an order of magnitude greater than CO rates measured in the same samples and production rates for both gases exhibited a linear correlation to photobleaching and initial absorbance of light at 350 nm [37]. Thus, DIC photoproduction constitutes the most rapid photochemical process of DOC that has been identified to date. It is interesting to note that rates and quantum yields for photochemical production of hydrogen peroxide in the sea are also about an order of magnitude greater than those for CO photoproduction [65]. The initial step for photochemical formation of hydrogen peroxide involves reduction of molecular oxygen to superoxide (or its conjugate acid hydroperoxyl) during DOC photooxidation reactions. Superoxide then dismutes to efficiently form hydrogen peroxide. Hydrogen peroxide, in turn, is decomposed enzymatically over periods ranging from days to weeks to regenerate molecular oxygen (via catalase-catalyzed reactions) or to form water and oxidized substrates (via peroxidase-catalyzed reactions) [66, 67]. The latter reaction, coupled with DOC photooxidation to reduce molecular oxygen to hydrogen peroxide, may provide a significant sink for molecular oxygen in the sea in addition to respiration.

Whatever the mechanism of DOC photooxidation, these recent studies have important implications for marine geochemists. As summarized by Hedges [68], marine scientists have long puzzled over the fate of riverine organic matter on entry to the ocean. Earlier in this paper it was noted that isotopic studies indicate that the DOC in the open ocean is primarily of marine origin although some terrestrial character would have been expected. The recent studies of DOC photooxidation induced by sunlight support Hedges' suggestion that DOC photodegradation likely accounts for the observed loss of riverine DOC.

How efficient is this mineralization process? Plots of fractional conversion of DOC to DIC versus fractional absorbance loss at 350 nm were close to linear with a slope of about 0.15 for both the Sapelo Island and Mississippi River plume water (Figure 10). This result indicates that possibly up to 85% of the DOC photoreaction yielded compounds that weakly absorb in the UV-A region. This fraction is almost two orders of magnitude greater than the combined yield of all known organic photoproducts of DOC photoreactions. It is likely, however, that the yield of these unidentified "bleached" compounds is somewhat lower, because a part of the original DOC in the samples likely was weakly absorbing organic matter such as polysaccharides.

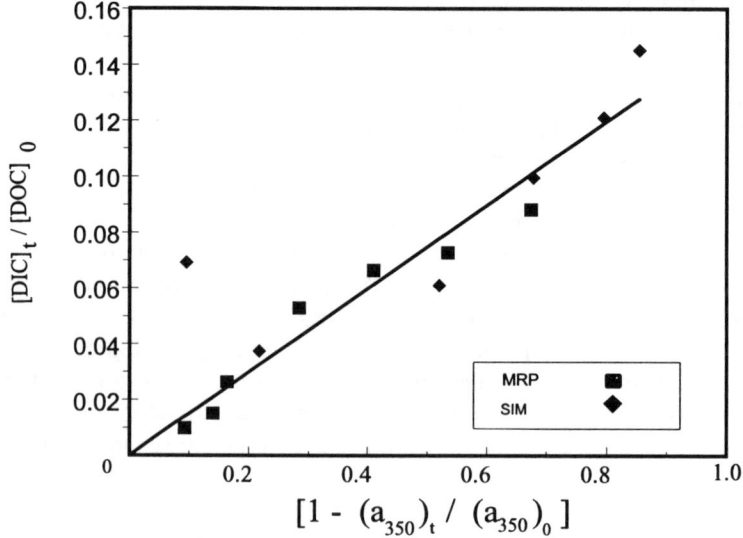

Figure 10. Comparison of fractional conversion of DOC to DIC with fractional loss of UV absorbance (350 nm) for water samples obtained from the Mississippi River plume and from Sapelo Island, Georgia. Taken from Miller and Zepp [37].

5. Photodegradation of Dissolved Organic Nitrogen (DON)

The majority of nitrogen that enters the ocean from land is in the form of DON [65]. Human activities on continents account for over half of the N that enters coastal ocean via rivers (about 41Tg N yr^{-1}), and DON constitutes the fastest growing component of riverine inputs of N to coastal regions [69].

It has been argued that denitrification in the sediments is the major fate of this riverine N [69] and that little N is transported to the open ocean. Others have suggested that the riverine N promotes coastal eutrophication and algal nuisance blooms in N-limited regions of the coastal ocean. Indeed, the riverine inputs of dissolved inorganic nitrogen (DIN) are biologically available and thus are rapidly taken up and cycled by phytoplankton and bacteria within estuaries and the inner shelf. The

DIN nitrogen eventually ends up in the sediments or, to a lesser extent, in the atmosphere after denitrification.

On the other hand, little is known about the fate of the riverine DON. The chemical composition of DON, as well as its degradation kinetics and bioavailability, are poorly defined, largely due to the heterogeneous nature of DON and consequent difficulties in structure elucidation. Most past research on total DON and individual components thereof has focused almost exclusively on the fraction that is readily assimilated by microorganisms. The persistent part of DON consists of predominantly higher molecular weight compounds, including humic substances. Although the persistent DON fraction constitutes the predominant component, its importance as a potential nitrogen source is generally assumed to be inconsequential.

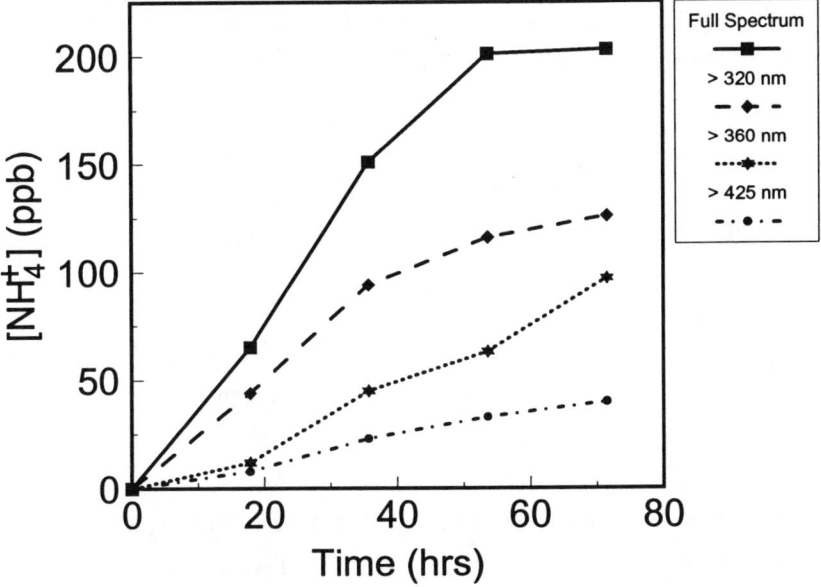

Figure 11. Photoproduction of ammonium in water from the Okefenokee Swamp, Georgia, USA on exposure to full-spectrum simulated solar radiation and to solar radiation that was filtered through various glass filters to remove varying portions of the UV component. The results show that solar UV radiation is most important in inducing photoammonification. [70]

Very recently, it was discovered that the photodegradation of bulk DON and DON fractions isolated from Georgia coastal rivers, wetlands and estuaries and boreal wetlands produces ammonium and low-molecular-weight organonitrogen compounds that are known substrates for microbial and algal growth [70]. Wavelengths in the UV region were most effective at inducing ammonium formation, as shown for the Okefenokee Swamp, the headwaters for two rivers that drain into the Atlantic Ocean and the Gulf of Mexico (Figure 11).

Early modeling results [70] indicate that most of the photoproduction of ammonium from riverine DON occurs in the clear, N-limited mid-shelf region off the eastern U.S. coast (photodegradation is much slower in the turbid estuaries and inner shelf); little riverine DIN reaches this mid-shelf region. These results suggest that the DON acts a nitrogen reservoir which is not significantly tapped until the nitrogen is released by the action of solar radiation. Under N-limiting conditions, the release of nitrogenous photoproducts from aquatic humic substances increased rates of bacterial growth 2-fold [70].

6. Effects of Photoreactions on Biological Availability of DOC

Photodegradation of DOC also produces biologically-labile, low-molecular weight organic compounds [26-35, 71-72] (Figure 12) that may stimulate bacterial growth and enhance carbon cycling rates in the upper ocean. A recent paper [72] reviews the photoproduction of biologically-available substrates by DOC and DON photoreactions.

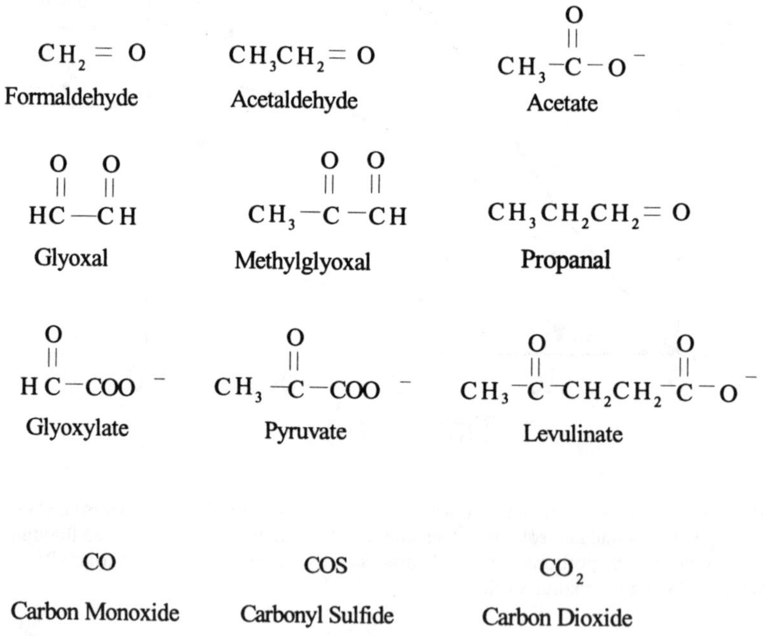

$CH_2 = O$ Formaldehyde

$CH_3CH_2 = O$ Acetaldehyde

$CH_3-\overset{O}{\underset{\|}{C}}-O^-$ Acetate

$\overset{O}{\underset{\|}{H C}}-\overset{O}{\underset{\|}{C H}}$ Glyoxal

$CH_3-\overset{O}{\underset{\|}{C}}-\overset{O}{\underset{\|}{C H}}$ Methylglyoxal

$CH_3CH_2CH_2 = O$ Propanal

$\overset{O}{\underset{\|}{H C}}-COO^-$ Glyoxylate

$CH_3-\overset{O}{\underset{\|}{C}}-COO^-$ Pyruvate

$CH_3-\overset{O}{\underset{\|}{C}}-CH_2CH_2-\overset{O}{\underset{\|}{C}}-O^-$ Levulinate

CO   Carbon Monoxide

COS   Carbonyl Sulfide

CO_2 Carbon Dioxide

Figure 12. Low-molecular-weight photoproducts from DOC photodegradation.

6. References

1. Zafiriou, O.C., Joussot-Dubien, J., Zepp, R.G. and Zika, R.G. (1984) Photochemistry of natural waters, *Environ. Sci. Technol.*, **18**, 358A-371A.
2. Goldberg, E.D. and Bard, A. (1988) CHEMRAWN IV. Modern chemistry and chemical technology applied to the ocean and its resources, *Applied Geochemistry*, **3**, 1-129.
3. Zepp, R.G. and Blough, N.V. (1990) Effects of solar ultraviolet radiation on biogeochemical dynamics in aquatic environments, Woods Hole Oceanographic Institution Report No. WHOI-90-09, January 1990.
4. Farman. J.C., Gardiner, B.G., and Shanklin, J.D. (1985) Large losses of total ozone in Antarctica reveal seasonal ClO_x/NO_x interaction, *Nature*, **315**, 207-210.
5. Butcher, S.S., Charlson, R.J., Orians, G.H. and Wolf, G.V. (eds.), *Global Biogeochemical Cycles*, Academic Press, Harcourt Brace Jovanovich, New York, 1992.
6. Schlesinger, W.H., *Biogeochemistry: An Analysis of Global Change*, Academic Press, San Diego, 1991.
7. Zepp, R.G., Callaghan, T.V., and Erickson, D.J. (1995) Effects of increased solar ultraviolet radiation on biogeochemical cycles, *Ambio*, **24**, 181-187.
8. Karentz, D., Bothwell, M.L., Coffin, R.B., Hanson, A., Herndl, G.J., Kilham, S.S., Lesser, M.P., Lindell, M., Moeller, R.E., Morris, D.P., Neale, P.J. , Sanders, R.W., Weiler, C.S. and Wetzel, R.G. (1994) Impact of UV-B radiation on pelagic freshwater ecosystems: Report of working group on bacteria and phytoplankton, *Arch. Hydrobiol., Beih. Ergebn. Limnol.,* 43, 31-69.
9. Williamson, C. E. (1995). UV impacts on lacustrine ecosystems, *Limnol. Oceanog.*, **40**, 386-392.
10. Harris, N. R. P. *et al.* 1995. Ozone measurements, in C. A. Ennis (ed.), *Scientific Assessment of Ozone Depletion: 1994*. World Meteorological Organization.
11. Madronich, S.; McKenzie, R.L.; Caldwell, M.M.; and Björn, L.O. (1995) Changes in ultraviolet radiation reaching the Earth's surface, *Ambio*, **24**, 143-152.
12. Baker, K.S. and Smith, R.C. (1982) Spectral irradiance penetration in natural waters, in J. Calkins (ed.), *The Role of Solar Ultraviolet Radiation in Marine Ecosystems*, Plenum Press, New York, pp. 79-91.
13. Kirk, J.T.O. (1994) *Light and Photosynthesis in Aquatic Ecosystems*, Cambridge University Press, New York, pp. 46-84.
14. Vodacek, A., DeGrandpre, M.D., Peltzer, E.D., Nelson, R.K., and Blough, N.V. (1996) Seasonal variation of CDOM and DOC in the Middle Atlantic Bight:Terrestrial inputs and photooxidation, *Limnol. Oceanog.*, in press.
15. Thurman, E. M. (1985) *Organic Geochemistry of Natural Waters*, Nijhoff/Junk, Boston, 497 p.
16. Zepp, R.G., Miller, W.L., Bourbonniere, R.A. and Tarr, M.A. (1995) Interactions of changing solar ultraviolet radiation and organic matter photooxidations in northern peatlands. *Preprint Extended Abstract, 210th National Meeting, American Chemical Society Division of Environmental Chemistry,* **35**, 394-398.
17. Blough, N. V. and Green, S. A. (1994) Spectroscopic characterization and remote sensing of non-living organic matter, in R.G. Zepp and Ch. Sonntag (eds.), *Role Of Non-Living Organic Matter in the Earth's Carbon Cycle*, Wiley, New York, pp. 42-57.
18. Blough, N.V., Zafiriou, O.C. and Bonilla, J. (1993) Optical absorption spectra of waters from the Orinoco River outflow: Terrestrial input of colored organic matter to the Caribbean. *J. Geophys. Res.*, **98(C2)**, 2271-78.
19. Zepp, R.G. and Schlotzhauer, P.F. (1981) Comparison of photochemical behavior of various humic substances in water. III. Spectroscopic properties of humic substances, *Chemosphere,* **10**, 479-86.
20. Bricaud, A., Morel, A. and Prieur, L. (1981) Absorption by dissolved organic matter of the sea (yellow substance) in the UV and visible domains, *Limnol. Oceanogr.*, **26**, 43-53.
21. Carder, K.L., Steward, R.G., Harvey, G.R. and Ortner, P. (1989) Marine humic and fulvic acids: Their effect on remote sensing of ocean chlorophyll, *Limnol. Oceanogr.*, **34**, 68-81.
22. Smith, R.C. and K.S. Baker (1981) Optical properties of the clearest natural waters (200-800 nm), *Applied Optics*, **20**, 177-184.
23. Whipple, G. C. (1914) *The Microscopy of Drinking Water*. Wiley, New York, 123 pp.

24. Gjessing, E.T. and Gjerdahl, T. (1970) Influence of ultra-violet radiation on aquatic humus, *Vatten*, **26**, 144-145.
25. Strome, D. J., and Miller, M.C. (1978) Photolytic changes in dissolved humic substances. *Verh. Internat. Verein. Limnol.*, **20**, 1248-1254.
26. Stewart, A. J., and R. G. Wetzel (1981) Dissolved humic materials: photodegradation, sediment effects, and reactivity with phosphate and calcium carbonate precipitation. *Arch. Hydrobiol.* **92**, 265-286.
27. Zepp, R. G. 1988. Environmental photoprocesses involving natural organic matter, in F. H. Frimmel and R. F. Christman (eds.). *Humic Substances and Their Role in the Enviroment*. Wiley, New York, pp. 193-214.
28. Kotsias, D., Herrmann, M., Zsolnay,A., Bayerle-Pfnur, R., Parlar, H. And Korte, F. (1987) Photochemical aging of humic substances, *Chemosphere*, 16, 1463-1468.
29. Kieber, D. J., McDaniel, J.and Mopper, K. (1989) Photochemical source of biological substrates in sea water: implications for carbon cycling. *Nature*, **341**, 637-639.
30. Kieber, R. J., X. Zhou, and K. Mopper (1990) Formation of carbonyl compounds from UV-induced photodegradation of humic substances in natural waters: fate of riverine carbon in the sea, *Limnol. Oceanogr.*, **35**, 1503-1515.
31. Mopper, K., Zhou, X., Kieber, R.J., Kieber, D.J., Sikorski, R.J. and Jones, R.D. (1991) Photochemical degradation of dissolved organic carbon and its impact on the oceanic carbon cycle, *Nature*, **353**, 60-62.
32. Kouassi, M., R.G. Zika, R.G. and Plane, J.M.C. (1990) Photochemical modeling of marine humus fluorescence in the ocean. *Neth. J. Sea Res.*, **27**, 33-41.
33. Kouassi, A.M. and R.G. Zika (1992) Light-induced destruction of the absorbance property of dissolved organic matter in seawater, *Toxicol. Environ. Chem*, **35**, 195-211.
34. De Haan, H. (1993) Solar UV-light penetration and photodegradation of humic substances in peaty lake water. *Limnol. Oceanogr.*, **38**, 1072-1076.
35. Hongve, D. (1994) Sunlight degradation of aquatic humic substances, *Acta Hydrochim. Hydrobiol.*, **3**, 117-120.
36. Frimmel, F. H. (1994) Photochemical aspects related to humic substances, *Environ. Internat.*, **20**, 373-385.
37. Miller, W. L., and Zepp, R.G. (1995) Photochemical production of dissolved inorganic carbon from terrestrial organic matter: significance to the oceanic organic carbon cycle, *Geophys. Res. Lett.*, **22**, 417-420.
38. Williams, P.M. and Druffel, E.R.M. (1987) Radiocarbon in dissolved organic matter in the central North Pacific Ocean, *Nature*, **330**, 246-248.
39. Williams, P.M. and Gordon, L.I. (1970) Carbon-13: carbon-12 ratios in dissolved and particulate organic matter in the sea, *Deep Sea Research*, **17**, 19-27.
40. Højerslev, N.K. (1982) Yellow substance in the sea,in J. Calkins (ed.), *The Role of Solar Ultraviolet Radiation in Marine Ecosystems*, Plenum Press, New York, pp. 263-281.
41. Che, R.F. and Bada, J.L. (1992) The fluorescence of dissolved organic matter in seawater, *Mar. Chem.*, 37, 191-197.
42. Hayase, K., Tsubota, H., Sunada, Goda, S. And Yamazaki, H. (1988) Vertical distribution of fluorescent organic matter in the North Pacific, *Mar. Chem.*, 25, 373-378.
43. Conrad, R., Seiler, W., Bunse, G. and Giehl, H. (1982) Carbon monoxide in seawater (Atlantic Ocean). *J. Geophys. Res.*, **87**, 8839-8852.
44. Valentine, R. and Zepp, R.G. (1993) Formation of carbon monoxide from photodegradation of terrestrial organic matter, *Environ. Sci. Technol.*, **27**, 409-412.
45. Erickson, D. J. III (1989) Ocean to atmosphere carbon monoxide flux:: Global inventory and climate implications, *Global Biogeochem. Cycles*, 3, 305-314.
46. Erickson, D.J. III and J.A. Taylor (1992) 3 D atmospheric CO modeling: The possible influence of the ocean, *Geophys. Res. Lett.*, 19, 1955-1958.
47. Zuo, Y., and Jones, R.D. (1995) Formation of carbon monoxide by photolysis of dissolved marine organic material and its significance in the carbon cycling of the oceans, *Naturwissenschaften*, **82**, 472-474.
48. Bates, T.S., Kelly, K.C., Johnson, J.E., and Gammon, R.H. (1995) Regional and seasonal variations in the flux of oceanic carbon monoxide to the atmosphere, *J. Geophys. Res. (D11)*, 100, 23,093-23,102.

49. Charlson, R. J.,Lovelock, J.E., Andreae, M.O., Warren, S.G. Oceanic phytoplankton, atmospheric sulfur, cloud albedo and climate, *Nature,* **326,** 655-661, 1987.
50. Keller, M. D., Bellows, W.K.and Guillard, R.R.L. (1989) A survey of dimethyl sulfide production in 12 classes of marine phytoplankton, in E. Saltzman and W. Cooper (eds.) *Biogenic Sulfur in the Environment,* American Chemical Society, Washington, D. C., pp. 167-182.
51. Kieber, D.J., Jiao, J., Kiene, R.P., and Bates, T.S. (1996) Impact of dimethylsulfide photochemistry on methyl sulfur cycling in the equatorial Pacific Ocean, *J. Geophys. Res.,* **101(C2),** 3715-3722.
52. Zepp, R.G. and Andreae, M.O. (1994) Factors affecting the photochemical production of carbonyl sulfide in sea water, *Geophys. Res. Lett.,* **21,** 281-2816.
53. Crutzen, P. (1976) The possible importance of COS for the sulfate layer of the stratosphere, *Geophys. Res. Lett.,* **3,** 73-76.
54. Andreae, M.O. and Ferek, R.J. (1992) Photochemical production of carbonyl sulfide in seawater and its emission to the atmosphere, *Global Biogeochem. Cycles,* **6,** 175-183.
55. Zafiriou, O. C., Blough, N. V., Micinski, E., Dister, B., Kieber, D. J. and Moffett, J. W. (1990) Molecular probe systems for reactive transients in natural waters, *Mar. Chem.,* **30,** 45-70.
56. Hoigné, J., Faust, B.C., Haag, W.R. and Zepp, R.G. (1989) Aquatic humic substances as sources and sinks of transients, in P. MacCarthy and I. H. Suffett (eds.), *Influence of Aquatic Humic Substances on Fate and Treatment of Pollutants.* ACS Symposium Series 327, Washington, DC. p. 363-384.
57. Blough, N.V.; Zepp, R.G. 1995.Reactive oxygen species in natural waters, in C.S. Foote and J.S. Valentine, J.S. (eds.), *Reactive Oxygen Species in Chemistry and Biochemistry,* Chapman & Hall: New York, pp. 280-333.
58. Miles, C. J., and Brezonik, P.L. Oxygen consumption in humic-colored waters by a photochemical ferrous-ferric catalytic cycle, *Environ. Sci. Technol.,* **15,** 1089-1095.
59. Miller, W.L. and Zepp, R.G. (1992) Photochemical carbon cycling in aquatic environments: Formation of atmospheric carbon dioxide and carbon monoxide. Preprint Extended Abstract, Environmental Chemistry Division, 203rd American Chemical Society National Meeting, San Francisco, CA.
60. Salonen, K., and Vähätalo, A. (1994) Photochemical mineralization of dissolved organic matter in lake Skjervatjern, *Environ. Internat.,* **20,** 307-312.
61. Granéli, W., Lindell, M., and Tranvik, L. (1996) Photo-oxidative production of dissolved inorganic carbon in lakes of different humic content, *Limnol. Oceanog.* , **41,** 698-706.
62. Faust, B.C. and Zepp, R.G. (1993) Photochemistry of aqueous iron(III)-polycarboxylate complexes: Roles in the chemistry of atmospheric and surface waters, *Environ. Sci. Technol.,* **27,** 2517-2522.
63. Voelker, B.M. and Sulzberger, B. (1996) Effects of fulvic acid on Fe(II) oxidation by hydrogen peroxide, *Environ. Sci. Technol.,* **30,** 1106-1114.
64. Budac, D., and Wan , P. (1992) Photodecarboxylation: mechanism and synthetic utility, *J. Photochem. Photobiol. A: Chem.,* **67,** 135-166.
65. Zafiriou, O.C., personal communication, 1996.
66. Moffet, J.W. and Zafiriou, O.C. (1990) An investigation of hydrogen peroxide chemistry in seawater by isotope ratio mass spectrometry using $^{18}O_2$ and $H_2^{18}O_2$, *Limnol. Oceanog.* , **35,** 1221-1229.
67. Cooper, W. J. and Zepp, R.G. (1990) Hydrogen peroxide decay in waters with suspended solils: Evidence for biologically mediated processes, *Can. J. Fish. Aquat. Sci.,* **47,** 888-893.
68. Hedges, J. I. (1992). Global biogeochemical cycles: progress and problems, *Mar. Chem.,* **39,** 67-93.
69. Galloway, J.N., Schlesinger, W.H., Levy, H., Michaels, A. and Schnoor,J.L..(1995) Nitrogen fixation: Anthropogenic enhancement - environmental response, *Global Biogeochem. Cycles,* **9,** 235-252.
70. Bushaw, K. L., Zepp, R.G.,Tarr, M.A., Schulz-Jander,D.,Bourbonniere, R.A., Hodson, R.E., Miller, W.L., Bronk, D.A. and Moran , M.A. (1996) Photochemical release of biologically labile nitrogen from dissolved organic matter, *Nature,* **381,** 404-407.
71. Wetzel, R. G., Hatcher, P. G. and Bianchi, T. S. (1995) Natural photolysis by ultraviolet irradiance of recalcitrant dissolved organic matter to simple substrates for rapid bacterial metabolism. *Limnol. Oceanogr.,* **40,** 1369-1380.
72. Moran, M.A. and Zepp, R.G. Role of photoreactions in the formation of biologically labile compounds from dissolved organic matter, *Limnol. Oceanog.,* in press.

EFFECTS OF LIGHT ON THE BIOLOGICAL AVAILABILITY OF TRACE METALS

BARBARA SULZBERGER
Swiss Federal Institute for Environmental Science and Tchnology
(EAWAG)
CH - 8600 Duebendorf
Switzerland

1. Introduction

Many metals are essential micronutrients, and they may limit or colimit phytoplankton growth in the euphotic zone of surface waters [1-3]. The hypothesis that iron is a limiting nutrient in some areas of the open ocean has been tested in mesoscale iron fertilization experiments [4, 5], and is an issue that is being discussed in connection with the global carbon cycle (see chapter entitled "The Influence of Iron on Carbon Dioxide in Surface Seawater" by Fank J. Millero, this volume).

The biological availability of metals depends on their speciation (see chapter entitled "Relationship between Metal Speciation and Bioavailability" by Margaret Farago, this volume). The speciation and thus the biological availability of redox-active metals is often strongly influenced by light. In this chapter, the discussion is limited to iron and manganese, although these are not the only trace metals whose speciation is affected by light (other examples are chromium, arsenic, selenium and mercury). In the euphotic zone of surface waters, the thermodynamically stable form of iron is colloidal or particulate Fe(III)-(hydr)oxide, and that of manganese is colloidal or particulate Mn(IV)-oxide. By most algae, however, iron and manganese are taken up as the dissolved species. Furthermore, not all dissolved iron and manganese species are biologically available to phytoplankton. The biological availability of these elements also depends on the chemical speciation of the dissolved species. Light-induced redox reactions play a pivotal role in the formation of dissolved and biologically available iron and manganese species, i.e., they control to a large extent the chemical speciation and thus the biological availability of iron and manganese in the euphotic zone of surface waters.

In this chapter the following issues are addressed: (i) Iron and manganese as phytoplankton nutrients (Subchapter 2); (ii) role of chemical speciation in controlling the biological availability of iron and manganese (Supchapter 3); (iii) speciation of iron and manganese in surface waters (Subchapter 4); (iv) role of light on the speciation of iron (Subchapter 5); and (v) role of light on the speciation of manganese (Subchapter 6). Finally, conclusions are drawn (Subchapter 7).

2. Iron and Manganese as Phytoplankton Nutrients

Both iron and manganese can exist in more than one oxidation state in cells, and catalysis of redox reactions and electron transport are two major functions of iron and manganese containing enzymes [6] (Table 1). Iron is present in the active centers of cytochromes and iron-sulfur proteins, e.g., ferredoxin, which are important components of the photosynthetic and respiratory electron transport chain (Figure 1). Iron, along with molybdenum, is also essential for nitrogen fixation and nitrate reduction and thus for the assimilation of the major nutrient nitrogen (Table 1). Nitrate assimilation increases the iron requirement for growth by about 60%, relative to NH_4^+ assimilation, and N_2 fixation increases this requirement by 100-fold [7].

Manganese serves as the primary electron acceptor in the oxidation of water in photosystem II (see Figure 1) and, along with iron, catalyzes the disproportionation of HO_2/O_2^- radicals to H_2O_2 (Table 1; reaction 4, below). Manganese has also been shown to compete with the uptake of cadmium and zinc by algae, and therefore to play an important role in the ecotoxicological effects of these elements [8, 9].

TABLE 1. Some enzymes and redox proteins containing iron and manganese cofactors (modified from Sunda [10])

Metal	Enzyme	Function
Fe	Cytochrome f	Photosynthetic electron transport
	Cytochromes b and c	Electron transport in photosynthesis and respiration
	Ferredoxin	Electron transport in photosynthesis and nitrogen fixation
	Iron-sulfur proteins	Photosynthetic and respiratory electron transport
	Catalase	H_2O_2 breakdown to H_2O and O_2
	Peroxidases	H_2O_2 reduction to H_2O
	Chelatase	Porphyrin and phycobiliprotein synthesis
Fe and Mo	Nitrogenase	Nitrogen fixation
	Nitrate and nitrite reductases	Nitrate reduction to ammonia
Mn or Fe	Superoxide dismutase	Disproportionation of O_2^- radicals to O_2 and H_2O_2
Mn	O_2 evolving enzyme	Oxidation of water to O_2 in photosynthesis

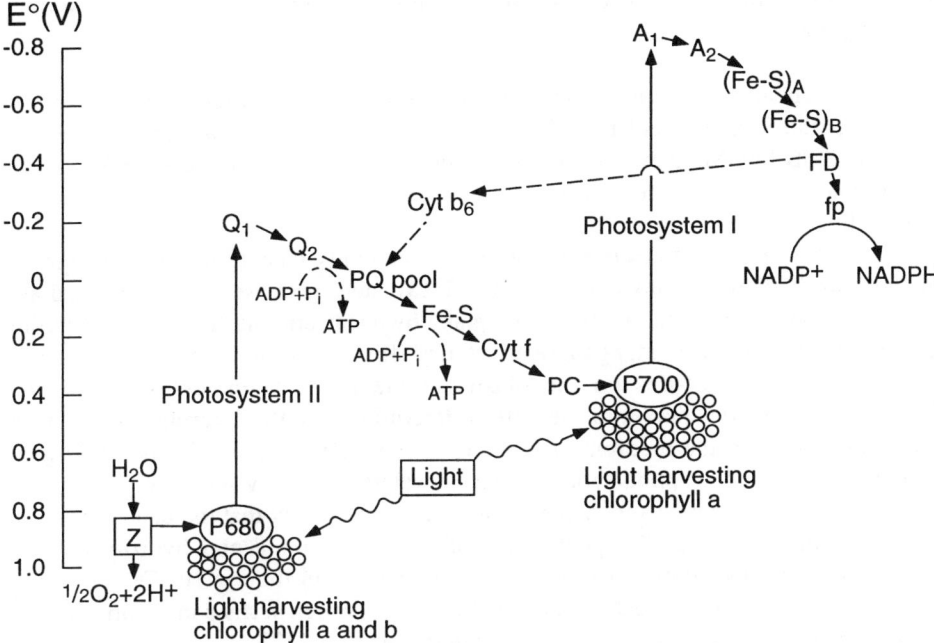

Figure 1. Z-scheme of biological photosynthesis (from Archer and Bolton [11]). E^o is the standard redox potential at pH 7. The essential features are the following: Photosystem II, which is associated with the oxidation of water, and photosystem I, which is associated with the reduction of $NADP^+$, are linked in series by an electron transport chain. Absorption of light by chlorophyll a and b of photosystem II mediates the transfer of an electron against the potential gradient from species Z (which is a manganese complex) to a species Q_1, to give Z^+ and Q^-_1. Z^+ is a strong oxidant that is capable of oxidizing water to oxygen. From the reduced species Q^-_1, an electron is transferred in spontaneous dark reactions to subsequent electron acceptors, until it reaches photosystem I. Absorption of light by chlorophyll a in photosystem I enables the second uphill electron transfer to the highly reducing component A_1. A second series of dark electron-transfer reactions occurs via the protein ferredoxin (FD) and other electron carriers and results finally in the reduction of $NADP^+$. NADPH acts as a reducing agent in dark reactions that reduce CO_2. These reactions occur in a series of catalyzed steps known as the Calvin cycle. The reactions of the Calvin cycle require ATP as well as NADPH. ATP is synthesized in reactions that proceed parallel to the Z-scheme and utilize some of the energy that would otherwise be wasted in the downhill electron transfer steps. For the assimilation of one CO_2 molecule according to the net reaction $CO_2 + H_2O \rightarrow [CH_2O] + O_2$, 4 electrons are transferred, which requires 8 photons, since 2 photons are needed to transfer an electron from Z to A_1. $(Fe-S)_A$, $(Fe-S)_B$, and Fe-S are different iron-sulphur proteins.

3. Role of Chemical Speciation in Controlling Biological Availability of Iron and Manganese

The role of chemical speciation for the biological availability of metals is discussed in detail in the chapter entitled "Relationship between Metal Speciation and Bioavailability" by Margaret Farago, this volume. In this chapter, some background is given which is important for the appreciation of the effects of light on the biological availability of iron and manganese.

Most marine photosynthetic organisms can only take up iron and manganese in the dissolved form. Several authors [12-15] have proposed a carrier-mediated transport model of iron and manganese uptake by eucariotic marine algae. According to this model, iron and manganese must first form coordination complexes with specialized transport sites on the plasmalemma to be taken up intracellularly. Therefore, metal transport into the cell is determined by the interplay between the metal speciation in the medium and ligand exchange reactions (Figure 2). Organic ligands present in the euphotic zone of marine waters form very stable complexes with dissolved Fe(III) (see next subchapter). Thus, it is most likely that the concentration of inorganic, dissolved Fe(III) species (including the hydrated and hydrolyzed ions) determines the upper limit for bioavailable Fe(III) [2]. In the case of Fe(II), organic complexes may be rather weak, and hence ligands at the cell surface may form stronger complexes with Fe(II) than organic ligands in the medium [12].

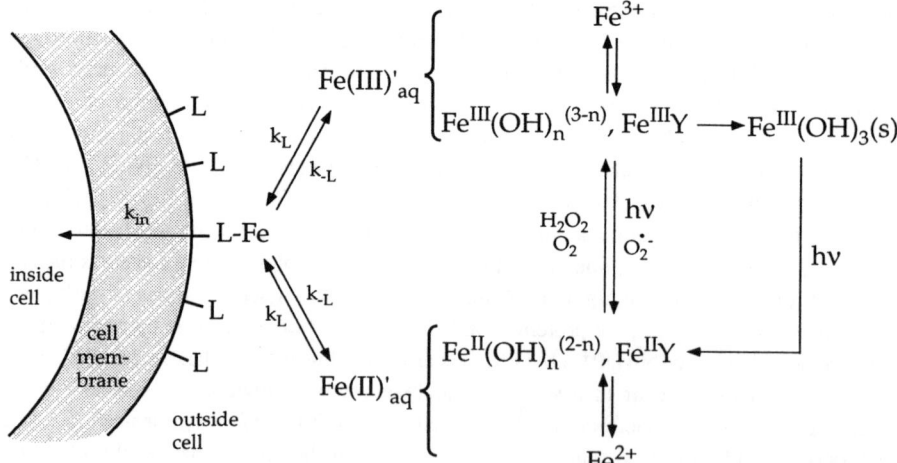

Figure 2. Schematic representation of the carrier-mediated transport model of iron, and of the (photo)redox cycling of iron in the euphotic zone of surface waters. $Fe^{III}Y$ and $Fe^{II}Y$ stand for dissolved organic Fe-complexes, and $Fe(III)'_{aq}$ and $Fe(II)'_{aq}$ for bioavailable iron species and designate the dissolved Fe(III) and Fe(II) coordination compounds with low stability constants and fast ligand exchange rates relative to the stability constants and ligand exchange rates of the membrane-bound iron-ligand complexes. The concentrations of the free iron ions, Fe^{3+} and Fe^{2+}, is a measure for the concentration of bioavailable iron species (modified from Anderson and Morel [12]).

A carrier-mediated transport model is consistent with classical Michaelis-Menten uptake kinetics [2, 10] for the uptake of both iron and manganese by marine phytoplankton [12-16]. The kinetic model is described by equations (1) and (2),

$$-L + M' \underset{k_{-L}}{\overset{k_L}{\rightleftarrows}} -LM \overset{k_{in}}{\rightarrow} -L + M_{in} \tag{1}$$

$$R = \frac{R_{max}[M']}{[M'] + K_s} \tag{2}$$

where -L is the transport ligand at the cell surface, -LM is the Fe- or Mn-ligand complex at the cell surface, M' and M_{in} are the bioavailable iron or manganese species outside the cell and the iron or manganese species inside the cell, respectively, R is the metal uptake rate, and R_{max} is the maximal rate achieved when the transport ligands at the cell surface are fully saturated. K_s is the half-saturation constant,

$$K_s = \frac{k_{-L} + k_{in}}{k_L} \tag{3}$$

It is noteworthy that the half-saturation constant, K_s, has been found to be invariant for a given trace metal and a given phytoplankton species, while the maximal uptake rate, R_{max}, can vary by a factor of 20-30, depending on the degree of limitation or sufficiency of the trace metal [13, 14]. The latter is achieved by increasing or decreasing the concentration of uptake ligands on the membrane [13, 14].

The biological availability of metals is often associated with the average steady-state concentration of the free metal species, e.g., $[Fe^{3+}]$. However, for many metals, the free metal concentration is too small (e.g., less than 10^{-22} M for Fe^{3+} in seawater, [17]) to represent the sole biologically available species. It is thus more likely that the concentration of free metal ions is a measure of the concentration of metal species, for which ligands at the cell surface can compete successfully with the ligands present in the surrounding medium (see Figure 2).

The binding of iron and manganese on cell surface sites previous to the transport into the cell may be ubiquitous among eucaryotic algae [12]. Another strategy which is used, e.g., by blue-green algae, is the excretion of siderophores. The production of siderophores by blue-green algae can inhibit the growth of green algae and has been proposed as a factor determining blue-green algal blooms in lakes [18]. Siderophores are low-molecular-weight organic molecules which specifically chelate Fe(III). Employing voltammetric methods, Lewis and coworkers [19] have estimated thermodynamic stability constants of approximately 10^{40} M^{-1} for the complexation of Fe^{3+} with catecholate-type siderophores isolated from the marine bacterium *Alteromonas luteoviolacea* and from the marine cyanobacterium *Synechoccus* sp. PCC 7002. Hence, such siderophores should be able to remove Fe(III) from other organic complexes, e.g. from Fe(III)-EDTA. Production of siderophores under iron-limiting conditions is considered to be a frequent strategy among both marine and freshwater cyanobacteria and eubacteria [20, 21].

Recently, an interesting link between biological processes and chemical speciation has been proposed: Digestion of colloidal iron in acidic food vacuoles of protozoan grazers may be a mechanism for the generation of bioavailable iron from biologically refractory iron phases [22].

4. Speciation of Iron and Manganese in Surface Waters

As discussed in the previous subchapter, the biological availability of iron and manganese depends on the speciation of the dissolved species. Therefore the chief questions are what are the steady-state concentrations of total dissolved iron and manganese, and what is their speciation in surface waters. The determination of dissolved iron and manganese concentrations and the assessment of their speciation in oceanic surface waters have become feasible thanks to the development of trace metal clean procedures and extremely sensitive analytical methods. Dissolved Mn can be extracted with organic ligands [23, 24]. Cathodic stripping voltammetry after addition of specific ligands has been used to determine low total dissolved iron concentrations [25], and to obtain insights into the iron speciation [26-29] (see also chapters by Marko Branica, this volume).

For the determination of dissolved Fe(II), a spectroscopic method with ferrozine as Fe(II) complexing agent has frequently been used [30-32] (Figure 3). When Fe(II) was preconcentrated prior to analysis, the detection limit was as low as 0.6 nM [32]. The preconcentration technique consists in passing water samples through a C_{18} Sep-Pak cartridge with adsorbed ferrozine, and subsequent extraction of the Fe(II)-ferrozine complex from the Sep-Pak cartridge with methanol [32]. The drawback of spectroscopic methods with Fe(II) complexing agents is the shift of the ambient redox and solubility equilibria by Fe(II) complexation. Therefore, Fe(II) determined with the ferrozine method is often referred to as "ferrozine-reactive iron" (see Figure 8, below).

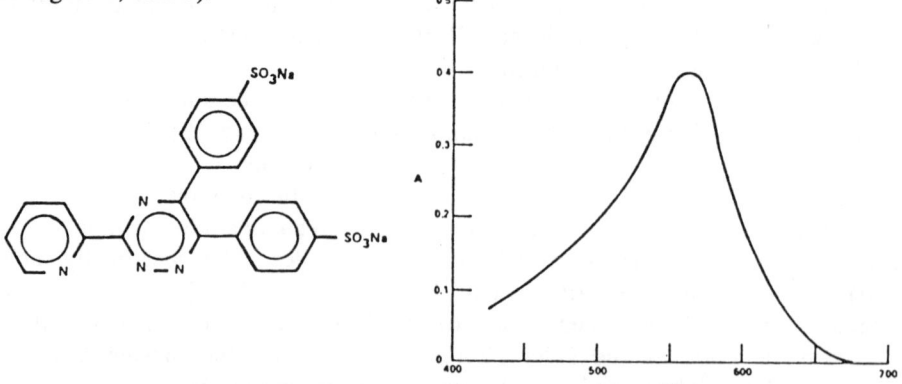

Figure 3. Stucture of ferrozine, and absorption spectrum of the Fe(II)-ferrozine complex (from Stookey [30]).

A less "intrusive" method for the determination of dissolved Fe(II) has recently been developed by O'Sullivan and coworkers [33] and King and coworkers [34]. It is a flow injection chemiluminescence method based on the Fe(II)-catalyzed reaction between luminol and oxygen (Figure 4). This technique allows for the selective detection of Fe(II) at nanomolar levels and, unlike the ferrozine method, does not require preconcentration of Fe(II) prior to analysis.

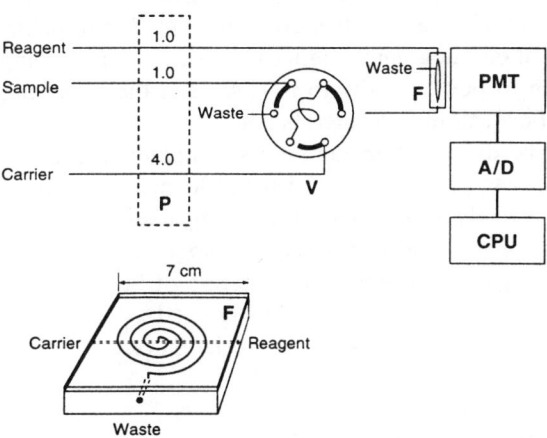

Figure 4. Schematic of Fe(II) analysis system with expanded view of flow cell: PMT, integrated photomultiplier; A/D, A/D converter and valve controller; CPU, Macintosh computer; P, pump with flow rates listed in mL min^{-1}; V, injection valve; F, flow cell (from King et al. [34]).

Under oxic conditions, the thermodynamically stable form of iron is Fe(III)-(hydr)oxide (e.g., lepidocrocite, γ-FeOOH, goethite, α-FeOOH or hematite, α-Fe$_2$O$_3$), and of manganese Mn(IV)-oxide (e.g., birnessite, γ-MnO$_2$). Under seawater conditions, the solubilities of Fe(III) and of Mn(IV) are extremely low (see chapter entitled "The Effect of Ionic Interactions on Thermodynamic and Kinetic Processes in Natural Waters" by Fank J. Millero, this volume). Careful analytical studies in the oceans have shown that the concentrations of total dissolved iron range from 0.02 to 10 nM; in some cases, nutrient-like profiles have been observed [23, 25, 28, 35, 36]. Concentrations of dissolved manganese in the oceans are also in the nanomolar range [24, 37-39]. One major difficulty in assessing the dissolved iron and manganese concentrations in environmental systems is the filtration process. Filtration is an opperational process defined by the filter pore size. Therefore, one obtains only the concentration of filtrable iron and manganese, and not that of truly, i.e., thermodynamically dissolved iron and manganese.

Interesting findings have recently been reported with regard to the speciation of dissolved iron in oceanic surface waters [26-29]. Rue and Bruland [28] combined the method of adsorptive cathodic stripping voltammetry with competitive ligand equilibration to examine the iron speciation in seawater. This new method revealed that 99.97% of the dissolved Fe(III) in central North Pacific surface waters is chelated by organic ligands [28]. Rue and Bruland [28] were able to model their data as consisting of two classes of iron-binding ligands, a strong ligand class with a conditional stability constant of $1.2 \cdot 10^{13}$ M^{-1}, and a weaker ligand class with a conditional stability constant of $3.0 \cdot 10^{11}$ M^{-1}. Also van den Berg [27] has reported that his measurements with a new cathodic stripping voltammetry method [26] indicates that 99% or more of iron is complexed by organic ligands throughout the water column of the Western Mediterranean. Employing also a competitive ligand equilibration / cathodic stripping voltammetry method, Wu and Luther III [29] were able to distinguish two classes of ligands in different oceanographic regimes. One class forms complexes with Ca^{2+} and Mg^{2+} in addition to Fe^{3+}, similar to EDTA, and the other class forms only strong complexes with Fe^{3+}, similar to enterobactin.

As will be discussed in the next subchapter, organically and inorganically complexed Fe(III) can be reduced under the influence of sunlight. Thereby, Fe(II)$_{aq}$ is produced which is likely to form less stable complexes with organic ligands present in the euphotic zone of seawater and hence is biologically more available than Fe(III)$_{aq}$. Although the average steady-state concentration of Fe(II)$_{aq}$ may be very low in the euphotic zone of seawater, considerably higher Fe(II)$_{aq}$ concentrations can be expected upon illumination with high intensity solar radiation (see Figure 8, below).

5. Role of Light on the Speciation of Iron

Photoreductive dissolution of Fe(III)-(hydr)oxides by suitable reductants (humic acids or simple carboxylic acids like oxalate) is the process of major importance for the formation of dissolved iron in oxic environmental systems [40-44] (see Figure 2). However, photoreductive dissolution of crystalline Fe(III)-(hydr)oxide phases, e.g., by humic substances, has been shown to occur only at acidic pH values [41, 44]. This is because both, the rate and the overall rate constant of photoreductive dissolution of Fe(III)-(hydr)oxides, e.g., lepidocrocite (γ-FeOOH), are strongly pH-dependent and decrease with increasing pH [42, 44, 45]. At near neutral pH, it thus appears that only amorphous Fe(III)-hydroxide phases undergo photoreductive dissolution [46, 47]. In surface waters, non-crystalline, colloidal Fe(III)-hydroxide phases are formed by oxidative precipitation of Fe(II) and by biological iron recycling. The question thus arises, in what environmental compartments the initially crystalline Fe(III)-(hydr)oxide phases undergo reductive dissolution.

Atmospheric waters often exhibit low pH values (between pH 3-6). Therefore, photoreductive dissolution of Fe(III)-(hydr)oxide phases is expected to occur much more readily in atmospheric waters than in seawater. (A study of photoreductive dissolution of various Fe(III)-(hydr)oxides under atmospheric water conditions is described in Section 5.1). Zhuang and coworkers [48, 49] have demonstrated that Fe(II) represents a large fraction of total iron in marine aerosols, e.g., 56±32% and 49±15% in marine aerosol samples collected in the central North Pacific and at Barbados in the North Atlantic, respectively (Figure 5). It is thought that atmospheric iron input is a significant source or iron to the surface waters in remote oceanic regions [50]. For example, the subarctic Pacific lies in the path of an extended aerosol plume originating in China [51]. Martin and coworkers [52] have estimated that atmospheric deposition accounted for 84 to 93% of the external iron input in these surface waters. In contrast, in the "high nutrient - low chlorophyll (HNLC)" regions (see chapter entitled "The Influence of Iron on Carbon Doxide in Surface Seawater" by Fank J. Millero, this volume) of the equatorial Pacific and Southern Ocean, iron input from upwelling and vertical mixing overwhelm by far atmospheric iron input [53].

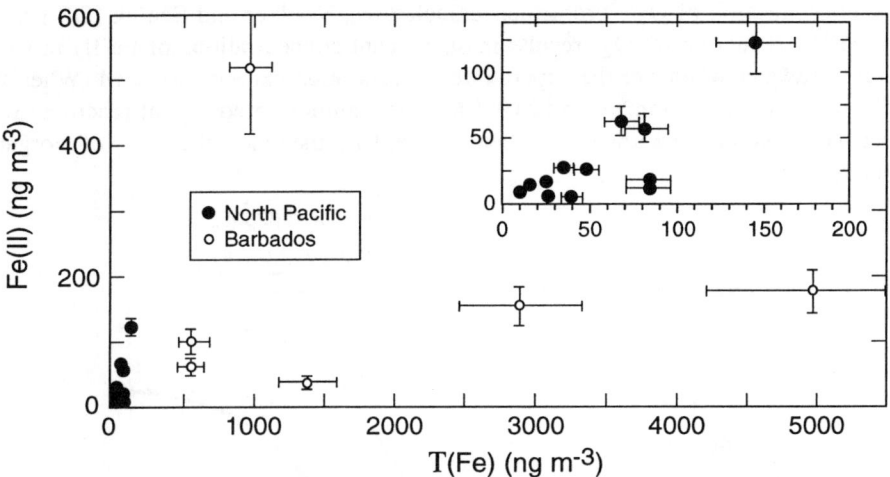

Figure 5. Fe(II) concentration as a function of total iron concentration, T(Fe), in remote marine aerosols from island in the North Pacific and from Barbados in the North Atlantic. Iron(II) was measured with a modified ferrozine method (from Zhuang et al. [48]).

Followed by photoreductive dissolution of particulate and colloidal iron, fast redox cycling between dissolved Fe(III) and dissolved Fe(II) occurs in oxic aquatic systems as long as suitable reductants such as humic substances or simple carboxylic acids like oxalate are in excess (Figure 2). In seawater, hydrogen peroxide (H_2O_2) is also an important oxidant of Fe(II) in addition to O_2 [34, 54-56]. Hydrogen peroxide is formed through dismutation of HO_2/O_2^- ($pK_a = 4.8$, [57]):

$$2\ HO_2/O_2^- \xrightarrow{H^+} H_2O_2 + O_2 \qquad (4)$$

Superoxide (O_2^-) is produced in photochemical reactions of colored dissolved organic matter (CDOM), e.g.:

$$^3CDOM^* + O_2 \rightarrow CDOM^+ + O_2^- \qquad (5)$$

In surface waters, reduction of dissolved Fe(III), which is in competition with precipitation (see Figure 2) can occur through various pathways: (i) Reduction of dissolved Fe(III) by superoxide [56, 58], (ii) photolysis of organic and inorganic Fe(III)-complexes [42, 59, 60], (iii) thermal, abiotic reduction of organically complexed Fe(III) [60, 61], and (iv) reduction of dissolved Fe(III) at cell surfaces [62]. Until the studies by Voelker and Sedlak [58] and Miller and coworkers [56], the role of O_2^- as a reductant of Fe(III) in seawater had not been recognized. Figure 6 shows the calculated percentage of Fe(II) (from total dissolved iron) as a function of the steady-state concentration of O_2^-. From their calculations, Voelker and Sedlak [58] predict that Fe(III) reduction by O_2^- results in significant concentrations of Fe(II) in sunlit surface seawater whenever the superoxide concentration exceeds 10^{-11} M. When the superoxide concentration exceeds 10^{-9} M, these authors predict that reactions with superoxide dominate the iron redox chemistry, at least the redox chemistry of inorganic dissolved iron.

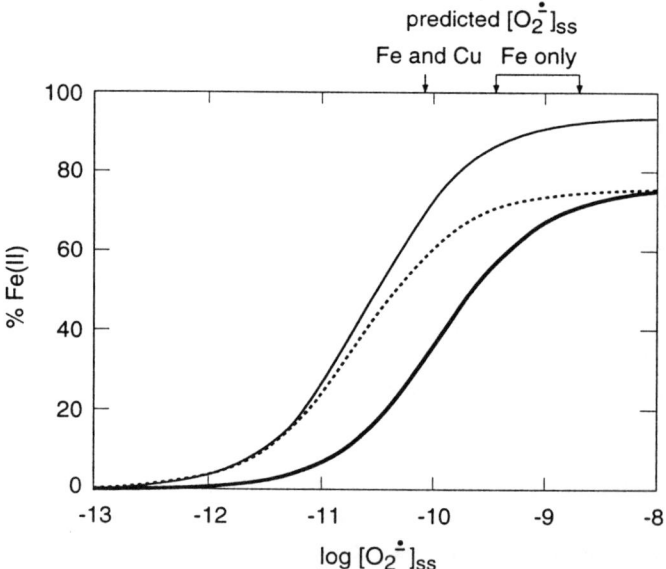

Figure 6. Estimated %Fe(II) as a function of $[O_2^-]_{ss}$ in the presence of 0.21 mM O_2 and 50 nM H_2O_2 (from Voelker and Sedlak [58]). The dashed line was calculated for these conditions using the following rate constants for reaction of O_2^- with Fe(II) and Fe(III), respectively (in $M^{-1} s^{-1}$): $k_{Fe(II)} = 1 \cdot 10^7$ [64], $k_{Fe(III)} = 1.5 \cdot 10^8$ [64]. The other lines represent the results of calculations with the following rate constants for reactions of iron with O_2^- in 0.7 M NaCl (in $M^{-1} s^{-1}$): solid line: $k_{Fe(II)} = 1 \cdot 10^7$, $k_{Fe(III)} = 3 \cdot 10^7$; dotted line: $k_{Fe(II)} = 5 \cdot 10^7$, $k_{Fe(III)} = 1.5 \cdot 10^8$. These rate constants were calculated from relative rates of Fe(III) reduction and Fe(II) oxidation at high concentrations of O_2^- (produced in continuous radiolysis experiments). For all three calculations, the rate constants of Fe(II) oxidation by O_2 and H_2O_2, k_{O_2} and $k_{H_2O_2}$, respectively, were assumed to be the same. The calculations are based on the following equation:

$$\%Fe(II) = \frac{100 \, k_{Fe(III)}[O_2^-]_{ss}}{k_{Fe(III)}[O_2^-]_{ss} + k_{Fe(II)}[O_2^-]_{ss} + k_{O_2}[O_2] + k_{H_2O_2}[H_2O_2]}$$

In the light of the recent reports [26-29] that most of the dissolved iron in seawater is in the form of organic complexes, the question arises as to the most efficient pathway of dissolved Fe(III) reduction, i.e., does photolysis of these organic Fe(III)-complexes occur more efficiently than reduction by O_2^-? Regarding this question, interesting findings have been reported by Miller and coworkers [56]. These authors measured time courses of dissolved Fe(II) and H_2O_2 in irradiated (with simulated sunlight), 0.2 μm filtered water samples from Narragansett Bay, and identified the most important processes determining Fe(II) steady-state concentrations in their systems with kinetic modeling. Their model showed that measured steady-state Fe(II) concentrations in irradiated seawater are controlled largely by pH dependent oxidation by O_2 and H_2O_2, and by both oxidation and reduction of iron by photochemically produced O_2^-. The kinetics of reduction and oxidation of dissolved iron species is probably crucial in determining the bioavailability of iron in those remote oceanic regions, where the main iron source is biological iron recycling and not atmospheric input or upwelling [63].

Because light affects to a larger extent the reduction of Fe(III) (through photoreductive dissolution of Fe(III)-(hydr)oxides and through light-induced reduction of dissolved Fe(III)) than the oxidation of Fe(II) [65] (although formation of H_2O_2 is a light-induced process, see reactions 4-5), the concentration of Fe(II) in the euphotic zone of surface waters is expected to depend on the light intensity and thus on the time of the day. A diurnal variation in the Fe(II) concentration has been observed by several research groups, not only in acidic lakes and rivers [66-68] (Figure 7), but also in shelf waters [69] (Figure 8). Note, that the concentrations of Fe(II)$_{aq}$ are much higher in acidic surface waters than in seawater. This is mainly because of the much slower oxidation of Fe(II) at low pH-values as compared to pH-values of seawater [56], the reason being that Fe(II) oxidation kinetics by O_2 and by H_2O_2 depends strongly on the Fe(II) speciation [60, 70, 71] (see also Figure 11, below).

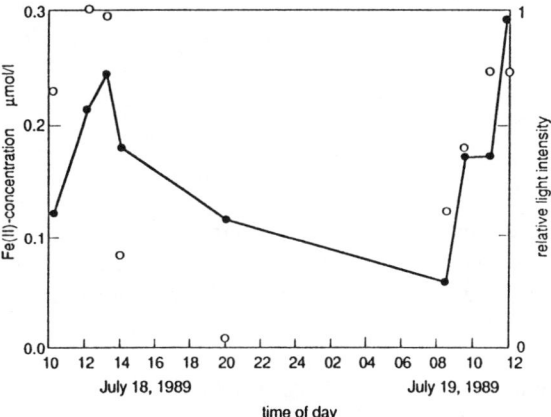

Figure 7. Diurnal variation of the concentration of dissolved Fe(II) ●, and of the incident light intensity ○ in Lake Cristallina, an acidic (pH ≈ 5.2) Swiss alpine lake. The maximal measured light intensity is arbitrary set to one. The dissolved Fe(II) concentration was determined spectrophotometrically using ferrozine as complexing agent of Fe(II) after fitration of the water samples through 0.1 μm cellulose nitrate filters (from Sulzberger et al. [68]).

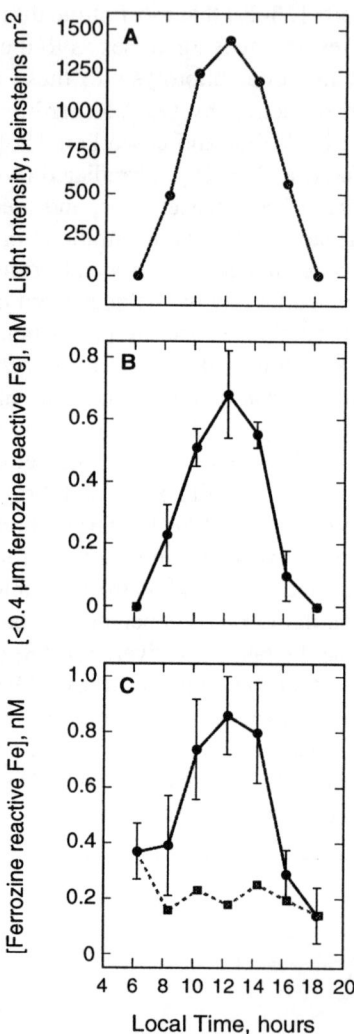

Figure 8. Results of a diel study of surface water ferrozine-reactive iron concentrations while following a body of water in the Gulf of Carpentaria on April 19, 1992. Plot (A) shows the light intensity measured over the day with the first point being at sunrise and the last at sunset. Plot (B) depicts filterable (< 0.4 μm) ferrozine-reactive iron concentrations for the same local time period. Plot (C) shows particulate (■) and total (●) ferrozine-reactive iron concentrations. Error bars indicate 98%-confidence limits (from Waite et al. [69]).

An interesting result of the field study by Waite and coworkers [69] is that no correlation was found between particulate iron concentrations and the concentration of ferrozine-reactive iron, which according to these authors suggests that the soluble rather than the particulate iron pool is most influenced by light in these shelf waters.

5.1. PHOTOREDUCTIVE DISSOLUTION OF Fe(III)-(HYDR)OXIDES UNDER ATMOSPHERIC WATER CONDITIONS

Photoreductive dissolution of particulate or colloidal Fe(III)-(hydr)oxide phases, e.g., with oxalate as reductant, can be described by the following steps [72] (see also chapter entitled "Light and Chemically Driven Reactions and Equilibria in the Presence of Organic and Inorganic Colloids in Sea Water" by Claudio Minero, this volume):

1. Surface complex formation:

$$\equiv Fe^{III}OH + HC_2O_4^- \rightleftarrows \equiv Fe^{III}C_2O_4^- + H_2O \tag{6}$$

2. Ligand-to-metal charge-transfer transition within the surface complex,

$$\equiv Fe^{III}C_2O_4^- \underset{k_{th}}{\overset{k_a(\lambda)}{\rightleftarrows}} \equiv Fe^{II}C_2O_4^{-*} \tag{7}$$

where $k_a(\lambda)$ is the specific rate of light absorption by the surface complex, creating a ligand-to-metal charge-transfer state, which is an electronically excited state (designated by an asterix), and k_{th} is the rate constant of thermal deactivation of the charge-transfer state.

3. Dissociation of the oxalate radical from the surface and decarboxylation,

$$\equiv Fe^{II}C_2O_4^{-*} \overset{k_p}{\rightarrow} \equiv Fe(II) + CO_2 + CO_2^- \tag{8}$$

where k_p is the rate constant of primary photoproduct formation.

4. Detachment of reduced surface iron from the crystal lattice, and reconstitution of the surface site,

$$\equiv Fe(II) \overset{k_{det}}{\rightarrow} Fe(II)_{aq} + \equiv rec \tag{9}$$

where k_{det} is the rate constant of detachment of Fe(II) from the Fe(III)-(hydr)oxide lattice, and $\equiv rec$ is the reconstituted surface site. In the presence of O_2 or other oxidants (Ox) of surface Fe(II), detachment of surface Fe(II) is in competition with its oxidation:

$$\equiv Fe(II) + Ox \overset{k_{ox}}{\rightarrow} \equiv Fe(III)^+ + Ox^- \tag{10}$$

Invoking the steady-state approximation for the charge-transfer state and surface Fe(II), the rate expression for Fe(II)$_{aq}$ formation is:

$$\frac{d[Fe(II)_{aq}]}{dt}(\lambda) = k_a(\lambda)[\equiv Fe^{III}C_2O_4^-] \cdot \frac{k_p}{k_p + k_{th}} \cdot \frac{k_{det}}{k_{det} + k_{ox}[Ox]} \tag{11}$$

The ratio $\frac{k_p}{k_p + k_{th}}$ is the quantum yield of primary photoproduct formation, Φ, and is expected to be wavelength-independent for a given surface complex. The ratio $\frac{k_{det}}{k_{det} + k_{ox}[Ox]}$ is the efficiency of detachment, η_{det}, of reduced surface iron from the crystal lattice. The rate expression of Fe(II)$_{aq}$ formation can then be written as:

$$\frac{d[Fe(II)_{aq}]}{dt}(\lambda) = k_a(\lambda)[\equiv Fe^{III}C_2O_4^-] \cdot \Phi \cdot \eta_{det} \tag{12}$$

Thus, the rate of Fe(II)$_{aq}$ formation through photoreductive dissolution of an Fe(III)-(hydr)oxide is expected to be linearly dependent on the surface concentration of an electron donor, e.g., oxalate, i.e., on the surface complex. Furthermore, according to the surface complexation model [73], the concentration of the surface complex is expected to be a Langmuir-function of the solution concentration of the electron donor, i.e., in the case of oxalate as electron donor,

$$[\equiv Fe^{III}C_2O_4^-] = \frac{K^{cond}[\equiv Fe(III)_T][C_2O_4^{2-}{}_{sol}]}{1 + K^{cond}[C_2O_4^{2-}{}_{sol}]} \tag{13}$$

where K^{cond} is the conditional formation constant for the surface complex, $[\equiv Fe(III)_T]$ is the total surface site concentration where oxalate can get adsorbed, $[\equiv Fe(III)_T] = [\equiv Fe^{III}OH] + [\equiv Fe^{III}C_2O_4^-]$, and $[C_2O_4^{2-}{}_{sol}]$ is the concentration of oxalate in solution.

Photolysis of a surface complex, formed from the specific adsorption of an electron donor at the surface of an Fe(III)-(hydr)oxide, is not the only pathway of photoreductive dissolution of the solid phase. Another possible pathway of photoreductive dissolution of semiconducting Fe(III)-(hydr)oxides is oxidation of an adsorbed electron donor by valence-band holes and reduction of surface Fe(III) by conduction-band electrons [74, 75].

Sulzberger and Laubscher [43] have studied the reactivity towards light-induced dissolution of three well defined Fe(III)-(hydr)oxide phases: γ-FeOOH (lepidocrocite), α-FeOOH (goethite), and α-Fe$_2$O$_3$ (hematite) with oxalate as reductant. Figure 9a shows the concentrations of oxalate and of dissolved Fe(II) as a function of time upon illumination of an aerated γ-FeOOH suspension at pH 3. Photoreductive dissolution under photooxidation of oxalate lead to net formation of one mol of dissolved Fe(II) per mole of oxidized oxalate. With the thermodynamically more stable phases α-FeOOH and α-Fe$_2$O$_3$, photooxidation of oxalate was not accompanied by net formation of appreciable concentrations of dissolved Fe(II) (Figure 9b,c). The most likely explanation of this phenomenon is reoxidation of reduced surface iron, i.e., of Fe(II) that has not been detached from the crystal lattice of α-FeOOH or α-Fe$_2$O$_3$ (reaction 10).

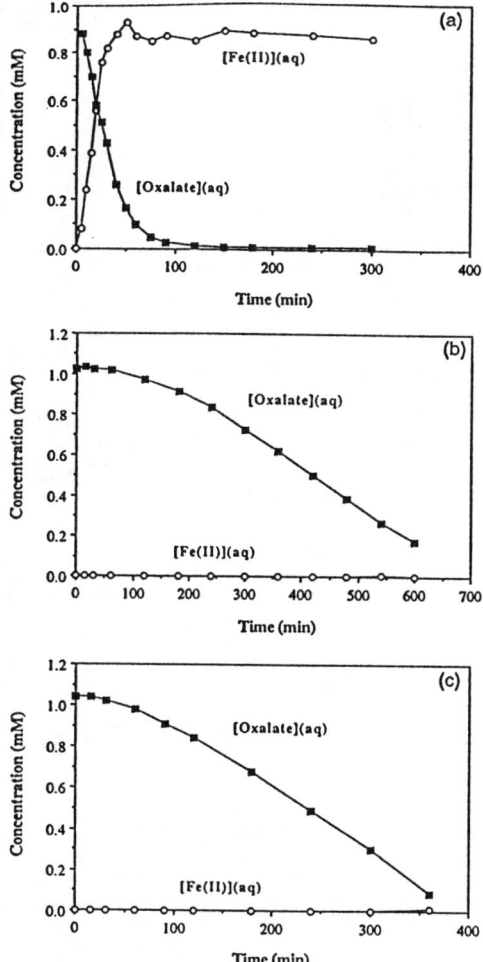

Figure 9. Concentrations of dissolved Fe(II) and of oxalate as a function of time upon illumination of **aerated** Fe(III)-(hydr)oxide suspensions at pH 3. Experimental conditions: Initial Fe(III)-(hydr)oxide concentration, 0.5 g L^{-1}; initial oxalate concentration, 1 mM; ionic strength, 5 mM (NaClO$_4$); temperature, 25 °C; light-source, white light from a 1000 W high-pressure xenon lamp that was filtered by the bottom window of the Pyrex glass vessel, which acts as cutoff filter ($\lambda_{1/2}$ = 305 nm), incident light intensity, I$_0 \approx$ 0.5 kW m^{-2}. Fe(III)-(hydr)oxide phases: (a) γ-FeOOH; (b) α-FeOOH; (c) α-Fe$_2$O$_3$ (from Sulzberger and Laubscher [43]).

In contrast, in deaerated suspensions, approximately two mol of dissolved Fe(II) were formed per mole of photooxidized oxalate with all three solid phases (Figure 10a-c), according to the following overall stoichiometry:

$$2\,\gamma\text{-FeOOH} + C_2O_4^{2-} + 6\,H^+ \xrightarrow{h\nu} 2\,Fe^{2+} + 2\,CO_2 + 4\,H_2O \tag{14}$$

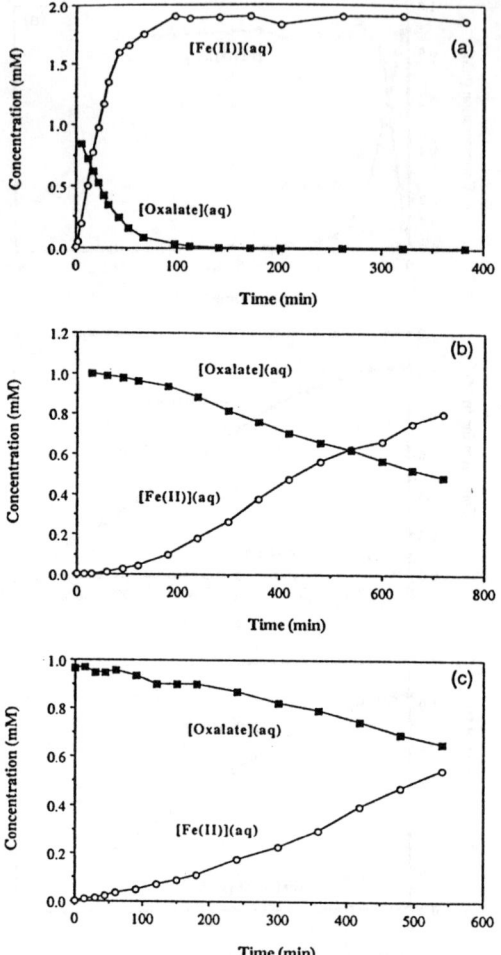

Figure 10. Concentrations of dissolved Fe(II) and of oxalate as a function of time upon illumination of **deaerated** Fe(III)-(hydr)oxide suspensions at pH 3. The experimental conditions were otherwise the same as in Figure 9. Fe(III)-(hydr)oxide phases: (a) γ-FeOOH; (b) α-FeOOH; (c) α-Fe$_2$O$_3$ (from Sulzberger and Laubscher [43]).

The observed 2:1 stoichiometry with regard to produced Fe(II) and consumed oxalate is explained in terms of reaction of the CO_2^- radical with Fe(III), either surface or dissolved species,

$$CO_2^- + Fe(III) \rightarrow CO_2 + Fe(II) \qquad (15)$$

where the speciation of Fe(III) and Fe(II) in Eq. (15) is not defined.

From the results shown in Figures 9 and 10, Sulzberger and Laubscher [43] hypothesize that with γ-FeOOH as solid phase, detachment of surface Fe(II) outcompetes to a large extent its reoxidation. Whereas, with the thermodynamically more stable phases α-FeOOH and α-Fe$_2$O$_3$, a large portion of reduced surface iron, i.e., of Fe(II) that has not been detached from the crystal lattice, is reoxidized (see reaction 10). Hence, the efficiency of detachment of reduced surface iron, η_{det}, is likely to be much smaller for α-FeOOH and α-Fe$_2$O$_3$ than for γ-FeOOH (see Eq. 12). Oxidation of surface Fe(II) is known to occur much faster than Fe^{2+} oxidation. This is because of the following two phenomena: (i) There exists a linear free energy relation between the redox potential of inorganic Fe(III)/Fe(II) species and the logarithm of the second-order rate constant of Fe(II) oxidation by O$_2$ (Figure 11), and (ii) the redox potential of the Fe(III)/Fe(II) redox pair depends on its speciation (Figure 12). [The redox potential of surface Fe(III)/Fe(II) probably lies between the redox potential of adsorbed Fe(III)/Fe(II) and of structural Fe(III)/Fe(II).]

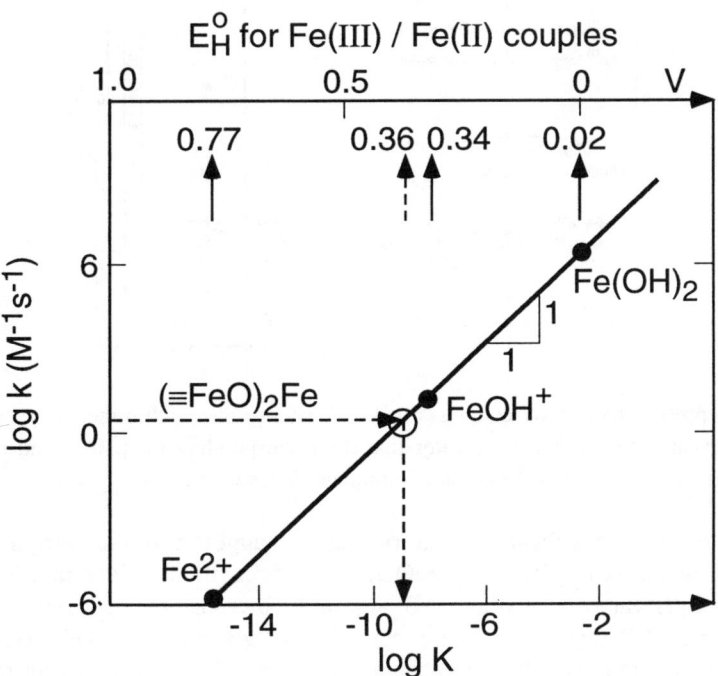

Figure 11. Logarithm of the reaction rate constant (M^{-1} s^{-1}) of oxidation of inorganic Fe(II) species by O$_2$ vs. log of the equilibrium constant and of the redox potential, E°$_H$ (vs. NHE), for the corresponding Fe^{III}/Fe^{II} redox couples (modified from Wehrli [76]). The redox potential of iron that is innerspherically adsorbed on a goethite surface (≡FeO)$_2$Fe$^{III/II}$ can be estimated from the experimentally determined rate constant [77]. Data for the rate constants of Fe^{2+} and Fe(OH)$^+$ oxidation by O$_2$ are from Singer and Stumm [78], and for Fe(OH)$_2$(aq) oxidation by O$_2$ from Millero et al. [55].

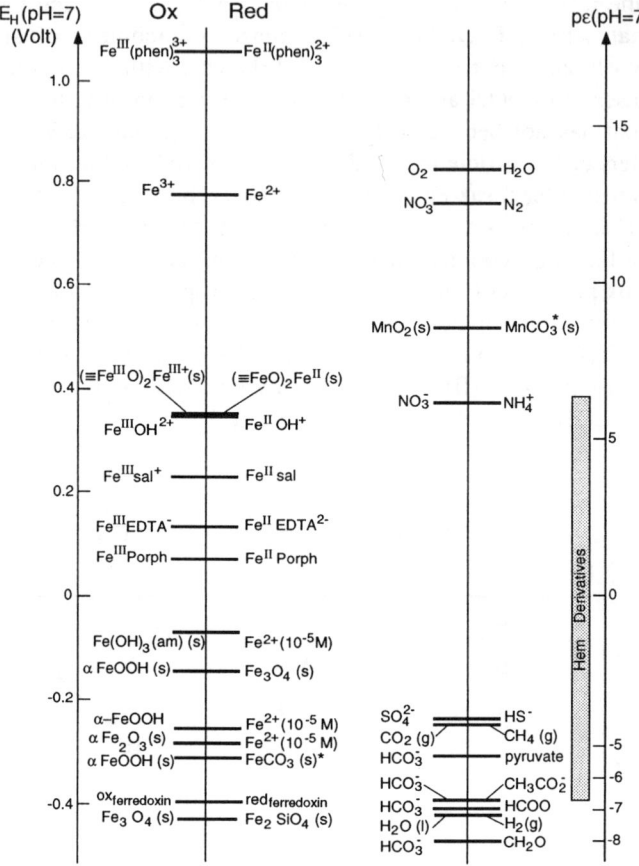

Figure 12. Representative redox couples at pH = 7. phen = phenanthroline, sal = salicylate, porph = porphyrin. The symbol * indicates that the corresponding $E_H(pH=7)$ and $p\varepsilon(pH=7)$ values are valid for $[HCO_3^-] = 10^{-3}$ M (from Stumm and Sulzberger [70]).

The results from these laboratory studies suggest that thermodynamically stable phases such as α-Fe_2O_3 and α-FeOOH are not readily dissolved in atmospheric waters, even at pH values as low as pH 3, and in the presence of reductants and light. Yet crystalline Fe(III)-(hydr)oxides (e.g. hematite and goethite) are likely to be major chemical forms of iron initially present in aeolian mineral dust, originating from soil surfaces and from desert dust [49]. For example, based on Mössbauer spectroscopy, Kopcewicz and Dzienis [79] have reported that aerosol samples collected in Warsaw and Zakopane, Poland, were predominantly in the form of α-Fe_2O_3 fine particles ($\approx$ 100 Å). The question then is how thermodynamically stable Fe(III)-(hydr)oxide phases are transformed into thermodynamically less stable, and thus more soluble phases, in the atmosphere. One possibility is redox cycling of iron at the surface of crystalline Fe(III)-(hydr)oxide phases, i.e., reduction and oxidation of surface iron without transfer into solution, resulting in more soluble Fe(III)-(hydr)oxide coatings on the surface of thermodynamically stable phases. Such surface transformation processes may take place in aerosols during their long-range transport. Zhuang et al. [49] found that the solubility of iron in remote marine aerosols, collected at four Pacific island stations, was 5-17 higher than that of iron in aerosols collected from an urban area near the Chinese loess plateau (at the same pH). Also reductive dissolution of Fe(III)-(hydr)oxides and subsequent oxidative precipitation of dissolved Fe(II) leads to a gradual increase in the solubility of colloidal iron phases [49], e.g., a crystalline phase like γ-FeOOH may be gradually transformed into ferrihydrite, an amorphous and hence much more soluble phase than γ-FeOOH [80]. Ferrihydrite undergoes readily reductive dissolution also at near neutral pH [61].

6. Role of Light on the Speciation of Manganese

Of any trace element in natural waters, manganese is possibly the most noticeably influenced by light. Both, the rates of reductive dissolution of Mn(IV)-oxides and oxidation of dissolved Mn(II) are now recognized to be affected by light [8, 24, 38, 81] (Figure 13).

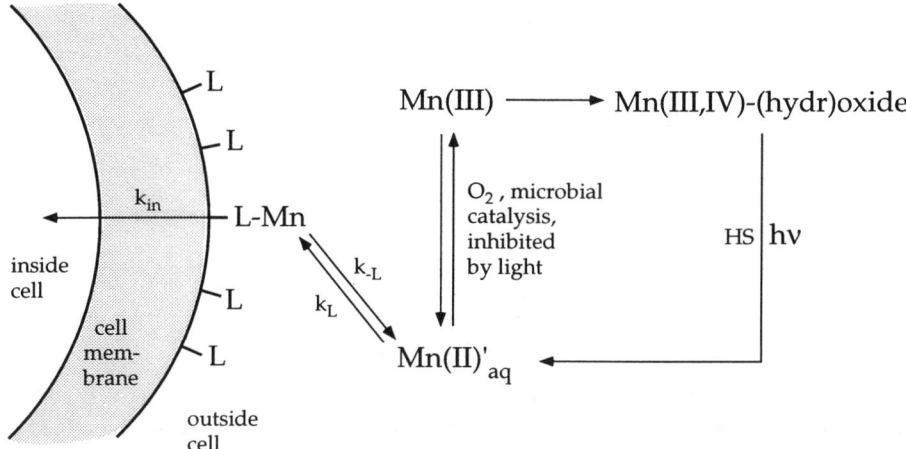

Figure 13. Schematic representation of the carrier-mediated transport model of manganese, and of the (photo)redox cycling of manganese in the euphotic zone of surface waters. The abbreviation "HS" stands for humic substances, and Mn(II)'$_{aq}$ stands for bioavailable manganeses species and designates the dissolved Mn(II) coordination compounds with low stability constants and fast ligand exchange rates relative to the stability constants and ligand exchange rates of the membrane-bound manganese-ligand complexes. Dissolved Mn(III) undergoes fast disproportionation to form Mn(II) and Mn(IV), unless it is stabilized by complexation. The existence of Mn(III)-pyrophosphate complexes has been demonstrated in surface waters, and it has been shown that these complexes are reduced chemically and biologically [86].

The rate of oxidation of Mn(II) is catalyzed by mineral and biological surfaces [76, 82], and is mediated by bacteria under natural water conditions [83-85]. The bacterially mediated Mn(II) oxidation appears to be significantly suppressed by solar radiation [38, 81]. Studies by Sunda and Huntsman [38] in coastal Bahamian waters revealed a diel variation in Mn(II) oxidation rates that was attributed to photoinhibitory effects. In another study [81], it was demonstrated that sunlight inhibited the oxidative precipitation of radiolabelled Mn(II) added to water samples from the Eastern Caribbean (Figure 14).

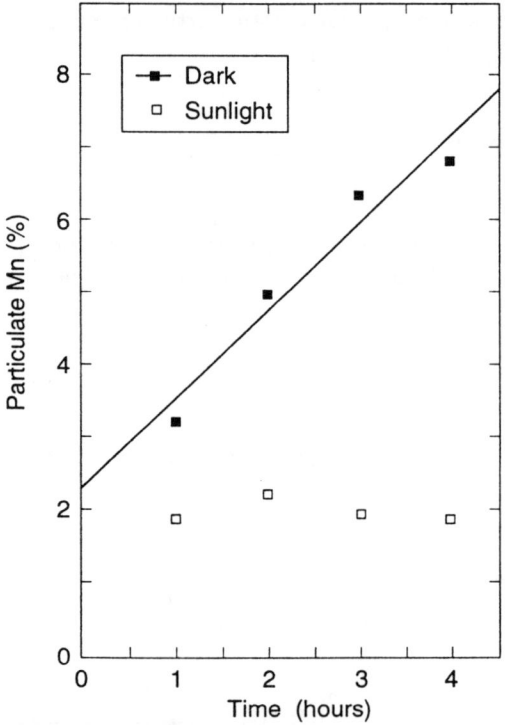

Figure 14. Effect of sunlight illumination on the percentage of radiolabelled Mn(II) added to water samples from surface waters in the Eastern Caribbean that was retained on a 0.22-μm pore-size Millipore filter (from Waite and Szymczak [81]).

While light acts to decrease the rate of bacterially-mediated Mn-oxide formation from dissolved Mn(II), it has also been shown to enhance the rate of reductive dissolution of particulate manganese in surface waters [8, 24, 39, 81]. It is believed that in surface waters, photoreductive dissolution of Mn-oxides occurs via reduction of Mn(III,IV) at the surface of Mn(III,IV)-(hydr)oxides by adsorbed humic substances [87]. This process is similar to photoreductive dissolution of Fe(III)-(hydr)oxides with adsorbed humic substances or simple carboxylic acids like oxalate or citrate acting as reductants. Assuming surface complex formation and subsequent ligand-to-metal charge-transfer reaction (in analogy to reactions 6-9), the following rate expression of dissolved Mn(II) formation through photoreductive dissolution can be expected [81],

$$\frac{Mn(II)_{aq}}{dt} = k_{red} \frac{K^{cond}[\equiv Mn_T][HA_{sol}]}{1 + K^{cond}[HA_{sol}]} \qquad (16)$$

where k_{red} is an overall rate coefficient of photoreductive dissolution of particulate manganese, K^{cond} is the conditional formation constant of the surface complex, $[\equiv Mn_T]$ is the total surface site concentration where humic acid can get adsorbed, and $[HA_{sol}]$ is the concentration of humic acid in solution (see also Eqs. 12 and 13). At low humic acid concentrations, only a small fraction of the total surface sites of particulate manganese is coordinated with humic acid and $K^{cond}[HA_{sol}] \ll 1$. Under these conditions, the rate of $Mn(II)_{aq}$ formation becomes linearly dependent on the humic acid concentration:

$$\frac{Mn(II)_{aq}}{dt} = k_{red}K^{cond}[\equiv Mn_T][HA_{sol}] \qquad (17)$$

Figure 15 shows that there exists a linear relationship between the rate of radiolabelled manganese that passed a 0.22 micron filter and the absorption coefficient at 365 nm of water samples from the Eastern Caribbean [81].

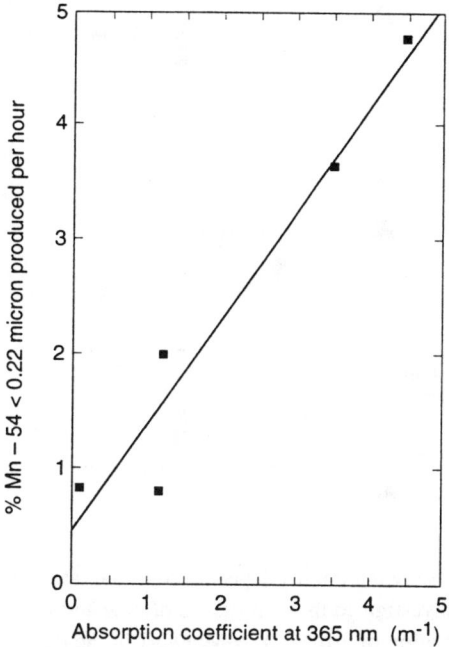

Figure 15. Correspondence between the absorption coefficient at 365 nm and the rate of photodissolution of added particulate manganese oxide to water samples from the surface waters of the Eastern Caribbean (from Waite and Szymczak [81]).

The absorption coefficient, $\alpha(\lambda)$, is a measure of the concentration of colored dissolved organic carbon (CDOM) and thus also of the humic acid concentration, $[HA_{sol}]$,

$$\alpha(\lambda) = \frac{A(\lambda)}{\ell} \approx \varepsilon'(\lambda)_{HA} \cdot [HA_{sol}] \qquad (18)$$

where $A(\lambda)$ is the absorption of the water sample (dimensionless), ℓ is the optical pathlength (in m), $[HA_{sol}]$ is the solution concentration of humic acids (in mg L^{-1}), and $\varepsilon'(\lambda)$ is the decadic extinction coefficient of humic acids (in L mg^{-1} m^{-1}), if the unit of the absorption coefficient, $\alpha(\lambda)$, is m^{-1}. This means that the rate of photoreductive dissolution of particulate manganese is linearly dependent on the concentration of humic substances (see Eq. 17).

In contrast to crystalline Fe(III)-(hydr)oxide phases, Mn(III,IV)-(hydr)oxides are also reductively dissolved in the absence of light by reductants such as humic substances or oxalate. Xyla et al. [88] have studied in detail the effects of pH and light on the reductive dissolution of two well defined manganese(III,IV)-(hydr)oxide phases with oxalate as reductant. Unlike Fe(III)-(hydr)oxides, γ-MnOOH and β-MnO$_2$ are reductively dissolved on a time scale of hours in the dark at low pH (Figure 16).

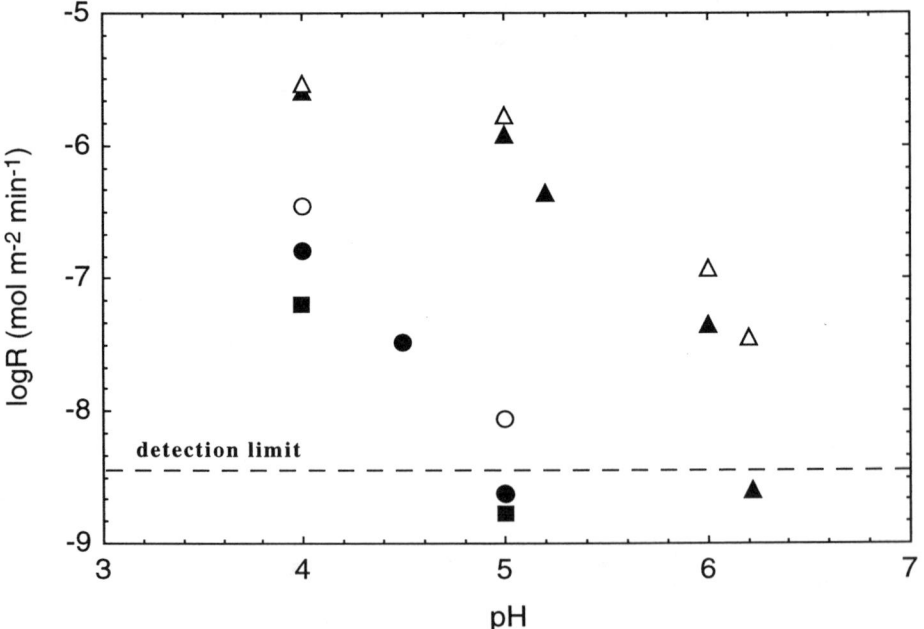

Figure 16. Summary plot showing the pH dependence of the rate of reductive dissolution of the solid manganese phases investigated in the presence of oxalate in the dark and in the light, with an initial oxalate concentration of 1 mM; ▲, γ-MnOOH in the dark; Δ, γ-MnOOH in the light; ●, β-MnO$_2$ in the dark; O, β-MnO$_2$ in the light; ■, Na$_4$Mn$_{14}$O$_{27}$·9H$_2$O in the dark (from Xyla et al. [88]).

Xyla and coworkers [88] observed a strong pH dependence of the rate of reductive dissolution of both manganese-(hydr)oxides; the rate of reductive dissolution decreased with increasing pH. This behavior is interpreted, in part, in terms of the pH dependence of oxalate adsorption on the surface of these Mn(III,IV)-(hydr)oxides. A plateau in the dissolution rate of γ-MnOOH was observed at high (total) oxalate concentrations, consistent with direct dependence of the rate of reductive dissolution on the concentration of surface-bound oxalate. At any given pH, the reductive dissolution rate per unit surface area was higher for γ-MnOOH than for β-MnO$_2$. Light had only a weak catalytic effect on the dissolution kinetics of the two Mn-(hydr)oxides studied, in contrast to Fe(III)-(hydr)oxides.

Another possible mechanism of dissolved Mn(II) formation would be reduction of Mn(III,IV) in Mn-(hydr)oxides by superoxide radicals (O_2^-) or by hydrogen peroxide (H_2O_2). Sunda and Huntsman [39] have reported that photoreductive dissolution of synthetic oxides resulted primarily from Mn(III,IV) reduction by H_2O_2. In contrast to the synthetic oxides, they found that the rates of photoreductive dissolution of natural oxides (formed in seawater from microbial oxidation of ^{54}Mn(II)) were not appreaciably altered by humic acid addition or by photochemical removal of natural organic matter. From these observations Sunda and Huntsman [39] concluded that H_2O_2 had only a minor effect on the reductive dissolution of natural Mn oxides, and that instead CDOM associated with the bacterial/Mn oxide aggregates acts as reductant.

7. Conclusions

Sunlight plays a crucial role in the formation of bioavailable iron and manganese, since it drives the redox cycling of these metals in oxic environmental systems. It is likely that the concentrations of dissolved Fe(II) species and of inorganic dissolved Fe(III) species (including the hydrated and hydolyzed species) determine the upper limit for bioavailable iron. The bioavailability of manganese is probably controlled by the concentration of Mn(II)$_{aq}$.

In oxygenated aquatic systems, sunlight enhances the reduction of Fe(III), through photoreductive dissolution of particulate and colloidal Fe(III)-(hydr)oxides and through light-induced reduction of dissolved Fe(III) species. Therefore, the Fe(II) concentration in sunlit surface waters tends to parallel the intensity of solar radiation and thus the time of day. In the euphotic zone of seawater, it is mainly the kinetics of light-induced redox cycling between dissolved or finely colloidal Fe(II) and Fe(III) species that controls the formation of biologically available iron. Thereby photochemically formed superoxide (O_2^-) plays probably a pivotal role as reductant of dissolved Fe(III). But also reduction of organically complexed Fe(III) through photolysis is likely to be an important process for the formation of Fe(II) in seawater.

The process that initiates redox cycling between dissolved Fe(II) and dissolved Fe(III) in oxygenated aquatic environments is photoreductive dissolution of Fe(III)-(hyrd)oxides. In seawater, this process seems to occur only with readily soluble, i.e., with amorphous Fe(III)-hydroxide phases. But in atmospheric waters which are usually acidic, also crystalline Fe(III)-(hydr)oxides undergo photoreductive dissolution in the presence of suitable reductants such as e.g. oxalate. Hence, atmospheric iron input is probably an important source of bioavailable iron to the surface waters in remote oceanic regions. Yet, thermodynamically very stable phases such as hematite (α-Fe_2O_3) and goethite (α-FeOOH) are not readily dissolved, also under acidic atmospheric water conditions. However, redox cycling of iron at the surface of α-Fe_2O_3 or α-FeOOH may take place, i.e., reduction and oxidation of surface iron without transfer into solution, resulting in formation of less crystalline surface coatings. As a result, initially crystalline, thermodynamically stable Fe(III)-(hydr)oxide phases could be gradually transformed into amorphous phases, which undergo more readily reductive dissolution. Such transformation processes may take place in aerosols during their long-range transport.

In the case of manganese, bacterially mediated Mn(II) oxidation is suppressed by light, and reduction of Mn(III,IV) is enhanced by light (through reductive dissolution of particulate Mn(III,IV)-(hydr)oxides). Therefore, the dependence of the Mn(II) concentration on the intensity of solar radiation should be even more pronounced than in the case of iron. Reductive dissolution of Mn(III,IV)-(hydr)oxides occurs at considerable rates also in the absence of light in oxic environments. Unlike Fe(III)-(hydr)oxides, not only finely colloidal but also particulate Mn(III,IV)-(hydr)oxide phases undergo reductive dissolution in surface seawater, both photochemical and thermal. Thereby humic acids associated with the bacterial/Mn oxide aggregates are probably the reductants of major importance.

8. References

1. Sunda, W.G., Swift, D.G., and Huntsman, S.A. (1991) Low iron requirement for growth in oceanic phytoplankton, *Nature* **351**, 55-57.
2. Morel, F.M.M., Hudson, R.J.M., and Price, N.M. (1991) Limitation of productivity by trace metals in the sea, *Limnol. Oceanogr.* **36**, 1742-1755.
3. Morel, F.M.M., Reinfelder, J.R., Roberts, S.B., Chamberlain, C.P., Lee, J.G., and Yee, D. (1994) Zinc and carbon co-limitation of marine phytoplankton, *Nature* **369**, 740-74.
4. Martin, J.H., et al., (1994) Testing the iron hypothesis in ecosystems of the equatorial Pacific Ocean, *Nature* **371**, 123-129.
5. Coale, K.H. et al. (1996) A massive phytoplankton bloom induced by an ecosystem-scale iron fertilization experiment in the equatorial Pacific Ocean, *Nature* **383**, 495-501.
6. Fraústo da Silva, J.J.R., and Williams, R.J.P. (1991) *The Biological Chemistry of the Elements*, Clarendon, Oxford.
7. Raven, J.A. (1988) The iron and molybdenum use efficiencies of plant growth with different energy, carbon and nitrogen sources, *New Phytol.* **109**, 279-287.
8. Sunda, W.G., Huntsman, S.A., and Harvey, G.R. (1983) Photoreduction of manganese oxides in seawater and its geochemical and biological implications, *Nature* **301**, 234-236.
9. Sunda, W.G. and Huntsman, S.A. (1996) Antagonisms between cadmium and zinc toxicity and manganese limitation in a coastal diatom, *Limnol. Oceanogr.* **41**, 373-387.
10. Sunda, W.G. (1991) Trace metal interactions with marine phytoplankton, *Biol. Oceanogr.* **6**, 411-442.
11. Archer, M.D. and Bolton J.R. (1991) *Photoconversion of Solar Energy: A Short Course*, Brigels, Switzerland, September 14-25.

12. Anderson, M.A. and. Morel, F.M.M. (1982) The influence of aqueous iron chemistry on the uptake of iron by the coastal diatom *Thalassiosira weissflogii*, *Limnol. Oceanogr.* **27**, 789-813.
13. Sunda, G.S. and Huntsman, S.A. (1985) Regulation of cellular manganeses and manganese transport rates in the unicellular alga *Chlamydomonas*, *Limnol. Oceanogr.* **30**, 71-80.
14. Harrison, G.I., and Morel, F.M.M. (1986) Response of the marine diatom *Thalassiosira weissflogii* to iron stress, *Limnol. Oceanogr.* **31**, 989-997.
15. Hudson, R.J.M., and Morel, F.M.M. (1990) Iron transport in marine phytoplankton: Kinetics of cellular and medium coordination reaction, *Limnol. Oceanogr.* **35**, 1002-1020.
16. Sunda, G.S. and Huntsman, S.A. (1986) Relationships among growth rate, cellular manganese concentrations and manganese transport kinetics in estuarine and oceanic species of the diatom *Thalassiosira*, *J. Phycol.* **22**, 259-270.
17. Bruland, K.W., Donat, J.R., and Hutchins, D.A. (1991) Interactive influences of bioactive trace metals on biological production in oceanic waters, *Limnol. Oceanogr.* **36**, 1555-1577.
18. Murphy, J.P., Lean, D.R., and Nalewajko, C. (1976) Blue-green algae: Their excretion of iron selective chelators enables them to dominate other algae, *Science* **192**, 900-902.
19. Lewis, B.L., Holt, P.D., Taylor, S.W., Wilhelm, S.W., Trick, C.G., Butler, A., and Luther III, G.W. (1995) Voltammetric estimation of iron(III) thermodynamic stability constants for catecholate siderophores isolated from marine bacteria and cyanobacteria, *Mar. Chem.* **50**, 179-188.
20. Reid, R.T. and Butler, A. (1991) Investigation of the mechanism of iron acquisition by the marine bacterium *Alteromonas luteoviolaceus*: Characterization of siderphore production, *Limnol. Oceanogr.* **36**, 1783-1792.
21. Wilhelm, S.W. and Trick, C.G. (1994) Iron-limited growth of cyanobacteria: siderophore production is a common response mechanism, *Limnol. Oceanogr.* **39**, 1979-1984.
22. Barbeau, K., Moffett, J.W., Caron, D.A., Croot, P.L., and Erdner, D.L. (1996) Role of protozoan grazing in relieving iron limitation of phytoplankton, *Nature* **380**, 61-64.
23. Landing, W.M. and Bruland, K.W. (1987) The contrasting biogeochemistry of iron and manganese in the Pacific Ocean, *Geochim. Cosmochim. Acta* **51**, 29-43.
24. Sunda, W.G. and Huntsman, S.A. (1988) Effect of sunlight on redox cycles of manganese in the southwestern Sargasso Sea, *Deep-Sea Res.* **35**, 1297-1317.
25. Wu, J. and Luther III, G.W. (1994) Size-fractionated iron concentrations in the water column of the western North Atlantic Ocean, *Limnol. Oceanogr.* **39**, 1119-1129.
26. Gledhill, M., and Van den Berg, C.M.G (1995) Measurement of the redox speciation of iron in seawater by catalytic cathodic stripping voltammetry, *Mar. Chem.* **50**, 51- 61.
27. Van den Berg, C.M.G. (1995) Evidence for organic complexation of iron in seawater, *Mar. Chem.* **50**, 139-157.
28. Rue, E.L. and Bruland, K.W. (1995) Complexation of iron(III) by natural organic ligands in the Central North Pacific as determined by a new competitive ligand equilibration / adsorptive cathodic stripping voltammetric method, *Mar. Chem.* **50**, 117-138.
29. Wu, J. and Luther III, G.W. (1995) Complexation of Fe(III) by natural organic ligands in the Northwest Atlantic Ocean by a competitive ligand equilibration and a kinetic approach, *Mar. Chem.* **50**, 159-177 (1995).
30. Stookey, L.L. (1970) Ferrozine - A new stectrophotometric reagent for iron, *Anal. Chem.* **42**, 779-781.
31. Landing, W.M. and Westerlund, S. (1988) The solution chemistry of iron(II) in Framvaren Fjord, *Mar. Chem.* **23**, 329-343.
32. King, D.W., Lin, J., and Kester, D.R. (1991) Spectroscopic determination of iron(II) in seawater at nanomolar concentrations, *Anal. Chim. Acta* **247**, 125-132.

33. O'Sullivan, D.W., Hanson, A.K., and Kester, D.R. (1995) Stopped flow luminol chemiluminescence dtermination of Fe(II) and reducible iron in seawater at subnanomolar levels, *Mar. Chem.* **49**, 65-77.
34. King, D.W., Lounsbury, H.A., and Millero, F.J. (1995) Rates and mechanism of Fe(II) oxidation at nanomolar total iron concentrations, *Environ. Sci. Technol.* **29**, 818-824.
35. Martin, J.H., Gordon, R.M., and Fitzwater, S.E. (1991) The case for iron, *Limnol. Oceanogr.* **36**, 1793-1802.
36. Powell, R.T., King, D.W., and Landing, W.M. (1995) Iron distibution in surface waters of the south Atlantic, *Mar. Chem.* **50**, 13-20.
37. Landing, W.M., and Bruland, K.W. (1980) Manganese in the North Pacific, *Earth Planet. Sci. Lett.* **49**, 45-56.
38. Sunda, W.G. and Huntsman, S.A. (1990) Diel cycles in microbial manganese oxidation and manganese redox speciation in coastal waters of the Bahama Islands, *Limnol. Oceanogr.* **35**, 325-338.
39. Sunda, W.G. and Huntsman, S.A. (1994) Photoreduction of manganese oxides in seawater, *Mar. Chem.* **46**, 133-152.
40. Finden, D.A.S., Tipping, E., Jaworski, G.H.M., and Reynolds, C.S. (1984) Light-induced reduction of natural iron(III) oxide and its relevance to phytoplankton, *Nature* **309**, 783-784.
41. Waite, T.D. and Morel, F.M.M. (1984) Photoreductive dissolution of colloidal iron oxides in natural waters, *Environ. Sci. Technol.* **18**, 860-868.
42. Waite, T.D. and Morel, F.M.M. (1984) Photoreductive dissolution of colloidal iron oxide: Effect of citrate, *J. Colloid Interface Sci.* **102**, 121-137.
43. Sulzberger, B. and Laubscher, H.-U. (1995) Reactivity of various types of iron(III) (hydr)oxides towards light-induced dissolution, *Marine Chemistry* **50**, 103-115.
44. Voelker, B.M., Morel, F.M.M., and Sulzberger, B. (in press) Iron redox cycling in surface waters: effects of humic substances and light, *Environ. Sci. Technol.*.
45. Sulzberger, B. and Laubscher, H.-U. (1995) Photochemical reductive dissolution of lepidocrocite: Effect of pH. In C. P. Huang, C. R. O'Melia, J. J. Morgan, Eds., *Aquatic Chemistry; Interfacial and Interspecies Processes*, Advances in Chemistry Series 244, American Chemical Society, Washington, DC, pp. 279-290.
46. Rich, H. and Morel, F.M.M. (1990) Availability of well-defined iron colloids to the marine diatom *Thalassiosira weissflogii*, *Limnol. Oceanogr.* **35**, 652-662.
47. Johnson, K.S., Coale, K.H., Elrod, V.A., and Tindale, N.W. (1994) Iron photochemistry in seawater from the equatorial Pacific, *Mar. Chem.* **46**, 319-334.
48. Zhuang, G., Yi, Z., Duce, R.A., and Brown, Ph. R. (1992) Link between iron and sulphur cycles suggested by detection of Fe(II) in remote marine aerosols, *Nature* **355**, 537-539.
49. Zhuang, G., Yi, Z., Duce, R.A., and Brown, Ph.R. (1992) Chemistry of iron in marine aerosols, *Global Biogeochem. Cycles* **6**, 161-173.
50. Duce, R.A. and Tindale, N.W. (1991) Atmospheric transport of iron and its deposition in the ocean, *Limnology and Oceanography* **36**, 1715-1726.
51. Young, R.W., Carder, K.L., Betzer, P.R., Costello, D.K., Duce, R.A., DiTullio, G.R., Tindale, N.W., Laws, E.A., Uematsu, M., Merrill, J.T., and Feely, R.A. (1991) Atmospheric iron inputs and primary productivity: Phytoplankton responses in the North Pacific, *Global Biogeochem. Cycles* **5**, 119-134.
52. Martin, J.H., Gordon, R.M., Fitzwater, S., and Broenkow, W.W. (1989) Vertex: phytoplankton/iron studies in the gulf of Alaska, *Deep-Sea Res.* **36**, 649-680.
53. Wells, M.K., Price, N.M., and Bruland, K.W. (1994) *Iron Chemistry in Seawater and its Relationship to Phytoplankton: A Workshop Report*, Bermuda, May 1994.
54. Moffett, J.W., and Zika, R.G. (1987) Reaction kinetics of hydrogen peroxide with copper and iron in seawater, *Environ. Sci. Technol.* **21**, 804-810.
55. Millero, F., Sotolongo, S., and Izaguirre, M. (1987) The oxidation kinetics of Fe(II) in seawater, *Geochim. Cosmochim. Acta* **51**, 547-553.

56. Miller, W.L., King, D.W., Lin, J., and Kester, D.R. (1995) Photochemical redox cycling of iron in coastal seawater, *Mar. Chem.* **50**, 63-77.
57. Bielski, B.H.J., Cabelli, D.E., Arudi, R.L., and Ross, A.B. (1985) Reactivity of O_2^-/HO_2 radicals in aqueous solution, *J. Phys. Chem. Ref. Data* **14**, 1941-1100.
58. Voelker, B.M. and Sedlak, D.L. (1995) Iron reduction by photoproduced superoxide in seawater, *Mar. Chem.* **50**, 93-102.
59. King, D.W., Aldrich, R.A., Charnecki, S.E. (1993) Photochemical redox cycling of iron in NaCl solution, *Mar. Chem.* **44**, 105-120.
60. Voelker, B.M. and Sulzberger, B. (1996) Effects of fulvic acid on Fe(II) oxidation by hydrogen peroxide, *Environ. Sci. Technol.* **30**, 1106-1114.
61. Deng, Y. and Stumm, W. (1994) Reactivity of aquatic iron(III) oxyhydroxides - implications for redox cycling of iron in natural waters, *Appl. Geochem.* **9**, 23-36.
62. Jones, G.J., Palenik, B.P., and Morel, F.M.M. (1987) Trace metal reduction by phytoplankton: the role of plasmalemma redox enzymes, *J. Phycol.* **23**, 237-244.
63. Morel, F.M.M. Cycling and availability of iron in the equatorial Pacific, Abstract Bermuda Workshop *Iron Chemistry in Seawater and its Relationship to Phytoplankton*, May 1994.
64. Rush, J.D. and Bielski, B.H.J. (1985) Pulse radiolysis studies of the reactions of HO_2/O_2^- with ferric ions and its implication on the occurrence of the Haber-Weiss reaction, *J. Phys. Chem.* **89**, 5062-5066.
65. Barry, R.C., Schnoor, J.L., Sulzberger, B., Sigg, L., and Stumm, W. (1994) Iron oxidation kinetics in an acidic alpine lake, *Wat. Res.* **28**, 323-333.
66. Collienne, R.H. (1983) Photoreduction of iron in the epilimnion of acidic lakes, *Limnol. Oceanogr.* **28**, 83-100.
67. McKnight, D.M., Kimball, B.A., and Bencala, K.E. (1988). Iron photoreduction and oxidation in an acidic mountain stream, *Science* **240**, 637-649.
68. Sulzberger, B., Schnoor, J.L., Giovanoli, R., Hering, J.G., and Zobrist, J. (1990) Biogeochemisty of iron in an acidic lake, *Aquatic Sci.* **52**, 56-74.
69. Waite, T.D., Szymczak, R., Espey, Q.I., and Furnas, M.J. (1995) Diel variations in iron speciation in northern Australian shelf waters, *Mar. Chem.* **50**, 79 - 91.
70. Stumm, W. and Sulzberger, B. (1992) The cycling of iron in natural environments: Considerations based on laboratory studies of heterogeneous redox porcesses, *Geochim. Cosmochim. Acta* **56**, 3233-3257.
71. Sedlak, D.L. and Hoigné, J. (1993) The role of copper and oxalate in the redox cycling of iron in atmospheric waters, *Atmospheric Environment* **27A**, 2173-2185.
72. Siffert, C. and Sulzberger, B. (1991) Light-induced dissolution of hematite in the presence of oxalate: A case study, *Langmuir* **7**, 1627-1634.
73. Stumm, W. (1992) *Chemistry of the Solid-Water Interface*, Wiley-Interscience, New York.
74. Kormann, C., Bahnemann, D.W., and Hoffmann, M.R. (1988) Photocatalytic production of H_2O_2 and organic peroxides in aqueous suspensions of TiO_2, ZnO, and desert sand, *Environ. Sci. Technol.* **22**, 798-806.
75. Faust, B.C., Hoffmann, M.R., and Bahnemann, D.W. (1989) Photocatalytic oxidation of sulfur-dioxide in aqueous suspensions of alpha-Fe_2O_3, *J. Phys. Chem.* **93**, 6371-6381.
76. Wehrli, B. (1990) Redox reactions of metal ions at mineral surfaces. In W. Stumm, Ed., *Aquatic Chemical Kinetics*, Wiley-Interscience, New York, pp. 311-336.
77. Tamura, H., Goto, K. and Nagayama, M. (1976) The effect of ferric hydroxide on the oxygenation of ferrous ions in neutral solutions, *Corrosion Sci.* **16**, 197-207.
78. Singer, Ph.C. and Stumm, W. (1970) Acidic mine drainage - the rate-determining step, *Science* **167**, 1121-1123.
79. Kopcewicz, M., and Dzienis, B. (1971) Mössbauer study of iron in atmospheric air, *Tellus* **23**, 176-182.

80. Pehkonen, S.O., Siefert, R.S., Erel, Y., Webb, S., and Hoffmann, M.R. (1993) Photoreduction of iron oxyhydroxides in the presence of important atmospheric organic compounds, *Environ. Sci. Technol.* **27**, 2056-2062.
81. Waite, T.D. and Szymczak, R. (1993) Manganese dynamics in surface waters of the Eastern Caribbean, *J. Geophys. Res.* **98**, 2361-2369.
82. Wehrli, B., Friedl, G., and Manceau, A. (1995) Reaction rates and products of manganese oxidation at the sediment-water interface, In C. P. Huang, C. R. O'Melia, J. J. Morgan, Eds., *Aqutic Chemistry; Interfacial and Interspecies Processes*, Advances in Chemistry Series 244, American Chemical Society, Washington, DC, pp. 111-134.
83. Tebo, B.M. and Emerson, S. (1986) Microbial manganese(II) oxidation in the marine environment: A quantitative study, *Biogeochemistry* **2**, 149-161.
84. Sunda, G.S. and Huntsman, S.A. (1987) Microbial oxidation of manganese in a North Carolina esturary, *Limnol. Oceanogr.* **32**, 552-564.
85. Mason, Y. (1995) Natural manganese oxides associated with Metallogenium-like particles as scavengers of metals in lakes. Ph. D. thesis ETH Zürich Nr. 11393.
86. Kostka, J.E., Luther III, G.W., and Nielson, K.H. (1995) Chemical and biological reduction of Mn(III)-pyrophosphate complexes: Potential importance of dissolved Mn(III) as an environmental oxidant, *Geochim. Cosmochim. Acta* **59**, 885-894.
87. Waite, T.D., Wrigley, I. C., and Szymczak, R. (1988) Photoassisted dissolution of a colloidal manganese oxide in the presence of fulvic acid, *Eviron. Sci. Technol.* **22**, 778-785.
88. Xyla, A.G., Sulzberger, B., Luther III, G.W., Hering, J.G., Van Cappellen, P., and Stumm, W. (1992) Reductive dissolution of manangese(III,IV) (hydr)oxides by oxalate: the effect of pH and light, *Langmuir* **8**, 95-103.

THE INFLUENCE OF IRON ON CARBON DIOXIDE IN SURFACE SEAWATER

DR. FRANK J. MILLERO
Rosenstiel School of Marine and Atmospheric Science
4600 Rickenbacker Causeway
Miami, FL 33149
Tel: 1 305 361 4707
Fax: 1 305 361 4144
Email: fmillero@rsmas.miami.edu

Abstract

The carbon dioxide produced due to the burning of fossil fuels has increased the concentration in the atmosphere from levels of 280 ppm before the industrial revolution to the present level of 360 ppm. Approximately 40% of the CO_2 added to the atmosphere is thought to be taken up by the oceans. This uptake of CO_2 can be from the solubility of the gas in cold sinking waters (the **Solubility Pump**) or the biological conversion to organic carbon that sinks to deep waters (the **Biological Pump**). For the Biological pump to work the phytoplankton growing in the surface waters need nutrients (nitrate, phosphate, and silicate). Most surface waters of the oceans are deplete of these nutrients, thus, the plants cannot grow without the upwelling of cold nutrient rich waters. In some parts of the ocean, however, the surface waters have high concentrations of nutrients and low concentrations of chlorophyll (HNLC) or phytoplankton. It has been postulated that the plants do not grow in these waters due to a lack of iron (the **Iron Hypothesis**). Recently there have been some open sea experiments that have been made to test this hypothesis. This was done by adding large amounts of iron enriched seawater to 64 km^2 of surface waters. The waters were then sampled for a number of days to examine the changes in the biological and chemical properties of the iron enriched patch. The results of these experiments will be discussed. The changes in the concentration of the carbonate parameters due to CO_2 uptake during the phytoplankton growth will be demonstrated.

1. The Carbon Dioxide System in the Ocean

Interest in the carbon dioxide system comes from its importance as a major greenhouse gas and its effect on the climate system (**Figure 1**). The sunlight in the visible range reaches the surface of the earth and is redirected to space as infrared energy. A number of gases (CO_2, CH_4, O_3, etc.) in the atmosphere are able to adsorb

GREENHOUSE GASES TRAP HEAT AT THE EARTH'S

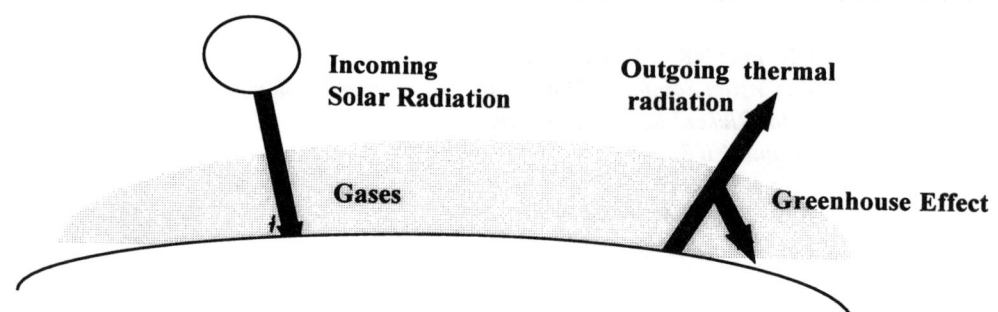

Figure 1. The Greenhouse Effect on the Energy Balance of the Earth.

this IR energy which results in a warming of the atmosphere (**Figure 2**). Although scientists agree that this greenhouse effect occurs and some of the greenhouse gases are increasing in the atmosphere, questions still are not settled on the effects this process has on the climate system.

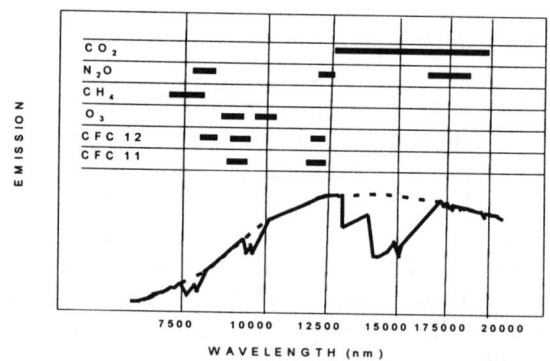

Figure 2. The absorption of the greenhouse gases of infrared energy.

Scientists have clearly demonstrated [1] that the CO_2 in the atmosphere has increased with time. This is shown in **Figure 3** where the direct atmospheric measurements have been made since 1958 started by Charles Keeling from U. Cal. San Diego. More recently the CO_2 in the past atmosphere have been obtained from measuring the concentration in the air trapped in ice cores. The pre-industrial level of CO_2 was about 280 ppm and has increase in an exponential manner to its present levels of 360 ppm. It is thought by many that this increase in CO_2 in the atmosphere will increase the global temperature. Although it is presently difficult to demonstrate that the temperature of the earth has increased, most global models support this assumption. The rate of the increase of CO_2 added to the atmosphere due to the burning of fossil fuels is increasing with time (See **Figure 4**). Although the actual amount of CO_2 measured in the atmosphere is also increasing its level is less that expected from fossil fuel emissions (**Figure 4**). Each year approximately 5.4 gigatons (Gt. = 10^{15} g) of CO_2 are released to the atmosphere due to the burning of fossil fuels (**Table 1**).

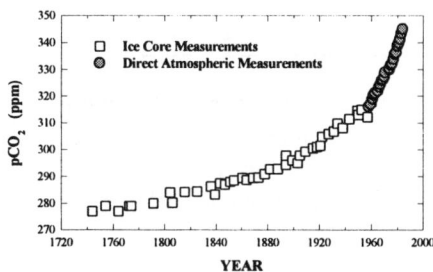

Figure 3. The increase of carbon dioxide in the atmosphere over time.

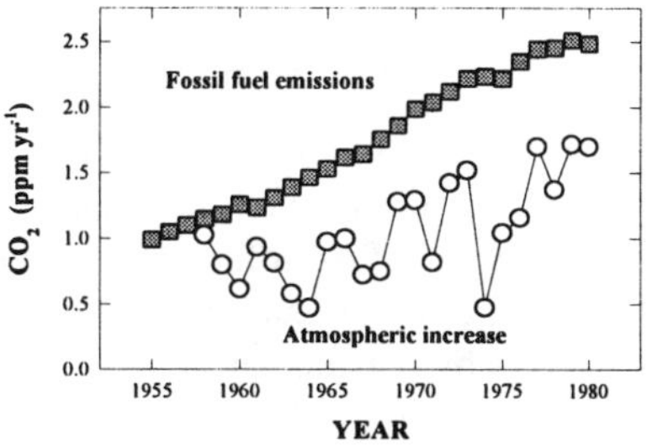

Figure 4. Changes in the rate of increase of carbon dioxide from the burning of fossil fuels and measured in the atmosphere.

Only 3.4 gigatons of the CO_2, however, are found in the atmosphere from direct measurements. This means that 2 gigatons or about 40% of the CO_2 released due to the burning of fossil fuels is being taken up by land plants or the oceans. Since the amount of CO_2 taken up by land plants is thought to be less or nearly the same as the amount released due to deforestation, most scientists feel that the oceans are the major sink for the missing CO_2. Ocean and atmospheric models support this contention as do some limited ocean measurements. At the present time a large scale international program is underway to examine the carbonate system in the oceans. The major goals of this program are

1. Quantify the uptake of CO_2 by the Oceans
2. Provide a global description of the CO_2 system to aid modelers
3. Characterize the transport of CO_2 across the air-sea interface

Table 1. Budget for The Global CO_2 System (1980-1989)

Sources	Average Perturbation (10^{15} g Carbon y^{-1})
Fossil Fuel Combustion	5.4 ± 0.5
Deforestation	1.6 ± 1.0
Total	7.0 ± 1.2
Sinks	
Atmosphere	3.4 ± 0.2
Oceans (Models)	2.0 ± 0.8
Land Plants	1.6 ± 1.4
Total	5.4 ± 0.8

To accomplish this goal a great deal of progress has been made in being able to make precise and accurate measurements of the carbonate parameters in ocean waters (pH; total alkalinity, TA; total carbon dioxide, TCO_2; and the partial pressure of carbon dioxide, pCO_2). A summary of the present precision and accuracy of measurements being made by marine carbonate chemist is given in **Table 2**.

Table 2. Estimates of the analytical precision and accuracy of measurements of carbonate parameters

Analysis	Precision	Accuracy	Reference
pH (spectrophotometric)	± 0.0004	± 0.002	Clayton and Byrne [2]
TA (potentiometric)	± 2 µmol kg^{-1}	± 4 µmol kg^{-1}	Millero et al.[3]
TCO_2 (coulometric)	± 1 µmol kg^{-1}	± 2 µmol kg^{-1}	Johnson et al.[4]
pCO_2 (Infrared)	± 0.5 µAtm	± 2 µAtm	Wanninkhof and Thoning[5]

It should be pointed out that the measurement of the carbonate parameters can be made with the above precision continuously on surface seawater and on individually collected samples. More recently it has also been demonstrated that our knowledge of the thermodynamics of the carbon dioxide system is nearly consistent with our ability to characterize the system [6,7].

The ocean can take up the CO_2 in two ways:
1. The physical chemical dissolution of CO_2 into ocean waters (**Solubility Pump**).

2. The uptake of CO_2 by plants and removal of organic carbon into the deep ocean (**Biological Pump**) where it is kept for 600 to 1000 years (the mixing time of the oceans).

The **solubility pump (Figure 5)** is related to the diffusion of CO_2 across the

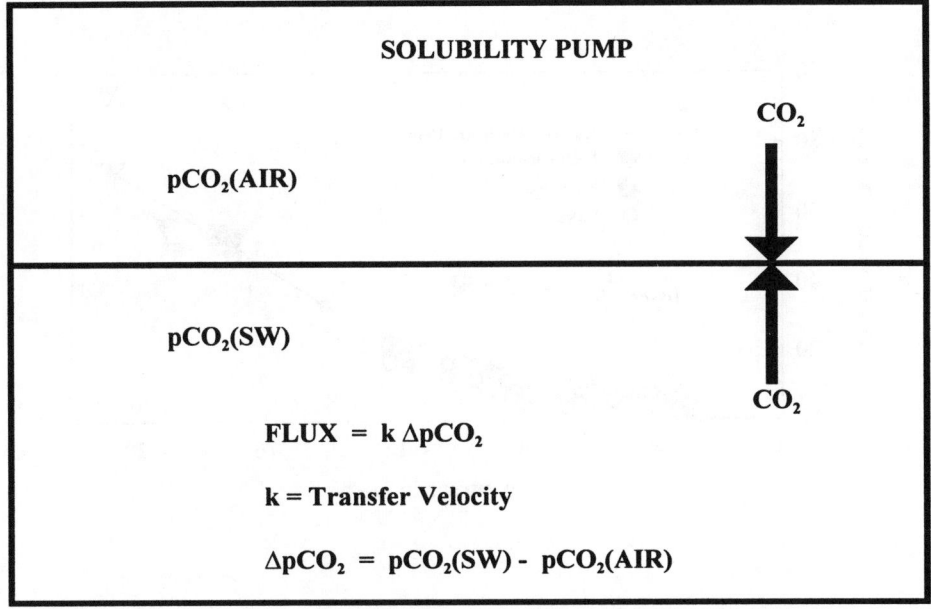

Figure 5. A sketch of the **Solubility Pump**.

air-sea interface. The driving force or flux of CO_2 across this interface is the differences between the concentrations in the atmosphere and oceans given by [8]

$$\text{Flux} = k \{pCO_2(SW) - pCO_2(ATM)\} = k \Delta pCO_2 \tag{1}$$

where the value k is called the transfer velocity. The transfer velocity can be attributed to two terms [8]

$$1/k = 1/ak_W + 1/Hk_A \tag{2}$$

where k_W and k_A are transfer velocities in the water and air, respectively; H is the Henry's law constant (a unitless value, the ratio of air to water concentrations at equilibrium) and a is a factor that accounts for any enhancement of the transfer on the water side due to chemical reactions between the gas and H_2O ($CO_2 + H_2O \rightarrow H^+ + HCO_3^-$). The value of a = 1.02 to 1.03 for CO_2. From ^{14}C measurements values of k_W average about 6 mol m^{-2} y^{-1} µatm^{-1} on a global basis. The transfer velocity[9] increases with increasing wind speed (see **Figure 6**). The higher the winds the greater mixing occurs and the flux of

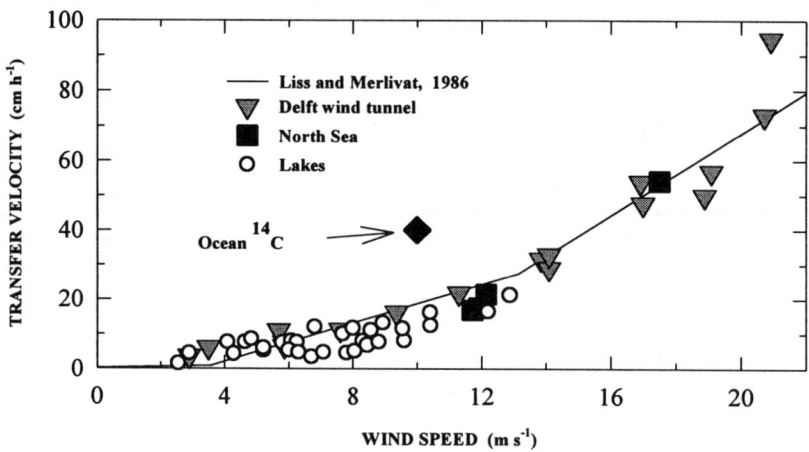

Figure 6. The transfer velocity of gases across the air-sea interface as a function of the wind speed.

CO_2 is increased. If the concentration of CO_2 is higher in the atmosphere than in the ocean, the CO_2 will dissolve in the oceans. The values of k determined in wind tunnel measurements are much smaller that the value estimated from ^{14}C measurements making it difficult to use equation (1) to calculate global fluxes of CO_2. When ΔpCO_2 is positive the oceans are a source of CO_2 and when it is negative the oceans are a sink for CO_2. For 2 gigatons of CO_2 to dissolve in the ocean the value of ΔpCO_2 globally would only need to be -8 ppm - which is only 2% of the amount in the atmosphere (360 ppm or 360 µatm). If rapid exchange takes place, one would expect the pCO_2 in the atmosphere to be equal to the values in the surface waters. If the exchange is sluggish, the pCO_2 in surface waters will be higher in upwelling areas and lower in colder waters than the values in the atmosphere. The variations of CO_2 in surface waters from place to place can be as large as 100 ppm and the differences in the values between the ocean and atmosphere can be positive or negative. This is shown in **Figure 7** from recent measurements of the pCO_2 in the surface waters of the Atlantic.

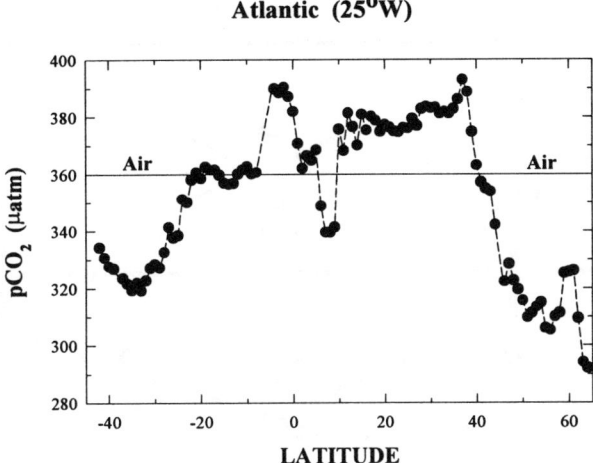

Figure 7. The differences in the partial pressure of carbon dioxide in the atmosphere and ocean in the North Atlantic (Dr. Rik Wanninkhof, NOAA/Miami).

The higher values of pCO_2 near the equator are due to equatorial upwelling and are a source of CO_2. The lower values of pCO_2 in the polar regions is due to waters being cold and where the oceans are a sink for CO_2. The complicated structure in the north and south Atlantic may be related to phytoplankton blooms. These large variations in ΔpCO_2 make it difficult to detect the sources and sinks of CO_2 in the oceans. Presently oceanographers are attempting to characterize the surface levels of CO_2 in ocean water throughout the world over different seasons. This information will allow one to characterize the sources and sinks of CO_2 in the world oceans.

The second method of removing CO_2 from surface waters is related to the growth of plants and is called the **biological pump** (see **Figure 8**). This so-called pump is simply the conversion of CO_2 into organic carbon and oxygen during photosynthesis in surface waters and the oxidation of this plant material by bacteria in deep waters. If the plant material sinks below the thermocline (that prevents mixing between surface and deep waters), the CO_2 is taken out of the surface waters and deposited into deep waters-where it is stored for 600 to 1000 years. For this biological pump to operate the plants need nutrients (nitrate, phosphate, and silica) and light to grow. Since the surface waters of the oceans are normally depleted of these nutrients, this process is limited by how much of the nutrients (at high concentrations) are upwelled to the surface. The fraction of organic carbon that is removed from the surface waters (or new production) is normally a small fraction (20 to 30%) of the amount fixed in surface waters.

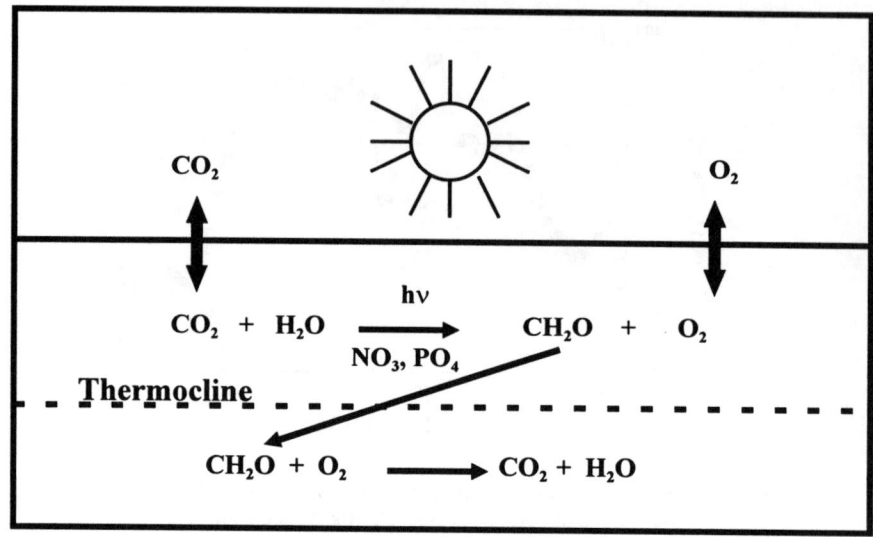

Figure 8. A sketch of the **Biological Pump**.

2. The Iron Hypothesis

The limiting nutrient in most ocean waters is thought to be nitrate. Over 10% of the world's ocean surface waters, however, are replete with the major plant nutrients (nitrate, phosphate and silicate) and light while standing stocks of phytoplankton remain low (see **Figure 9**). These areas include the surface waters in the North Pacific, the Equatorial Pacific, and the Southern Ocean waters around Antarctica. The factors that limit phytoplankton growth and biomass in these high nutrient, low chlorophyll (HNLC) areas have been vigorously debated [10]. The suggestion that increasing the growth of plants in HNLC areas could remove significant amounts of carbon dioxide from the atmosphere has stimulated the interest in this topic [11]. Some have proposed that if plants can made to grow in these areas that it may be possible to control the uptake of CO_2 by the oceans.

The cause of the lack of growth of plants in some of the HNLC areas, has been attributed to:
1. Zooplankton grazing that may contribute to the maintenance of low chlorophyll levels [12].
2. Strong turbulence at high latitudes might mix phytoplankton below the critical depth and result in light limitation [13].

THE INFLUENCE OF IRON ON CARBON DIOXIDE IN SURFACE SEAWATER

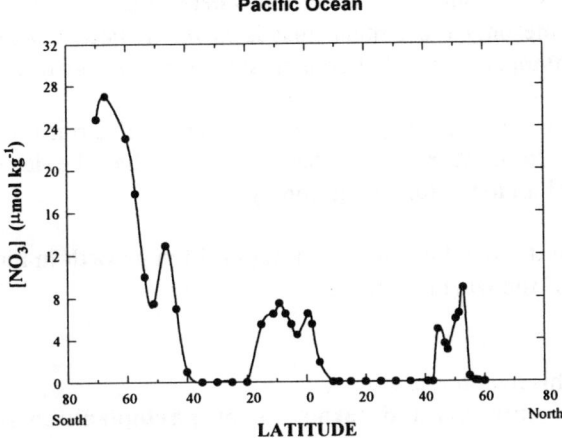

Figure 9. The concentration of nitrate in surface waters in the Pacific Ocean.

3. Lack of needed micronutrient elements, such as **iron** may limit the growth [14].

The limitation of metals for the growth of phytoplankton can be demonstrated by examining the relative concentrations of metals to phosphate in ocean waters compared to the ratio in phytoplankton (**Figure 10**). The relative concentrations of iron in

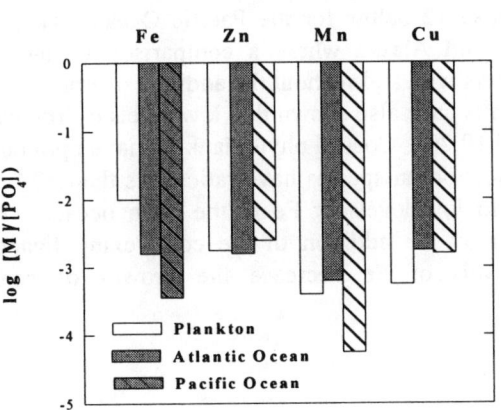

Figure 10. The ratio of metals to phosphate in phytoplankton and ocean waters.

phytoplankton are higher than the levels in the waters suggesting that it may be limiting to growth. Mn is the only other metal that is lower in Pacific waters relative to the concentration in phytoplankton. Laboratory studies on the growth by phytoplankton support this contention.

Since iron is an essential trace element used in electron transport and enzyme systems, the late **John Martin** [15] has suggested that the latter reason is most important. This led him to the following iron hypothesis:

Iron availability limits specific rates of phytoplankton growth in high nutrient low chlorophyll areas of the world's oceans.

Two corollaries to this hypothesis are:
1. **The uptake of nutrients and carbonate by phytoplankton is limited by iron availability.**
2. **Since iron limitation controls the growth of phytoplankton removes CO_2 from the atmosphere, the levels of atmospheric CO_2 varies as a function of iron transport to the surface of the oceans.**

A number of preliminary studies have been made that support of the iron hypothesis including:
1. Measured iron levels are quite low (**Figure 11**) and may not support high phytoplankton biomass at maximum growth rates [16].
2. The addition of Fe to uncontaminated seawater from the HNLC areas has been shown to stimulate the growth of phytoplankton, especially diatoms [17,18]. This is demonstrated in **Figure 12** below for the Pacific Ocean waters collected near the Equator, Antarctica, and Alaska where a comparison of the doubling times of phytoplankton in waters with and without the addition of iron.
3. Laboratory experiments have also shown that low levels of iron can limit the growth of phytoplankton [14,19,20]. Coastal phytoplankton have optimum ratios of Fe:P of 10^{-2} to 10^{-3}, while most ocean species have ratios less than 10^{-4}, indication that the latter have adapted to low levels of Fe in the open oceans. By controlling the concentration of iron by the addition of the complexing ligand EDTA, it can be shown that low levels of Fe decrease the growth of coastal and oceanic cyanobacteria.

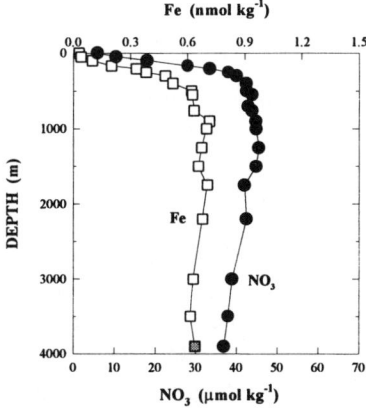
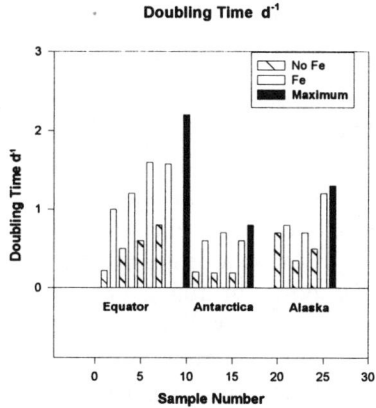

Figure 11. Profiles of nitrate and iron in the North Pacific.

Figure 12. The effect of the addition of iron on the doubling time of the growth of phytoplankton in Pacific Ocean waters.

4. Fast repetition rate fluorometry has shown that the photochemical energy conversion efficiency in phytoplankton is less than maximum in HNLC areas and is significantly increased with the addition of nM levels of iron [21].

These experiments suggest iron availability and supply may regulate ocean production in HNLC areas. The historical record of atmospheric CO_2, dust deposition and non sea-salt aerosols (from the growth of phytoplankton) in the Vostok ice core suggest that iron deposition may be related to ocean productivity. The supply of iron-rich dust to the oceans may cause an increase in ocean productivity and decrease the atmospheric concentration of CO_2 [22]. The extrapolation of these shipboard and laboratory experiments to whole ecosystems has, however, been strongly criticized [12]. Bottle experiments by design, do not accurately represent the community response to nutrients since some components of the community are excluded. The effects of iron on phytoplankton growth in HNLC areas can only be resolved by actual mesoscale enrichments of ocean waters to see the effect on the entire ecosystems.

In 1994 a natural iron enrichment study called **IRONEX-I** was conducted in the waters ~500 km south of the Galapagos Islands in the Equatorial Pacific [23,24,25]. This location was selected as a favorable spot since it is a HNLC area and the Chlorophyll plume off the western coast of the Galapagos island which provides a proximal comparison as a possible natural iron enrichment experiment. These experiments were undertaken to examine the connection between iron availability, phytoplankton productivity, and atmospheric carbon dioxide.

The initial iron concentrations in the area of the experiment were quite low (0.1 nM). To be sure that the changes in the patch could be attribute to the addition of iron, the area was surveyed over a period of two days to determine its biological, chemical and physical heterogeneity. The surface waters to about 35 m had quite uniform properties (t = 22°C and S = 35.36, pH = 8.008, NTA = 2309 µM, $NTCO_2$ = 2044 µM, fCO_2 = 408 µatm compared to the atmospheric levels 360 µatm, Oxygen = 216 µM, NO_3 = 10.8 µM, PO_4 = 0.92 µM and SiO_2 = 3.9 µM, chlorophyll = 0.24 µg L^{-1}) concentrations typical of the equatorial HNLC region.

An area of approximately 64 km^2 (8 x 8 km) was enriched by adding 455 kg (7,800 moles) of iron as $FeSO_4$ (0.5M) in acidic seawater (pH = 2) to the prop wash of the ship. Along with the iron, sulfur hexafluoride, SF_6, was added as an inert tracer. The SF_6 solution was prepared by dissolving 0.35 moles of SF_6 in seawater in a 2500 L steel tank. The SF_6 was pumped together with the iron solution into the prop wash as the ship steamed at ~ 9 km hr^{-1}. The goal was to increase the iron concentration to 4 nM (a level, in bottle experiments, that was sufficient to cause large increases in chlorophyll and complete depletion of the available major nutrients within 5 to 7 days). This enriched patch was tracked for 10 days and monitored for changes in biological, chemical, and physical parameters. The iron hypothesis required that we detect increased rates of growth in the patch, relative to water outside the patch. The increased growth in the patch should reduce major nutrient concentrations and decrease the fugacity of carbon dioxide. The inert chemical tracer, SF_6, added to the patch was detected continuously by electron capture gas chromatography at detection limits of less than 10^{-16} M. Dissolvable iron concentrations (DFe) as high as 6.2 nM were determined in the core of the patch 4 hours after the fertilization. These values decreased rapidly when the water column mixed, with the highest values on the subsequent day being 3.6 nM.

The changes in the chlorophyll in the patch became apparent within a day after the addition of Fe. A profile of Chlorophyll and primary production three days after the addition of Fe are shown in **Figure 13**. Photosynthetic energy conversion efficiency

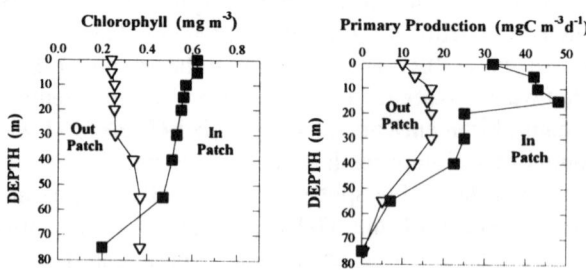

Figure 13. Changes in the chlorophyll and primary production in waters three days after the addition of iron.

(F_v/F_m) was the first biological response detected (**Figure 14**). The distribution of F_v/F_m

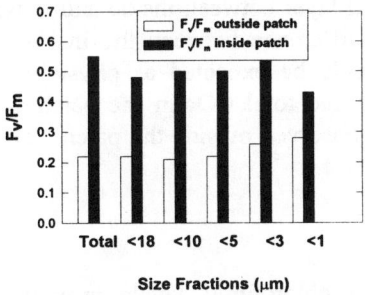 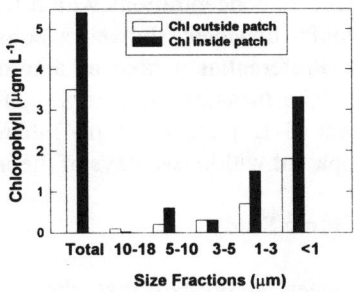

Figure 14. Changes in the photosynthetic energy conversion efficiency (F_v/F_m) and chlorophyll inside and outside the iron patch.

was also coincident with the area enriched in iron. The rapid increase in photosynthetic energy conversion efficiency in all sizes of phytoplankton in the patch confirm that the ambient populations in all size classes are physiologically limited by lack of available iron. The changes in the chemical constituents that occurred in the surface waters is shown in **Figure 15**. The nutrients showed little or no systematic differences of nitrate,

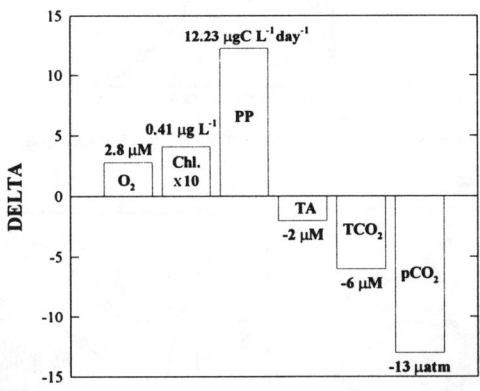

Figure 15. Changes in the chemical parameters in the surface waters three days after the addition of iron.

phosphate and silicate concentrations within the mixed layer between inside and outside stations. Ammonia, however, showed a consistent difference between the inside and outside stations. Preferential uptake of ammonia would be expected as physiological rates increased. The measurements of CO_2 fugacity and total CO_2 in the patch were significantly lower (3-12 µatm and 6 µM) than those observed outside the patch. These changes were apparent within two days of the iron release.

3. Galapagos Plume Study

It has been suggested that these elevated chlorophyll levels around the Galapagos islands are produced by the addition of iron derived from the island platform. As part of the IRONEX study a series of stations were occupied [26,27] both upstream (to the east) and downstream (to the west) of the Galapagos Islands (**Figure 16**). The surface values of nitrate and fCO_2 around the islands show [27] the strong upwelling that occurs off the west coast (see **Figure 17**).

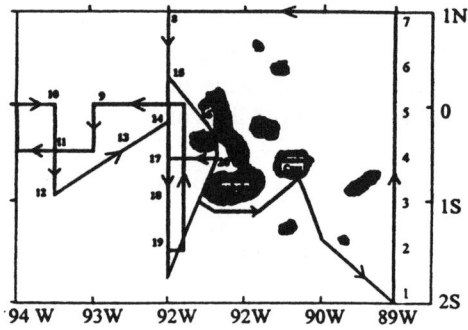

Figure 16. Hydrographic stations occupied around the Galapagos Islands.

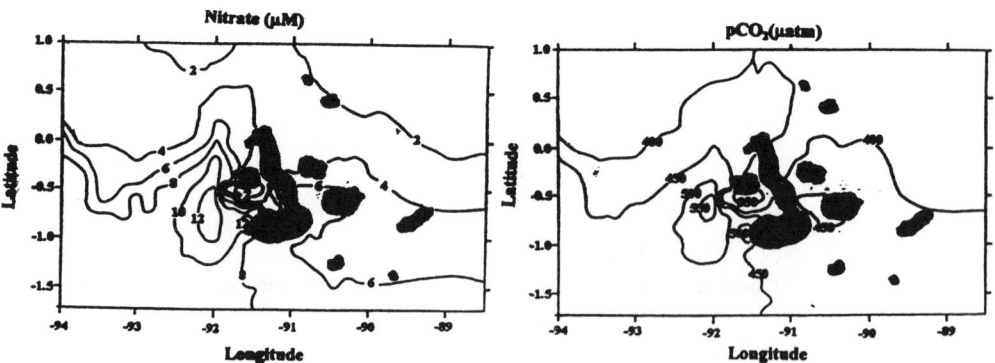

Figure 17. Surface values of temperature, nitrate and partial pressure of carbon dioxide around the Galapagos Islands.

In the Bolivar Canal we [26,27] saw a nearly complete depletion of nitrate and decreases of pCO_2 of as much as 200 µAtm (**Figure 18**).

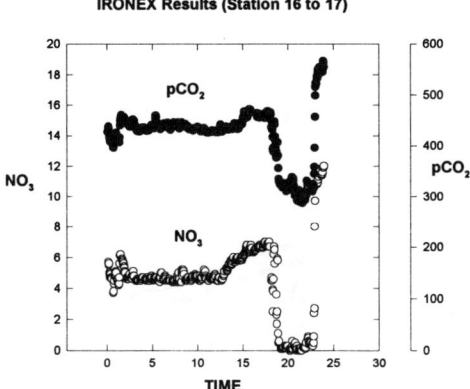

Figure 18. Changes in the surface nitrate and partial pressure of carbon dioxide between the Galapagos Islands (between Stations 16 and 17).

This suggests that nutrient consumption and phytoplankton growth is limited in the patch and downstream of the Galapagos Islands because of a loss term that occurs in open waters, but not in bottles or in shallow shelf regions. This loss term might be sinking of large diatoms. Iron would also be lost much more rapidly from open ocean systems than from bottles or shallow waters. Iron cannot be lost from a bottle. In shallow waters, sinking iron is trapped in a nepheloid layer near the bottom where concentrations of Fe can be greater than 10 nM. These enriched waters from this layer will continually resupply the euphotic zone with iron. This must account for the iron concentrations greater than 1 nM observed near the Islands. The biological similarity of shipboard iron enrichment experiments and stations over the Galapagos platform and Bolivar Canal suggest that it is the elimination of the loss term for iron that may be responsible for complete utilization of nutrients.

The iron that comes from the island platform must have a short residence time in surface waters. Iron was not detected far from the source region downstream of the islands. This indicates that both systems reflect a transient addition of iron rather than a sustained addition characteristic of bottle enrichments and shallow shelf stations. With a transient addition, only a few cell divisions are possible before the iron is removed from the system. Continual supply must occur to sustain production. The factors controlling the cycling of Fe in ocean waters is shown in **Figure 19**.

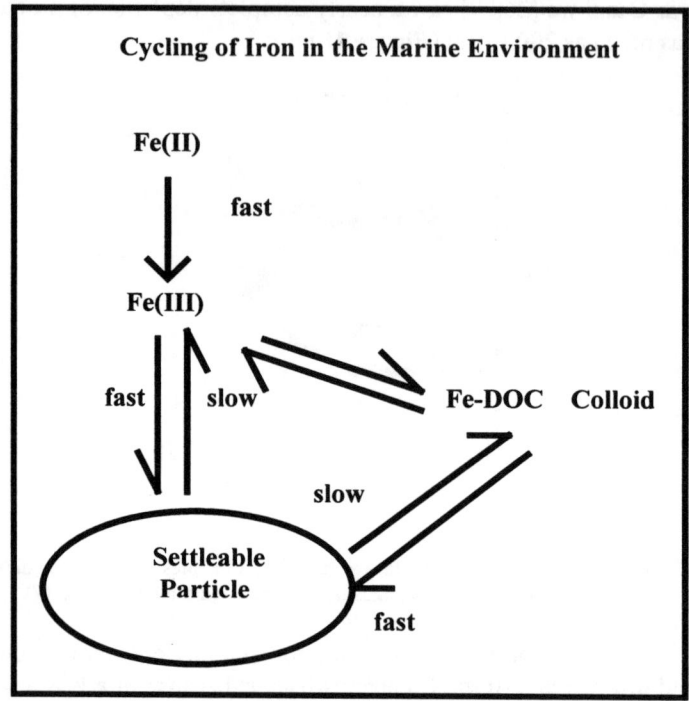

Figure 19. The processes controlling the state of iron in natural waters.

Although the results of the **IRONEX-I** experiment were encouraging, many questions still remain however after this study was completed. It is not clear that the modest increase in productivity and negligible loss of nitrogen and phosphate resulted from the loss of Fe or an increase in Zooplankton grazing. Last Summer (1995) the **IRONEX-II** was conducted in the Equatorial Pacific. During this experiment the patch was repeatedly fertilized with Fe. The results led to a significant increase in the growth of diatoms and the resultant decrease in nitrate and carbon dioxide [28]. The changes in the concentration of the pCO_2, nitrate, and fluorescence in and out of the fertilized patch shown in **Figure 20** clearly demonstrate that large changes in the CO_2 system can be affected if Fe does not leave the system [28]. Since most of the organic carbon formed is not delivered to the deep ocean this will not be a method of relieving the atmosphere of its increasing burden of CO_2. The experiments however have led to a better understanding of the interactions between the effect of nutrients and metals on the growth of phytoplankton and the resultant grazing of Zooplankton. The issue of whether iron additions can sequester substantial quantities of atmospheric CO_2 will lead to further debates. These experiments, however, demonstrate that open ocean manipulative experiments are possible and ecological studies in the open ocean are no longer limited to

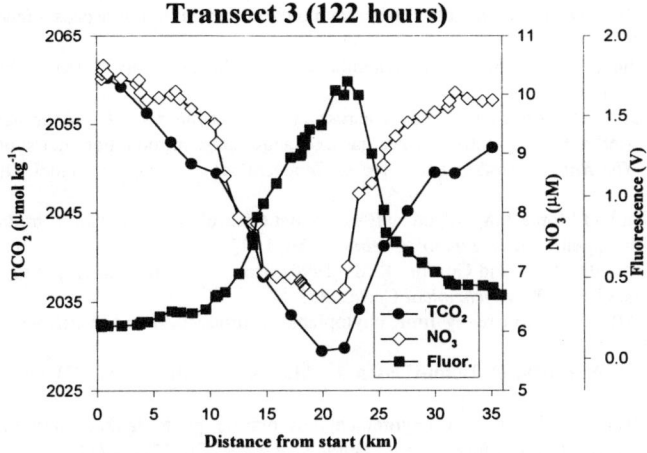

Figure 20. The values of nitrate, fluorescence and partial pressure after nine days across the iron enriched waters during IRONEX-II study in 1996.

passive observations. This will change significantly the way geochemical and ecological experiments are done in the future.

Acknowledgments

The authors wish to acknowledge the support of the National Science Foundation, The Department of Energy, The National Oceanic and Atmospheric Association, and the Office of Naval Research for this study.

References

1. Millero, F.J. (1996) *Chemical Oceanography*, 2nd Ed., CRC Press, Inc., Boca Raton, FL, 469 pp.
2. Clayton, T. and Byrne, R.H. (1993) Calibration of m-cresol purple on the total hydrogen ion concentration scale and its application to CO_2-system characteristics in seawater, *Deep-Sea Res.* **40**, 2115-2129.
3. Millero, F.J., Zhang, J.Z., Lee, K, and Campbell, D.M. (1993) Titration alkalinity of seawater, *Mar. Chem.*, **44**, 153-165.
4. Johnson, K.M., Wills, K.D., Butler, D.B., Johnson, W.K., and Wong, C.S. (1993) Coulometric total carbon dioxide analysis for marine studies: maximizing the performance of an automated gas extraction system and coulometric detector *Mar. Chem.*, **44**, 167-187.
5. Wanninkhof, R.H. and Thoning, K. (1993) Measurement of fugacity of CO_2 in surface water using continuous and discrete sampling methods *Mar. Chem.*, **44**, 189-204.

6. Millero, F.J. (1995) Thermodynamics of the carbon dioxide system in the oceans, *Geochim. Cosmochim. Acta.*, **59**, 661-677.
7. Lee, K. and Millero, F.J. (1995) Thermodynamic studies of the carbonate system in seawater, *Deep Sea Res.*, **42**, 2035-2061.
8. Liss, P.S. (1973) Processes of gas exchange across an air-water interface, *Deep-Sea Res.*, **20**, 221-238.
9. Liss, P.S. and Merlivat, L. (1986) Air-sea gas exchange rates: introduction and synthesis, in P. Buat-Menard, (ed.), *The Role of Air-Sea Exchange in Geochemical Cycling*, D. Reidel, Dordrecht, Holland, pp. 113-127.
10. Chisholm, S.W., and Morel, F.M.M., eds. (1991) What Controls Phytoplankton Production in Nutrient-Rich Areas of the Open Ocean? *Limnol. Oceanogr.* **36**, 1507-1970.
11. Martin, J.H, Fitzwater, S.E., and Gordon, R.M. (1990) Iron deficiency limits phytoplankton growth in Antarctic waters, *Global Biogeochemical Cycles*, **4**, 5-12.
12. Banse, K. (1990) Does iron really limit phytoplankton production in the offshore Pacific? *Limnol. Oceanogr.*, **35**, 772-775.
13. Mitchell, B.G., E.A. Brody, O. Holm-Hansen, C. McClain, J. Bishop, (1991) *Limnol. Oceanogr.* **36**, 1662-1677.
14. Brand, L. (1991) Minimum iron requirements of marine phytoplankton and the implication for biogeochemical control of new production, *Limnol. Oceanogr.*, **36**, 1756-1771.
15. Martin, J.H. (1992) Iron as a limiting factor in oceanic productivity, in P.G. Falkowski and A.D. Woodhead (eds.), ***Primary Productivity and Biogeochemical Cycles in the Sea*** Plenum, New York, 123-127.
16. Martin, J.H. and Gordon, R.M. (1988) Northeast Pacific iron distributions in relation to phytoplankton productivity, Deep-Sea Res., **35**, 177-196.
17. Martin, J.H. and Fitzwater, S.E. (1989) Iron deficiency limits phytoplankton growth in the north-east Pacific subartic, *Nature*, **331**, 341-343.
18. Martin, J.H., Gordon, R.M., and Fitzwater, S.E. (1991) The case for iron, *Limnol. Oceanogr.*, **36**, 1793-1802.
19. Sunda, W., Swift, D.H., and Huntsman, S. (1991) Low iron requirement for growth in oceanic phytoplankton, *Nature* **351**, 55-57.
20. Hudson, J.M. and Morel, F.M.M. (1990) Iron transport in marine phytoplankton: Kinetics of cellular and medium coordination reactions, *Limnol. Oceanogr.*, **35**, 1002-1020.
21. Greene, R.M., Geider, R.J., and Falkowski, P.G. (1991) *Limnol. Oceanogr.*, **36**, 1772-1782.
22. Martin, J.H., (1990) *Paleoceanogr.*, **5**, 1-13.
23. Martin, J.H et al. (1994) Testing the iron hypothesis in ecosystems of the equatorial Pacific Ocean, *Nature*, **371**: 123-129.
24. Kolber, Z.H., Barber, R.T., Coale, K.H., Fitzwater, S.E., Greene, R.M., Johnson, K.S., Lindley, S. and Falkowski, P.G. (1994) Iron limitation of phytoplankton photosynthesis in the equatorial Pacific Ocean, *Nature*, **371**, 145.
25. Watson A.J., Law, C.S., Van Scoy, K.A., Millero, F.J., Yao, W., Friederich, G.E., Liddicoat, M.I., Wanninkhof, R.H., Barber, R.T., and Coale, K.H. (1994) Minimal effect of Iron fertilization on sea-surface carbon dioxide, *Nature*, **371**, 143-145.
26. Millero, F.J., Yao, W., Lee, K., Zhang, J.Z., and Campbell, D.M. (1996) Carbonate system in the waters near the Galapagos Islands, *Deep Sea Res.*, submitted.
27. Sakamoto, C.M., Millero, F.J., Yao, W., Friederich, G.E., and Chavez, F.P. (1996) Surface seawater distributions of inorganic carbon and nutrients around the Galapagos Islands: Results from the PlumEx experiment using automated chemical mapping, Deep Sea Res., submitted.
28. Coale, K.H. et al. (1996) The IronEx-II mesoscale experiment produces massive phytoplankton bloom in the equatorial Pacific, *Nature*, submitted.

SUBJECT INDEX

A

AAS 133,151,167
Acid-base equilibria
        colloid effect on 40,48
Activity coefficients
        model for 17,61
Adsorption 40,42,212,244
Air pollution... 315
AES 151, 155, 169
Algae 259
Alkenones 201
Alkyl lead.. 85
        analysis 157
Amines 65
Amino acids 67
Ammonium ions... 347
Analytical quality control 173,225
Antimonium... 87
Arsenic 87,89,275
        analysis 157
        cycling 284
        sediments, in... 284
        toxicity.. 286
Atomic absorption spectroscopy, see AAS
Atomic emission spectroscopy, see AES
Azarenes 293

B

Bacteria 259
Bioaccumulation.. 287
Bioavailability 149,348,353
Biological pump 381
Biomarkers 201
Biomass 101
Butyl tin 85

C

Carbon cycle 3
Carbon dioxide 329,355,381
       uptake 383
Carbon monoxide 329,339
Carbonate 15,384
Carbonyl sulfide 329,343
Carboxylic ligands 63
Certification 229
Chloride 11,34
Chlorophylls 101
Chromatography 150,173,179
Chromium
       analysis 156
       speciation 156
Colloids 2,39,41
Complex formation 11
       ability 59,211
       colloid effect on 40
       model for 62
Concentration, see Preconcentration
Coprostanol 271

D

Data evaluation 173
DDT 294
Debye-Huckel equation 46,63,81
Derivatization 165
Dimethyl sulfide 325
Dissolved inorganic carbon (DIC) 345
Dissolved gases (see also specific gases) 4,128
Dissolved organic carbon (DOC) 3,329
Dissolved organic matter (DOM) 3,41,54,59,361
Dissolved organic nitrogen (DON) 329,346
Dissolved oxygen, see Oxygen
DOM, see Dissolved organic matter
DON, see Dissolved organic nitrogen

SUBJECT INDEX

Dynamic structures 107,244

Electrons, aquated... 40
Electrochemical preconcentration 143
Eutrophication... 268
Eutrophic conditions 259
Extraction 136,163,222

Filtration... 211
Flotation 135
Flow injection 133
Fluoride 29
Freundlich isotherm... 45
Fulvic substances, see also Humic 203,211

Gas chromatography... 150,165,179,190,300
Gases (see also specific gases) 4,128
Global changes 329
Greenhouse effect... 382
Guy-Chapman equation... 47

High performance liquid chromatography (HPLC)........ 149,179,190
Humic substances... 39,59,203,211
Hydrocarbons... 309
Hydrogen peroxide 12,40,313,361
Hydroxide ion 69
Hydroxyl radical... 40,313,343
Hyphenated techniques 149,167,190

ICP-AES 133,155
ICP-MS 156,167
Inductively coupled plasma, see ICP
Interfaces... 39,118,246,383
Ion adsorption... 40
Ion pairs 12,62

Ionic interactions 11
       model for 13
Ionic strength 11,40,59,62
Iron 353,381
       activity coefficient 20
       complexes 19,24,365
       cycling 396
       hydrolysis 20
       hypothesis 5,381
       photoreduction 365
       reduction 26
       solubility 21,23
       speciation 21,358
       thermodynamics of 14

Kinetics (see also Reaction rate)
       colloid effect on 40
       of light induced processes 40
       of removal 245

Langmuir equation 45
Lead (see also alkyl lead) 85
Ligands
       amine 65
       amino acids 67
       carboxylic 63
       chloride 12
       hydroxide 68
       inorganic 11
       organic 24
       phenols 67
Light
       absorption 335
       driven reactions 53
       on bioavailability 353

SUBJECT INDEX

Manganese 353
    speciation 371
Marine pollution (see also specific pollutants) 5,173,191,225,259,293
    chlorinated hydrocarbons, see also PCB
    dioxins, see PCDD
    distribution model 237
    oestrogenic chemicals 268
    oil pollution 203
    PCB, see PCB
    pesticides... 191
    polynuclear aromatic hydrocarbons, see PAH
    sewage, see Sewage
    volatile organic compounds, see VOC
Mass spectrometry (see also ICP-MS) 170,189,300
Matrix removal 131
Mercury 85
    analysis 157,164
    cycle 91
Metal ions (see also specific metal ions) 85,389
    analysis 131,149
    distribution model 237
    reference materials... 231
    speciation model 60
    transfer from atmosphere 94
Methylation 86
Methyl mercury 85,157,164
    reference material 230
Methyl tin 88
    reference material 230
Monitoring programs 6

Nitrate 101
Nitrogen oxides cycle... 316
Nutrients 99,249,259,353

SUBJECT INDEX

Oestrogenic chemicals.. 268
Oil spills 203,309
Organic carbon (see also DOC).. 3
Organic matter (see also DOM).. 3
       classification 3
Organometallic compounds.. 85,157,161
       reference materials.. 231
       toxicity.. 162
Oxygen
       depletion 265
       dissolved.. 55,109,267
       singlet.. 40,344
Ozone 313,324,331

PAH 173,196,293,309
       reference material 230
Particle sedimentation 40
Partitioning 296
Particulate, inorganic... 2,39,244
Particulate, organic.. 2,39,41,212,336
PCB 173,191
       reference material 230
PCDD 194,294
Peroxyl radicals 40,343
Petroleum hydrocarbons, see Oil spills
Phenols 67
Phosphate 102,389
Photosynthetic available radiation.. 104
Phytoplankton 104,353,389
Pitzer equation 16,63,81
PIXE 252
Plankton, see Phytoplankton
Point of zero charge (PZC) 41
Poisson-Boltzmann equation 46
Pollution, see Marine Pollution and specific pollutants
Polychlorinated biphenyls, see PCB

SUBJECT INDEX

Polychlorinated dioxins, see PCDD
Precipitation
    of iron hydroxide 21
    separation technique 152
Preconcentration 131,213
Pseudophase model 52

Quality control, see Analytical quality control

Radicals (see also specific radicals) 54
Radionuclides 5
Reaction rates
    colloid effect on 51
Redox reactions 11
    rates of 12
    ionic interactions in 12
    colloid effect on 40
Reference materials 225

Salinity 338
Sample treatment 131,163,173
Sampling 115,173,299
Seafloor 126
Seawater
    composition 93,119
    extraction from, see Extraction
    sampling 119
    synthetic 78
Sediments 85,248,272
    metals in 85
    organometallic compounds in 85
    sample preparation, see Sample treatment
    sampling 115,196
Selenium 87
Separation techniques 131,166,211

Sewage 259
    composition of 261
    metal ions 265
    treatment 261
Solar radiation (see also Light) 40,104,331
Solubility
    of iron hydroxide 23
Solubility pump 381
Speciation 4,21,59,94,149,156,161,279
    calculation program for 71
Stability constants 19
    calcium 63
    calculation program 71
    iron 19,353
    manganese 353
    sodium 63
Standard reference materials, see Reference materials
Sulfite
    autooxidation 33
    iron reduction 27
Supercritical fluid chromatography 180
Superoxide 12,343,361
Surface
    catalyzed reactions 40
    complexation model 41
    microlayers 122
    water 248
Suspended matter 125
Synthetic seawater 78

Tandem mass spectrometry 207
Tin 85
    organocompounds 158,162
Toxicity 162
Toxins 205